Ginkgo

Yale UNIVERSITY PRESS

New Haven and London

Ginkgo

THE TREE THAT TIME FORGOT

Peter Crane

WITH A FOREWORD BY PETER RAVEN

Published with assistance from the foundation established in memory of
Philip Hamilton McMillan of the Class of 1894, Yale College.

Yale University Press books may be purchased in quantity for educational, business,
or promotional use. For information, please e-mail sales.press@yale.edu (U.S. office)
or sales@yaleup.co.uk (U.K. office).

Designed by Nancy Ovedovitz and set in Minion type by Tseng Information Systems, Inc.
Printed in the United States of America.

The Library of Congress has cataloged the hardcover edition as follows:
Crane, Peter R.
Ginkgo : the tree that time forgot / Peter Crane.
p. cm.
Includes bibliographical references and index.
ISBN 978-0-300-18751-9 (cloth : alk. paper)
1. Ginkgo. I. Title.
QK494.5.G48C73 2013
585′.7—dc23   2012032831

ISBN 978-0-300-21382-9 (pbk.)

A catalogue record for this book is available from the British Library.

10 9 8 7 6 5 4 3 2 1

Ginkgo biloba

Dieses Baum's Blatt, der von Osten
Meinem Garten anvertraut,
Giebt geheimen Sinn zu kosten,
Wie's den Wissenden erbaut.

Ist es ein lebendig Wesen?
Das sich in sich selbst getrennt;
Sind es zwey? die sich erlesen,
Dass man sie als eines kennt.

Solche Frage zu erwiedern
Fand ich wohl den rechten Sinn;
Fühlst du nicht an meinen Liedern,
Dass ich Eins und doppelt bin?

—Johann Wolfgang von Goethe, September 15, 1815

Ginkgo biloba

This tree's leaf which from the Orient
Is entrusted to my garden
Lets us savor a secret meaning
As to how it edifies the learned man.

Is it one living being?
That divides itself into itself
Are there two who have chosen each other,
So that they are known as one?

To reply to such a question
I found, I think, the condign sense.
Do you not feel that in my poems
I am single and twofold?

—English translation by Kenneth Northcott, 2006

Contents

CONTENTS

Foreword

Perhaps the best known and most easily recognized of the world's 100,000 kinds of trees, ginkgo stands out by virtue of its unique features, amazing history, and long association with people. With their distinctive fan-shaped leaves and tall trunks, ginkgo trees adorn parks, streets, and recreational areas throughout the temperate regions of the world. When the weather turns sharp, all of the leaves suddenly turn a brilliant yellow, dropping soon after to lay a lovely, bright yellow carpet under each tree. Ginkgo trees are bisexual, some producing seeds and others only pollen-bearing organs. The seeds are naked, as in other gymnosperms (such as pines, cycads, and cedars); the fleshy outer coat of the ginkgo seed smells strongly of rancid butter (butyric acid). The inner "kernel" of the seed is nutlike and eaten widely in the Orient, once the smelly outer layer and the stony inner one are removed. Among the seed plants, only ginkgo and cycads form motile sperm within their pollen tubes, a fascinating example of the survival of an archaic characteristic.

A particularly fascinating feature of ginkgo's history is that it is extremely rare as an uncultivated native tree. Once widespread throughout the Northern Hemisphere, it disappeared at different times in different regions, including most of Eurasia and all of North America, as the climate changed and new plant communities came into being. In the Northern Hemisphere, China represents the zone where the most ancient lineages have survived, including many that were once widespread. Only in two mountainous areas of southern China do the varied genetic patterns of the standing trees make it likely that they or their immediate ancestors were native to the forests where they still occur. In addition, however, individual trees here and there in southern China may represent originally wild individuals that survived and eventually were protected by the people who lived near them, as is the case in the dawn redwood, *Metasequoia*. That ginkgo has survived essentially unchanged for as much as 200 million years is a miracle: virtually all other kinds of plants and animals that occurred with it more than 150 million years ago have become extinct. Those that disappeared included the relatives of the surviving evolutionary lineage to which *Ginkgo biloba* belongs. Although defining species from fossil material alone is difficult, ginkgo may legitimately be regarded as the oldest surviving kind of plant: the characteristics of the living species closely resemble those of its Jurassic ancestors!

Starting about a thousand years ago, ginkgo was brought in from the wild, or simply allowed to survive where native, in the temple gardens and protected forests of China; it has been nurtured and valued by human beings over the past thousand years or more. Ginkgo evidently was spread from there to Korea and Japan over the course of the past 800 years or so. After Europeans discovered it in Japan in the late seventeenth century, it was brought into cultivation in Europe over the next several decades, eventually spreading around the world in areas with a suitable climate. Ginkgo is resistant to air pollution and to pests and diseases, doing very well as a city tree through temperate zones of both Northern and Southern Hemispheres. Without the attention of human society, ginkgo would doubtless have become extinct by now, or at best reduced to a few individuals. In that sense, its history affords an excellent example of how we must deal with many plant species in this age of change and extinction if we wish to save them for the future.

As Peter Crane has emphasized in the closing pages of this fine book, human beings have had a short span of existence on this 4.5 billion–year–old planet, and we live only a short time as individuals. That we share the Earth with such a venerable organism, with a history where hundreds and tens of millions of years are relevant, should help to give us a better perspective from which to think of our own lives and existence here and thus prepare as well as we are able for the future. Our short-term actions are ruining the world in which we live much faster than we can imagine, with the world's sustainable capacity sufficient to provide less than two-thirds of what we consume each year, even though billions of us live hungry and in extreme poverty. The two billion or more additional people who will join our numbers in the next several decades will almost all be poor, entering a world that we, with our largely short-term view of progress, are destroying rapidly. Might not we be able to learn from the deep past and then redouble our efforts to sustainably use our planet's resources and thus live within its productive capacity while we still have time to do so?

Peter H. Raven, President Emeritus, Missouri Botanical Garden, Saint Louis

Preface

The origins of this book trace to an early fondness for the wonderful accounts of economically important plants written by the late Charlie Heiser in the 1970s and 1980s. These stand in a long tradition of popular science writing, but it was their particular blend of science and culture, leavened with personal experience, that provided the inspiration for this book. In focusing on the ginkgo, a tree with such a long and varied life story, my reach may have exceeded my grasp, but the challenge of trying to balance my scientific inclination for depth with the need for breadth has brought its own rewards.

From time to time researching and writing this book has impinged on my other responsibilities. Therefore, I am especially indebted to the three organizations that have been my professional homes over the last decade. Work began on this book while I was director of the Royal Botanic Gardens, Kew; continued at the University of Chicago; and was finally completed at Yale. I am grateful to all three institutions for allowing me to devote time to this project, which is not quite a normal piece of scientific research, had little do with my professional responsibilities, and was always something of an indulgence. Much of the writing was done in Seoul in the summers of 2009, 2010, and 2011, surrounded by ginkgo trees, during my time as a visiting professor in the World Class Universities program of the National Research Foundation of Korea. I am especially grateful to EWHA University, and my colleagues there, Jae Choe and Yikweon Jang, for their hospitality and support.

This book would not have happened without the unwavering kindness of Fumiko Ishizuna in Tokyo, who led me on several incomparable expeditions to key sites for ginkgo in Japan. Those expeditions, and the opportunities they provided to see truly spectacular great ginkgos, convinced me that this book was worth writing. I am also deeply grateful to my friend Zhou Zhiyan in Nanjing for his patience and thoughtful guidance in many aspects of this work, and not only in those chapters dealing with the fossil record. Finally, this manuscript probably would have languished forever had it not been for the dedication and commitment of Ashley DuVal, who more than anyone else helped drive it through to completion.

Among the many other individuals who have helped me with different aspects of this story I would especially like to thank Heidi Anderson and Andrew Drinnan in Australia; Johanna Eder and Michael Kiehn in Austria; Kevin Aulenback and Spencer Barrett in Canada; Cheng Quan, He Shan-an, Hu Yonghong, and Yungpeng Zhao in China; Branko M. Begović Bego in Croatia; Kaj Raunsgaard Pedersen in Denmark; Hans Kerp in Germany; Mitsuyasu Hasebe, Toshiyushi Nagata, Tetsuo Ohi-Toma, and Masamichi Takahashi in Japan; Gerard Thijsse in the Netherlands; Jolanta Kalisz in Poland; Adrian Patrut in Romania; Lye Lin Heng in Singapore; John Anderson and Brian Huntley in South Africa; Hyosig Won in South Korea; Else Marie Friis in Sweden; Julia Buckley, Eleanor Bunnell, Martin Hamilton, Liz Jaeger, Stephen Jury, Tony Kirkham, Christine Leon, Brian Mathew, Andrew McRobb, Mark Nesbitt, John Parker, Martin Postle, Hugh Prendergast, Malin Rivers, Moctar Sacande, Anna Saltmarsh, Wolfgang Stuppy, and Fiona Wild in the United Kingdom; and Selena Ahmed, Mark Ashton, Bruce Baldwin, Alona Banai, Jeremy Beaulieu, Graeme Berlyn, Kevin Boyce, Eric Brooks, Gary Brudvig, Bret Buskirk, Ed Buyarski, Bill Carvell, Jeff Courtney, David Dilcher, Laura Donnelley, Gerry Donnelly, Michael Donoghue, Ian Glasspool, Chris Haufler, Dave Hayes, David Heidler, Christie Henry, Pat Herendeen, Nancy Hines, Michelle Holbrook, Kirk Johnson, Bill LeFevre, Andrew Leslie, Stefan Little, Chris Liu, Marie Long, Steve Manchester, Greg McPherson, Herb Meyer, Rachel Meyer, Colleen Murphy-Dunning, Andrew Newman, Peter Purdue, John Rashford, Laurel Ross, Kemba Shakur, Pamela Soltis, Leroy Squires, Scott Strobel, Gregory Tarver, Douglas Trainor, Mary Evelyn Tucker, Warren Wagner, Marianne Welch, Elisabeth Wheeler, Mimi Yiengpruksawan, and Qingfu Xiao in the United States. All contributed ideas, information, or experiences that found their way into the book. I am especially indebted to Bill Chaloner, Pat Horn, Charles Jarvis, Toshiyushi Nagata, Peter Raven, and Scott Wing for their sound advice and review of the manuscript, and to Jean Thomson Black, Sara Hoover, and Dan Heaton at Yale University Press and Al Zuckerman at Writers House for their kind guidance. The drawings that introduce each section are by Pollyanna von Knorring, who has illustrated my research for almost thirty years.

Research for this book benefited greatly from the availability of a few key resources that had already drawn together much scattered material on ginkgo, most important the wonderful Ginkgo Pages Web site by Cor Kwant, the various writings of Peter Del Tredici, and especially for Western readers the important book edited by Terumitsu Hori, Robert Ridge, Walter Tulecke, Peter Del Tredici, Jocelyne Trémouillaux-Guiller, and Hiroshi Tobe, which was produced as part of the celebrations on the centenary of Hirase's remarkable discovery of swimming sperm in ginkgo.

Finally, I am deeply grateful to my longtime friend Peter Raven for writing the foreword to this book, and to my family, Elinor, Emily, and Sam, for their infinite patience during my multiyear obsession with this very special tree.

Prologue

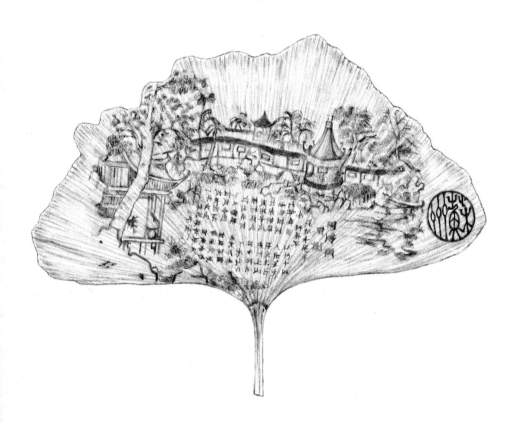

1
Time

One huge ginkgo tree, topping all the others, shot its great limbs
and maidenhair foliage over the fort which we had constructed.
In its shade we continued our discussion . . .
—Sir Arthur Conan Doyle, *The Lost World*

To most people ginkgo is either the tree with the smelly "fruits" or the plant that is
good for your memory, but that unmistakable acrid aroma, or the leaf extract in the
health food store, is only part of what makes ginkgo unique. Common on city streets
from Beijing to London and Tokyo to New York, ginkgo is an increasingly common
backdrop to the bustle of modern city life. It is hard to imagine that these trees, now
towering above cars and commuters, grew up with the dinosaurs and have come down
to us almost unchanged for 200 million years. Ginkgo is one of the world's most distinctive plants and has one of the longest botanical pedigrees; there is no other living
tree with a prehistory so deeply intertwined with that of our planet. This book, an
abridged global biography, sets out to tell ginkgo's evolutionary and cultural life story.[1]

My interest in ginkgo stretches back more than thirty-five years, but it grew during
my time, from 1999 to 2006, as director of the Royal Botanic Gardens, Kew. During
those seven years, the oldest and perhaps the most important ginkgo in the United

Kingdom flourished just steps from our family's home in the garden. We passed it almost every day; it was a regular stop on tours for distinguished visitors. It was one of the most beloved of the Old Lions, that handful of trees remaining from the mid-eighteenth century, when Kew was a royal estate. We watched this tree as it changed through the seasons, and we worried when a big storm came through. In 2002 it was designated one of fifty Great British Trees by the Tree Council in its celebration of the Queen's Golden Jubilee. It was among the most precious of the nearly fifteen thousand trees on the Kew landscape.[2]

I first came across Siegfried Unseld's delightful little book *Goethe und der Ginkgo: Ein Baum und ein Gedicht* through my longtime colleague Else Marie Friis at the Swedish Museum of Natural History in Stockholm. It led me to track down an English translation of the famous poem at the heart of Unseld's book. Then, upon my return to the United States in late 2006, Christie Henry at the University of Chicago Press presented me with a copy of Kenneth Northcott's rendering of *Goethe and the Ginkgo*. The idea that those who enjoy Unseld's book might like to know more about ginkgo was part of the stimulus for this biographical sketch of a singular tree.

Ginkgo is a botanical oddity, a single species with no close living relatives. Once regarded as a cousin of pines, yews, and cypresses, it was later recognized as something quite different. It was first distinguished from conifers in plant classifications of the early nineteenth century. The evidence that has come to light since—particularly an astonishing discovery made in Japan in 1896 about the intimate details of its reproduction—has reinforced the isolated position of ginkgo among living plants. So in the twentieth century, as the world of plants has come into sharper evolutionary focus, ginkgo has assumed new scientific importance. To borrow a phrase from Darwin, ginkgo has become a platypus for the plant kingdom, and plant paleontologists have traced its lineage millions of years into prehistory. Ginkgo is now the most widely recognized of all botanical "living fossils": a tree that time forgot and an increasingly familiar living link to landscapes of the distant past.[3]

For almost all of its long tenure on our planet ginkgo inhabited a world without people, and for much of that time, a world very different from that of today. For tens of millions of years it lived alongside plants and animals that are long since extinct. Several different kinds of ginkgolike trees watched as our ancestors transformed from reptiles to mammals. Fossil ginkgo leaves are known from every continent. The prehistory of ginkgo goes back to before the Atlantic Ocean existed and before the southern continents broke from Antarctica and went their own ways.

As our planet has changed over the past 200 million years, ginkgo has proved remarkably resilient. It felt the shock as new kinds of plants came to dominate the Earth about 100 million years ago, but it was oblivious to the traumatic events that eliminated the dinosaurs a few tens of millions of years later. Soon afterward ginkgo lost its foothold in the southern continents, but in Asia, Europe, and North America it continued to prosper. It flourished during the great warmth of 50 million years ago and once grew close to the North Pole. But as the Earth grew cooler, ginkgo found itself in retreat and was driven from its home in the Arctic.[4]

For the next forty million years, ginkgo was widespread across the northern continents. Its distinctive leaves are common in the fossil record. Eventually, though, and for reasons that are not fully clear, ginkgo began to suffer. By the time our ancestors diverged from those of other living apes, five million to seven million years ago, ginkgo was probably already in decline. It was nearly extinguished by the great Ice Ages that gave birth to our own species. When the last great southward push of the ice had retreated, ginkgo was barely hanging on, perhaps only in protected valleys scattered across eastern and south-central China. By the time modern people arrived in that part of Asia, perhaps fifty thousand years ago, ginkgo was already a relic.[5]

Human dominance on our planet could have meant the end for ginkgo, but unlike many other trees, it has flourished alongside people. In one way or another, it has proved useful; more unusually, it has become revered. In many cultures, in many different ways, ginkgo commands unusual respect. Together, these qualities earned ginkgo a reprieve. The nuts became a delicacy and were used for oil and in medicine. The tree, with its distinctive leaves and great longevity, also took on symbolic meaning in Buddhism, Daoism, and Confucianism. From China and Korea, ginkgo spread to Japan and became incorporated in the indigenous religion of Shintoism. Many of the great ginkgos of China, Japan, and Korea are in the grounds of Buddhist temples or Shinto shrines.

From Asia, again with the assistance of people, ginkgo began its renewal. In the eighteenth century, through the Dutch trading colony of Deshima in southern Japan, ginkgo became known in Europe. Soon after, seeds found their way back to the Low Countries and Britain, probably from both China and Japan, but perhaps also from

*A small Buddhist shrine at the base of an ancient ginkgo
at the Heungju Temple, in Taean, Chungnam, South Korea.*

Korea. Ginkgo then spread quickly to North America and elsewhere as a horticultural novelty. The Old Lion at Kew is one of several early ginkgo trees growing in Europe. By the time of Goethe, and later in the nineteenth century, ginkgo was widely grown as an unusual and striking tree: a symbol of the East. In just a hundred years, ginkgo returned to many of those places from which it had been extinguished millions of years before.

In the past fifty years ginkgo has been resurgent; interest in growing ginkgo, what it stands for scientifically, and the ways in which it might be useful has never been higher. Ginkgo has become recognized as a valuable street tree that grows well in tough places. Resistant to disease, tolerant of pollution, and able to withstand extremes of heat and cold, it is now familiar in urban landscapes over much of the world. Parts of Seoul are a near ginkgo forest, and ginkgo is among the most common street trees in Manhattan.

You can see ginkgo in San Francisco's Golden Gate Park, in Monet's garden at Giverny, and in parks and gardens in all but the warmest and coldest places on our planet. In recent decades ginkgo has also found its way into the pharmacy. It is among the most popular of herbal remedies, and its medicinal properties are the subject of advanced biomedical research. Extracts from ginkgo leaves are the source of a multibillion-dollar pharmaceutical industry.

Beginning with an introduction to the living tree, this book follows the prehistory of ginkgo over the past 250 million years from its origin, proliferation, and eventual decline, to its reprieve, renewal, and resurgence through its association with people. The dominant theme in the biography of ginkgo is survival, and its resilient life story offers hope for other botanical biographies that are still being written.

2
Trees

The best time to plant a tree was twenty years ago. The second best time is now.

—Chinese proverb

Among my clearest memories as a student beginning in botany are the giant Atlantic cedars that dominated the center of campus at the University of Reading. They were part of my life for nearly a decade, and still today every time I see one of these majestic trees, with their sparse flattened branching and truncated crowns, it takes my mind back to what now seems like a different life. Together with the giant sequoias scattered across campus, and the gnarled strawberry tree near Whiteknights House, the Atlantic cedars were planted at the end of the eighteenth century by the Marquess of Blandford, an ancestor of Winston Churchill. They were his living legacy to thousands of students who came two hundred years later.[1]

From the summer of 1974, when I was studying at Kew, it is the magnolias and monkey puzzles that stick in my mind. A vivid memory from my time as a new lecturer at Reading in the late 1970s is of a classically proportioned English oak, long since swept away, which grew near the Plant Science Laboratories. When I was newly arrived in Chicago in the early 1980s, the massive American beeches and sugar maples at Warren Woods, Michigan, a rare tract of midwestern forest spared by the loggers, made an indelible impression. And everyone in our family remembers the statuesque stone pine that grew by our garden gate when we lived at Kew; it has watched all fifteen directors come and go. These encounters with trees punctuate the chapters of my life. They are the touchstones to many other memories and connect me to the lives of people from times long before I was born.[2]

I have been fortunate; a life lived studying plants has given me the chance to interview some of the most spectacular trees on the planet. In 1975 I explored forests of Atlantic cedars in the Middle Atlas Mountains of Morocco and, in 1981, my first summer in the United States, I visited the coast redwoods just a few miles north of San Francisco. It is not hard to see how such miracles of nature, with their cousins the giant sequoias, galvanized the passion of John Muir and have stirred the hearts of many others.[3]

At Humboldt Redwoods State Park, two hundred miles up the coast from San Francisco, more than a hundred coast redwoods reach 350 feet or more. Few trees can match them for sheer size, but on a family holiday in Australia in the early 1990s, we visited equally immense swamp gums in the temperate rain forest of Tasmania. Several of those monumental eucalypts top out at more than 330 feet. Thomas Pakenham's spectacular photographs go a long way toward summoning up their grandeur, but there is no substitute for seeing these miracles of nature close up; you need to smell them, feel them, and be humbled by the cathedral-like spaces they create.[4]

For many of us, trees of all kinds, not just the largest and most spectacular, often take on special meaning. My affinity for the grove of ginkgo trees where we walked with our baby daughter, and for the black tupelo planted in our garden around the time our son was born, could not be more different from my feelings for the bricks and concrete that otherwise define the places where we have lived. Such connections probably go back deep into our evolutionary past. Our bodies reflect the arboreal lives of our ancestors, and in a world without skyscrapers it was trees, like mountains, that reached up into the sky. It was their roots, like caves, that penetrated deep into the Earth. Trees are part of us, and they connect heaven and the underworld with primeval power. Our urge to climb them may not be so different from our urge to climb mountains.[5]

Until recently my own tree climbing expertise had been pretty much confined to the small lilac that grew behind my father's garden shed fifty years ago. Because I have a tendency to vertigo, scaling tall trees is something I have mainly avoided. But that phobia was temporarily banished early one morning in the rain forest of northeastern Borneo. The promise of sightings of tropical birds, combined with just a hint of peer pressure, coaxed me onto a ladder strapped to the trunk of an enormous koompassia. It took me 120 feet, straight up, to the roof of some of Sabah's last remaining virgin lowland rain forest. It is hard to grasp the scale of such a giant until you have inched your way up its trunk and been enveloped by its canopy.[6]

Trees have been adopted as important symbols that are firmly entwined in human culture. In Genesis we learn that "out of the ground made the Lord God to grow every tree that is pleasant to the sight, and good for food; the tree of life also in the midst of the garden." It was the fruit of the "tree of the knowledge of good and evil" that Adam and Eve struggled to resist. The ancient Norse saw their world organized around Yggdrasil, the World Ash.[7]

Different cultures in different parts of the world have also been attracted to similar kinds of trees. Among figs, the banyan and peepal have special significance in the Hindu faith, and Buddha found enlightenment under one special peepal, the Bodhi tree. The Mediterranean fig was the tree of life in ancient Egypt, and five species of fig are at the center of the Afro-Brazilian religion of Candomblé, part of the legacy of slavery in Brazil. These fig trees have special meaning in the *terreiros*, the houses of worship.[8]

Trees are also living links to our cultural history. Transplanted people carried the seeds of baobabs from Africa to the New World. Brazil takes its name from Pau Brasil, the once common tree of the Atlantic rain forest that produced a rich red dye much prized by early Portuguese traders. The cedars of Lebanon, mentioned many times in the Bible, are a symbol of nationhood and appear on the Lebanese flag. The palmetto has similar significance in the state of South Carolina. The bearded fig is part of the national identity of Barbados, and in China there have been efforts to have ginkgo adopted officially as the national tree.[9]

We also connect trees with key people and important moments in history. In Britain the Royal Oak at Boscobel is a descendant of the tree in which King Charles II reportedly hid after defeat at the battle of Worcester. At Hatfield House, an English oak marks the spot where Queen Elizabeth I is said to have learned of her ascent to the throne. In Sherwood Forest, more than half a million visitors every year flock to see the Major Oak, the massive tree linked to the legend of Robin Hood.

A stylized ginkgo leaf, carved into the planks of a wooden bridge and painted yellow,
marks the way to the great ginkgo at the Yongmunsa Temple, South Korea.

Oaks have assumed cultural importance on the other side of the Atlantic, too. In Austin, Texas, the Treaty Oak is the sole survivor of the fourteen Council Oaks, a sacred meeting place for Native Americans weighing matters of war and peace. The Emancipation Oak at Hampton, Virginia, marks the site of the first southern reading of Lincoln's Emancipation Proclamation. In Chico, California, the Hooker Oak, named for Joseph Dalton Hooker, one of my predecessors at Kew, is featured in the Errol Flynn version of the Robin Hood legend. Renamed the Gallows Oak, it is the tree under whose shade the "merry band" first come together.[10]

There are many such legends related to the great ginkgo trees of eastern Asia. An hour and a half drive east of Seoul, South Korea, the massive ginkgo at the Yongmunsa Temple is one of the largest and most visited in the world. Established late in the first millennium A.D. in the foothills of Mount Yongmun, the temple is approached by a gentle walk uphill through the tranquil mountain forest. As the temple comes into view, the colossal ginkgo is perfectly placed on a broad plinth just above a rushing mountain stream. It is approached from below and leaves a strong impression.[11]

Legends about the origin of the Yongmunsa Ginkgo emphasize its place in Korean history. According to one, it grew from the staff of Ūisang, the great Buddhist priest of the Silla Dynasty. Another has it planted in sorrow by Maui, the crown prince of Gyeongsun, the last Silla king, as he lamented the fall of his nation. It hardly matters whether either account is true; more important is that this great tree is of deep significance to the Yongmunsa Temple and its monks, and has special meaning for the people of Korea.[12]

Cultural prominence is just one sign of the importance of the roughly 100,000 different kinds of trees on our planet. From the earliest civilizations trees have been used for fuel, timber, and food, as well as to make objects of all kinds. They also provide rubber, oils, and medicines, and are indispensable in the day-to-day livelihoods of poor people in many parts of the world. At Kew we worked on tree conservation with the Forestry Department in Burkina Faso. Every year Burkina Faso loses tens of thousands of acres of its scrubby forest to agriculture and the cutting of trees for fuel. At the same time, its population depends on fuel wood for 90 percent of their household energy. Understanding how people use trees, and how they can continue to use them sustainably, has practical value.[13]

Trees are also crucially important in the developed world. The average American uses a ton of wood every year, equivalent to about forty-three cubic feet of lumber. It gets turned into everything from baseball bats and baby cribs to xylophones and yo-yos. A huge amount of wood is also used in buildings, and trees are the main source of pulp for making card and paper. America consumes around ninety million tons of paper products every year.[14]

The OneTree Project took a symbolic approach to try to capture our debt to trees. An ailing 170-year-old English oak about sixty feet tall was felled in Tatton Park, Cheshire, England, and was turned over to more than seventy designers, artists, and artisans. Sawdust was burned to make a glaze for pottery, bark was used for tanning, and wood fibers were used to make paper. The wood itself was turned into an astonishing range of objects: toys, sculpture, furniture, window frames, ladders, yurts, bowls, and even a fetal stethoscope.[15]

Trees are also crucial participants in the ecological processes that determine the

The great ginkgo at the Yongmunsa Temple, South Korea.

environment on our planet. The fossilized remains of ancient trees are the main component of coal, and the carbon of which they were made is an important source of the carbon dioxide now accumulating in our atmosphere. Forests also affect local and regional climate, and regulate the quantity and quality of water in streams and rivers. Both above- and belowground, trees help structure different kinds of vegetation and the communities of animals and microbes that depend on them.

Yet despite the importance of trees, we often remain curiously ambivalent about them and their future. Trees are easy to take for granted. They seem ubiquitous, and when they stand in the way of something we really want—a site for a new parking lot, perhaps—they are easily pushed aside. Bill Vaughan, a columnist for the *Kansas City Star,* summed it up with a memorable line: "Suburbia is where the developer bulldozes out the trees and then names the streets after them."[16]

Nevertheless, threats to particular trees can bring out strong emotions. The Washington Oak in New Jersey survived two battles of Princeton: the first in 1777, when George Washington rallied his troops to defeat the British; the second in the 1980s, when it was threatened by a local developer. The citizens of Charleston annexed the massive Angel Oak to save it from similar threats. Through 2007 and much of 2008 the University of California, Berkeley, became the site of the longest running "urban tree sit," a protest to prevent the removal of ninety trees in the campus's Oak Grove to make way for a new athletic facility. Despite a lengthy lawsuit, numerous appeals, and, at one point, a hundred nude protesters perched in the boughs of the threatened oaks, the Alameda County Superior Court eventually ruled in favor of athletics.[17]

In the same way, controversies over the logging of old-growth forest or expanding agriculture into the rain forest can attract national and international political attention. U.S. President Bill Clinton must have known that he had crafted a fine compromise when neither loggers nor conservationists were happy with his plan for the management of old-growth forest in the Pacific Northwest. More recently, trees and forests have featured with increasing prominence in discussions of climate change policy.[18]

However, as always, it is the immediate and the local that get most of the attention; longer-term consequences and connections to what is happening elsewhere are too easily overlooked. Trees, and the ecological systems of which they are part, are assailed all over the world, all the time. The threats come from all directions; some are direct and rapid, others are indirect and insidious. Land conversion that removes forest and replaces it with agriculture, including tree monocultures, such as oil palm or rubber plantations, is one threat; urbanization is another. Overexploitation for fuel or timber is also a problem. For commercially valuable trees, such as mahogany and teak, selective, and sometimes illegal, logging—the botanical equivalent of hunting whales or poaching tigers—is a persistent threat.

Trees are also caught up in broader environmental changes, such as water or air pollution, as well as invasions of species, including diseases, from elsewhere. They are also in trouble from changes in soils, water, drainage, and the frequency of fire; rapid climate change and extreme climate events pose an increasing challenge. The stately life cycles of most trees are not well suited to the frenetic world that we are creating. The time it takes for a seed to grow into a mature tree, combined with fragmented landscapes that hamper the natural movement of plants and animals, makes it difficult for new generations to adapt or migrate.[19]

A large part of what attracts us to trees is their timelessness. Compared to us, and to much else in our modern world, trees have comforting longevity. They change slowly, almost imperceptibly. They are beacons of stability that can cross generations. To quote a well-known proverb, "A society grows great when old men plant trees whose shade they know that they shall never sit in." Ecologically speaking, trees find a place to live and put down roots; in the parlance of ecology they are site occupiers. They arrive, courtesy of the wind, or perhaps a bird or squirrel, and there they stay. In a densely populated world filled by restless people with ever escalating demands, such a ponderous lifestyle may not be a winning strategy. Our habits disturb the world on a massive scale. We create environments that favor weeds: plants that live fast, reproduce early, and die young. The soothing calm of long-lived trees is easily lost in the turmoil.

With its resilience and longevity ginkgo epitomizes much of what we admire about trees, and its strong cultural associations underline the strength of that bond. But the biography of ginkgo also has the dimension of deep time; the link that it provides to worlds gone by. Many of our familiar trees—ash, hornbeam, magnolia, oak, or walnut—have prehistories measured in millions or even tens of millions of years, but ginkgo is extreme; it has been ever present for 200 million years or more. And like the dawn redwood of China and the Wollemi pine of Australia, two other charismatic trees with biographies that stretch far into the past, ginkgo also has a further attraction; it very nearly went extinct. Ginkgo is a good news story: a tree that people saved.

3
Identity

> . . . just plant for me a funny looking ginkgo.
>
> —LuEsther Mertz, commemorative plaque, New York Botanical Garden

William Jackson Hooker, the first director of the Royal Botanic Gardens, Kew, was born in 1785. By then the ginkgo that stands in the historic heart of Kew Gardens was becoming mature. Half a century later, when Hooker came to Kew to create a national botanic garden, it was nearly a hundred years old; it had already outlived King George III, Sir Joseph Banks, and the others from the eighteenth century who had presided over its planting. Hooker would have seen this tree nearly every day and he would have recognized it as one of the more unusual and important in his care, just as it was for me 150 years later.[1]

William Hooker was the right man, in the right place, at the right time. The son of a brewer who became professor of botany at the University of Glasgow, he was a superb artist, gifted scientist, and remarkable visionary. He was tall and thin, with boundless energy, as well as patience, tact, and charm; his manners were "easy and urbane." Hooker was also an industrious and effective administrator: "He deemed nothing too small for his notice."[2]

Hooker came to Kew in the mid-nineteenth century to take charge of a small part of a country estate in decline. In the eighteenth century Kew had been a royal favorite, with the landscape of Richmond Gardens created by Capability Brown, and the adjacent estate of Frederick and Augusta, Prince and Princess of Wales, studded with magnificent buildings by Sir William Chambers. However, with the death of George III and also Sir Joseph Banks in 1820, Kew and its collection of living plants fell into decline.[3]

Appointed director in 1841, William Hooker set Kew on a new trajectory and brought it back to life. In a little more than two decades he built astonishing greenhouses, planted grand vistas, unified the eighteenth-century landscapes, and introduced exotic plants from all around the world. Behind the scenes, using his personal collection as the core, he began the development of the library and preserved plant collections that are the foundation of Kew's global work in plant conservation and science. By the time Hooker died, through his energy and political genius, he had greatly expanded the parts of the estate under his control and created a magnificent spectacle for the public.[4]

Hooker was a scientist, like most directors of Kew have been, but he was also a teacher. He understood that the way to interest students and others in plants is to show how they are useful to people. At Kew the collections of useful plants that he used to illustrate his lectures in Glasgow became the nucleus of what he called the Museum of Economic Botany, and for the national botanic garden this interest had another dimension. Hooker knew that the uses of plants were important to his political masters and to the commercial interests of Britain's Victorian Empire.[5]

With more than seventy-six thousand specimens, the Economic Botany collection at Kew is now the largest of its kind in the world. Tucked away in an air-conditioned vault, it is a vast miscellany of food plants, medicinal plants, dye plants, and timbers. There are endless artifacts made from different plant parts: fish traps and a dugout canoe alongside beautiful necklaces and exquisite fabrics. There is a shirt made from pineapple fibers and a bowler hat made out of cork. Among the spices, the psychoactive plants, the tools, the musical instruments, and the many other samples that illustrate the interdependence of plants and people, there are some that document the uses of ginkgo.

There are multiple accessions of ginkgo seeds. In China, ginkgo has been cultivated for a thousand years for its edible nuts. There are also samples of ginkgo wood, collected by John Quin, one of the first British diplomats in Japan. Between 1867 and 1896, at the instigation of Kew, he made meticulous observations of the ancient Japanese art of creating lacquerware. He recorded how the work was done, and he collected the materials and tools that were used. Among them is a piece of ginkgo wood. Quin lists it as used for "such articles as are turned in a lathe, as bowls, rice cups, round trays, etc."[6]

Among the other delights of the Economic Botany collection is a single dried ginkgo leaf acquired not long ago by a Kew botanist visiting China. It is decorated with a scene from a Chinese garden drawn in fine black lines with an accompanying poem in minute Chinese characters. To the left a tree, perhaps a pine, stands in the foreground. In the center is a group of rocks around a small pavilion. Behind, and off in the distance, are more trees, some of them perhaps willows. On the right a red stamp bears the name Suzhou in old Chinese characters from the Qin Dynasty.[7]

To those familiar with Chinese gardens the scene on the ginkgo leaf is well known. It is from one of the most famous gardens in Suzhou, the Chinese City of Gardens. It might also be familiar to those who know New York; it is partly re-created in the Metropolitan Museum of Art. Wang Shi Yuan, the Master of the Nets Garden, is one of the best examples of a small late-eighteenth-century Chinese private garden. Established by a retired bureaucrat on the site of a twelfth-century garden from the southern Song Dynasty, it is one of nine classical Chinese gardens in Suzhou that are recognized on the UNESCO list of World Heritage Sites. Several contain ancient ginkgo trees, as does the oldest of all botanic gardens at the University of Padua in northern Italy.[8]

The unique fan-shaped leaf has given ginkgo a strong presence all over the world. The seeds probably first attracted people to this tree, and perhaps saved it from extinction, but its distinctive leaf has played a large part in transforming ginkgo from a minor food plant into cultural icon. Ginkgo has the most memorable leaves of any tree, and its unusual form lies behind much of ginkgo's rich cultural history.[9]

The ginkgo leaf has been taken up repeatedly as an instantly recognizable motif. Ginkgo is native to China and has its longest history of cultivation there. Guo Moruo, a Chinese revolutionary, a contemporary of Mao, and the first president of the modern Chinese Academy of Sciences, wrote many poems about plants and flowers, but he reserved his highest praise for ginkgo. In the depth of the communist struggle against Chiang Kai-shek he called ginkgo "the Holy One of the East," a worthy and resounding symbol of Chinese nationalism.

In South Korea, many famous old ginkgo trees are preserved as natural monuments, including the ginkgo in Nongso, Seonsan, which villagers honor on the fifteenth day

The ginkgo motif on the gate of the
Sung Kyun Kwan University, which
incorporates the Munmyo Confucian
Shrine, in Seoul, South Korea.

of the lunar New Year. Nearly a hundred feet tall and sixteen feet in diameter, it is imposing even in its winter nakedness. It was planted more than four hundred years ago near a temple and marketplace that are now just a ruin. According to local legend it is so sacred that birds will not land on it.[10]

Ginkgo is also deeply embedded in Japanese culture. The ubiquitous *T* that serves as the symbol for the prefecture of Tokyo, home to thirteen million people, looks suspiciously like a stylized ginkgo leaf. The best sumo wrestlers have their hair tied at the top in shapes that take the ginkgo name; and when in 1958 two Japanese scientists described a new species of beaked whale from the warm waters of the Pacific and Indian Oceans, they named it *Mesoplodon ginkgodens,* the "ginkgo-toothed beaked whale." In Japan there are also ginkgo crabs, ginkgo mushrooms, ginkgo potatoes, and ginkgo sharks. Not to mention cupboards, tables, flower vases, farm tools, and musical instruments that all incorporate ginkgo as a part of their name.[11]

Around the world, governments, businesses, and organizations of all kinds have taken the ginkgo leaf as part of their identity; Zhejiang Forestry University in China, Osaka University in Japan, and Sung Kyun Kwan University in South Korea all have the ginkgo leaf as part of their logo. In the West, when the artist Larry Kirkland was asked to design the entrance to the new building of the National Academy of Sciences in Washington, D.C., and to convey the development of human knowledge of the natural world, he included a sprig of ginkgo leaves and their seeds. Along with Darwin's finches, Mendel's pea pods, and Morgan's fruit flies, ginkgo was among the nine images Kirkland chose to superimpose on the molecular structure of DNA. Appropriately for researchers working close to the site of the old Cavendish Laboratory, where the structure of DNA was first worked out, the Department of Plant Sciences at Cambridge University chose a similar juxtaposition; its logo surrounds a ginkgo leaf with a stylized double helix of DNA.[12]

The ginkgo identity also extends beyond academia into commerce. You can visit a Ginkgo café in Melbourne, Australia, or Frankfurt, Germany, and many places in between. Around the world there are chic Ginkgo spas and Ginkgo restaurants. Almost all advertise themselves by flourishing the distinctive ginkgo leaf. Marketers and brand developers link the word and the leaf with an image that is contemporary but timeless, exotic yet familiar, and, most of all, sublimely elegant.

In the Chicago suburb of Oak Park, a large ginkgo stands outside the former home and studio of Frank Lloyd Wright. It was probably a young tree when the architect bought the land. He could easily have cast it aside, but instead he worked around it as he extended his home. Certainly he would have seen ginkgo on his visits to Japan. Today, the tens of thousands of visitors each year who come to this shrine of twentieth-century architecture pass under his great ginkgo and buy their tickets at the Ginkgo Bookshop. Inside there is ginkgo merchandise of all kinds, from ginkgo dishes to ginkgo jewelry boxes. An even bigger selection can be found through a few keystrokes on the Web. The variety of ginkgo products is almost overwhelming, and they continue a long tradition. For hundreds of years artists and artisans of all kinds, in China, Japan, and Korea, have incorporated ginkgo into their work.[13]

The connection between ginkgo and Frank Lloyd Wright seems a natural one. The influence of the Arts and Crafts movement is especially strong in many contemporary products that incorporate the ginkgo leaf motif. Its elegance and clean curves connect easily to an aesthetic that began as a reaction against the machine. Ginkgo was also taken up in Art Nouveau; there are spectacular renderings of ginkgo twigs and leaves in the Art Nouveau architecture of Nancy and Prague.[14]

Ginkgo also remains a source of artistic inspiration. In 2005, in the British Pavilion at the Fifty-first Venice Biennale, Gilbert and George, artists from the East End of London, showed their Ginkgo Pictures: a striking exhibition of twenty-five photo images created from ginkgo leaves collected in Gramercy Park, New York. Brightly colored, framed in a black grid, and arranged with portraits of Gilbert and George in repeated, often symmetrical, patterns, each of the fourteen-foot-high panels is a kind of surreal-istic stained glass window.[15]

Ginkgo has made its deepest cultural mark in China, Japan, and Korea, where the most ancient of all ginkgo trees are beloved. Two large ginkgos stand on either side of the Temple of the Reclining Buddha in Beijing, in what is now the Beijing Botanic Garden. There are ginkgos at the Confucius Temple in QuFu: just as the Buddha sat beneath the Bodhi tree, legend has it that Confucius spent time reading, reflecting, and teaching beneath a ginkgo tree. A massive old ginkgo in Shanxi Province is said to have been planted by Lao Tzu, according to legend the founder of Daoism.[16]

In Korea many of the largest and oldest ginkgos are associated with Buddhist temples, but there are two large ginkgos at the Munmyo Confucian Shrine on the campus of Sung Kyun Kwan University in Seoul. My son and I found them there one hot July day a few years ago. They were in good health, planted symmetrically in the oldest courtyard of the university. They had stood there for hundreds of years, through good times and bad.

In Japan there are large ginkgos at some of the country's most important Buddhist temples, including the Tamba Kokubunji Temple in Kyoto Prefecture, the Jonichiji Temple in Toyama Prefecture, the Hida-Kokubunji in Gifu Prefecture, and the Zenpukuji Temple in Tokyo. There are also ginkgos at some of the most important Shinto shrines: the Ubagami Shrine in Miyagi Prefecture, the Katsushika Hachiman Shrine in Chiba Prefecture, and many others, including the controversial Yasukuni Shrine in Tokyo.[17]

In the West, ginkgo is widely used as a commemorative tree. There are several, with their memorial plaques, in the park close to my home. In Hoboken, New Jersey, just across the Hudson from where the twin towers of the World Trade Center once stood, a grove of ginkgo trees provides a living memorial to the victims of the 9/11 terror attack. In Detroit there is a ginkgo planted by Yoko Ono, and in Caen another planted by the Dalai Lama. At Kew in 1916, when Sir William Chamber's eighteenth-century Temple of the Sun was crushed by a massive cedar of Lebanon that came down in a storm, Queen Mary marked the site by planting a ginkgo just a few feet from the Old Lion placed there about 150 years earlier, in the time of her husband's great-great-great-grandmother. In Morgantown, Indiana, there is a single large ginkgo planted around 1880, soon after the end of the American Civil War. It commemorates Union prisoners who survived the notorious Andersonville Prison of the Confederacy.[18]

Ginkgo also turns up in celebrated locations around the world. There are ginkgos on the grounds of the White House in Washington, D.C., and the Imperial Palace in Tokyo, as well as in Tiananmen Square in Beijing and at the Alamo in San Antonio. In Ottawa, on the grounds of the residence of the governor general, a ginkgo commemorates the 1985 visit of Chinese President Li Xiannian. Most famous is the ginkgo at Hiroshima that survived the blast of the first atomic bomb on August 6, 1945. It became a symbol of endurance in the midst of great destruction and human suffering. On the other side of the world, there are ginkgos in Missouri near the home of President Harry Truman, who ordered the bomb to be dropped, and at the University of Chicago there are ginkgos on Ellis Avenue, a stone's throw from where Enrico Fermi and his team ushered in the nuclear age.[19]

Ginkgo has also become an icon of singularity. Joy Morton, who created the Morton Salt Company and founded the Morton Arboretum, just west of Chicago, is reputed to have said: "The Morton Arboretum is a ginkgo and a ginkgo it shall remain." His vision was of an organization that was unique and memorable. The Morton Arboretum has a Ginkgo Way, and its Ginkgo Restaurant is decorated with ginkgo-inspired furniture. More than seventy ginkgo trees from forty different sources are planted in its beautiful wooded landscape.[20]

Joy Morton was right: the inverted deltoid shape and distinctive fan of fine veins distinguish the leaf of ginkgo from those of every other plant. Engelbert Kaempfer, the first westerner to take note of the tree during his time in Japan at the end of the seventeenth century, saw similarities with leaflets of the maidenhair fern. From Kaempfer's comparison comes one of the English common names of ginkgo, the maidenhair tree. But while the leaves of the maidenhair fern do have a hint of the ginkgoesque, there is nothing with which the leaf of a real ginkgo is easily confused.

The Swedish naturalist Linnaeus, who sought to name and catalogue all the world's plants in the eighteenth century, gave ginkgo its formal scientific name. He took it from the name that Kaempfer transliterated from the Japanese. Despite the awkward combination of consonants, Linnaeus was evidently content; he simply added the epithet *biloba*, referring to a characteristic feature of some ginkgo leaves: a blade that is notched along its leading edge, or sometimes deeply divided into two.

Only a few decades later Goethe took Linnaeus's designation, *Ginkgo biloba*, for his famous poem in the West-East Divan. In the middle stanza he asks:

Is it one living being?
That divides itself into itself
Are there two who have chosen each other,
So that they are known as one?

Goethe turned to ginkgo to express his feelings for Marianne Willemer, his muse and the young wife of a close friend, but he also knew that his question had deeper significance. He struggled not only with his attraction to Marianne but also to find meaning behind the structure of plants. The scientific study of plant form begins with Goethe. He was the first to use the term *morphology,* as the scientific study of biological form is now called.[21]

Modern science provides no clear answer to Goethe's question about the ginkgo leaf, but his broader point has been taken up. The overwhelming variety of plant life, of which ginkgo is a key part, still challenges us to search for a fundamental organizational theme, a *Bauplan,* through which it all can be linked. Goethe once said: "From top to bottom a plant is all leaf." How leaves arose, and how they have been modified over thousands of millennia, is the key to much that we would like to understand about the evolution of plants.[22]

The Living Tree

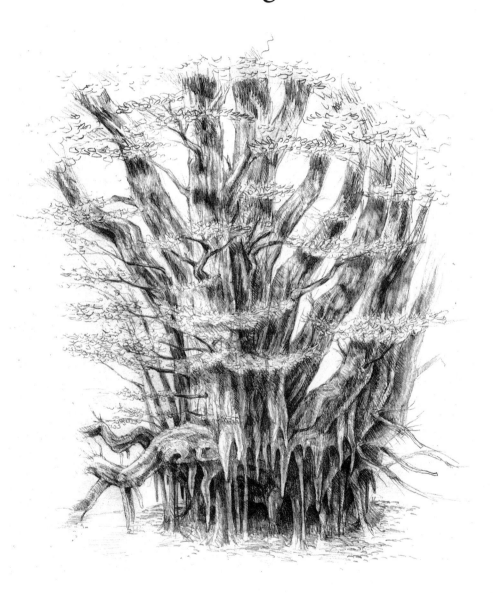

4
Energy

All growth depends upon activity. There is no development physically
or intellectually without effort, and effort means work.
—Calvin Coolidge, "Have Faith in Massachusetts"

The elegance of a ginkgo leaf begins with its stalk; it is long, sometimes a little longer than seems quite right for the length of the blade, but the two flow easily together. As the blade expands, the two nerves, which pass concealed through the leaf stalk, emerge as fine veins: each supplies one half of the leaf. The veins divide and diverge relentlessly within the blade; only rarely is the pattern disturbed by veins that rejoin.[1]

Leaves are of vital importance to a living ginkgo tree; they provide energy independence. Leaves are clean-energy factories; natural, mass-produced solar panels, each packed with sophisticated biochemical machinery capable of transforming the sun's energy into chemical energy that plants and animals can use. This miracle of natural alchemy, the process of photosynthesis, has been probed by some of the world's greatest scientific minds. We understand a great deal about how it works, down to the level of molecules and atoms, but much still remains elusive. My Yale colleague Gary Brudvig investigates the subatomic transfers of energy that lie at the heart of photosynthesis. Yet this process, first accomplished by simple organisms more than two billion years

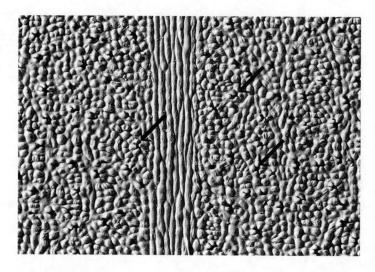

Scattered stomata, the breathing pores on the underside of a ginkgo leaf, seen from the outside, showing the projections from nearby cells that arch over the entrance. Carbon dioxide enters the leaf through the stomata, and the density of stomata on fossil ginkgo leaves has been used to estimate the amount of carbon dioxide in the Earth's atmosphere in the distant past: more stomata are needed to sustain the same supply of carbon dioxide when the concentration of carbon dioxide in the atmosphere is low.

ago, is one we still struggle to re-create in even the world's most advanced chemistry laboratories.[2]

Leaves are where the chemical building blocks that make and sustain a ginkgo tree are created, and they are assembled from the simplest of materials. Water comes from the soil, drawn up the trunk from the roots, and brings with it key nutrients, especially nitrogen and phosphorous. It is distributed across each fan-shaped leaf by the diverging and dichotomizing veins. Carbon dioxide comes from the air, taken into the leaf through breathing pores, the stomata, adjustable valves that are concentrated on the lower leaf surface. The carbon that it carries is combined with hydrogen from water to make sugars. These sugars, and the energy they contain, can be converted into many different kinds of carbohydrates and can combine and interact with other molecules to make proteins and the great variety of chemicals needed for life. Oxygen, essential for us, is a mere by-product.[3]

This basic function of leaves is fundamental to all plants. The ability to harvest light and to use it to create chemical energy is the ultimate accomplishment in green chemistry, and it could scarcely be more important. Ancient photosynthesis created the reserves of coal and oil on which much of modern society is based. Today photosynthesis builds forests and grasslands. Notwithstanding the other sources of energy available to us—nuclear energy, wind energy, geothermal energy, and others—photosynthesis remains the energetic foundation on which most ecosystems, all of agriculture, and almost all of civilization rest. As ancient people recognized more clearly than we do today, we all depend on the sun; photosynthesis is what sustains us and most other living organisms on our planet.[4]

Over the short term most of the carbon-containing molecules produced by photosynthesis are transformed back into carbon dioxide and water as they react with oxygen. This happens in our own bodies, and the bodies of most animals and plants, as energy in carbohydrates and other compounds is used for a myriad of life processes. The results of photosynthesis are also reversed over geologic time as temporarily buried carbon-containing molecules, produced by once living organisms, come to the surface and react with oxygen in the air. However, if those carbon compounds are separated from contact with oxygen, literally taken out of the equation—as, for example, when they are buried and deeply locked away in oceans and swamps—then the oxygen left behind builds up.

Two billion years ago, oxygen produced by ancient photosynthesizing bacteria accumulated in this way, to such a point that it changed the fundamental conditions for life. Poisonous to the earliest microbes, oxygen now became essential and channeled evolution in new directions. It created opportunities for new kinds of organisms, and it allowed new kinds of biological processes to emerge. The long-term impact of photosynthesis on our world has been vast. It was one of the key contingent steps in the history of life to which the origin of ginkgo, and eventually the origin of our own species, can be traced.[5]

The magic of photosynthesis resides within chlorophyll, the molecule that makes plants green. Chlorophyll molecules are packaged together inside tiny green light-gathering discs, chloroplasts, that are vital to the internal workings of most of the millions of plant cells that make up a ginkgo leaf. Each cell may contain up to a hundred chloroplasts; there are hundreds of millions in a single leaf. Inside each chloroplast, chlorophyll molecules and their complex associated biochemistry are embedded in membranes less than one–five millionth of an inch thick.[6]

The clever trick of the chlorophyll molecule, with help from other light-gathering pigments, is to release an electron for every photon of light it absorbs. That electron is passed through complex intermediaries in less than a thousandth of a second. It provides the energy that drives the slower process by which carbon dioxide is captured from the air, combined with hydrogen, and transformed into simple sugars.[7]

Water is a basic raw material for photosynthesis; the chemical breaking apart of water molecules provides the hydrogen and replenishes the electrons lost by chlorophyll molecules in the light-harvesting step. Water is also indispensable to many other essential life processes. It was readily available to the earliest plants that lived in the sea, lakes, ponds, or rivers, but as evolution created opportunities to move on to land, it also presented a fundamental challenge: the reality of water in short supply. In large part, the success of plants on land, and everything that has followed from it, including ourselves, has been a triumph of biological water management.

A fresh ginkgo leaf is stiff, full of water; the leaf stalk is wiry. However, plucked from the tree, it soon changes character. The leaf blade goes limp. What was once crisp and firm becomes soft, almost membranous; it is the pressure of water in countless individual cells that keeps the leaf stiff. When water is lost and not replenished, the leaf wilts. For plants, as for animals, a shortage of water can be fatal; to fulfill their energy-generating and other functions, leaves need to stay hydrated.

In the leaves of most trees big veins supply smaller ones in a hierarchy, and the veins are connected into a network as they rejoin. Such reticulation builds in redundancy, a safeguard against damage, and differences in vein sizes allow a single large vein to supply many smaller ones. However, ginkgo does it differently; the architecture of its water supply is unique among the world's trees. The hierarchy of veins is much less pronounced, and all can be traced back to the two slender veins that enter the leaf base from the leaf stalk. Only one leaf in ten shows any fusion among the veins, and such reticulations are always sparse. It is an idiosyncratic system, a water supply more reminiscent of drip irrigation than a hosepipe; the water leaks from the elongated cells that make up the veins into the tissues between.[8]

Keeping the water content of leaves just right is a challenge at the best of times; it is still more difficult when the soil dries out and water is scarce. Water evaporates from the leaves of plants and is also lost through the breathing pores. When water is plentiful and temperatures are warm, this is a good thing; it is the botanical equivalent of sweating, one way by which the leaves stay cool. Water loss from the leaves also draws water up the trunk from the roots and distributes it to every branch, every branchlet, every twig, and every leaf in the crown. But keeping the whole process in balance is crucial; if too much water is lost and not replenished, then the leaves suffer, and soon the tree will be in trouble.

In most plants that live on land, the difficulty of staying hydrated is compounded by the fact that when the breathing pores open to take in carbon dioxide, they also lose water. Driven by this awkward physiological reality, the upper leaf surface in ginkgo, and most trees, has a well-developed waterproof covering, the cuticle, which reduces water loss. Breathing pores are present only on the lower leaf surface, away from the direct heat and glare of the sun, and they close when water is in short supply.

In mature ginkgo leaves the upper surface is smooth and dark green; the cuticle is tough, colorless, and almost impermeable. The epidermis, the outermost layer of cells beneath the upper leaf cuticle, also lacks color, and sunlight streams straight through to the chlorophyll-rich cells below. Of all the cells in a living ginkgo leaf, these are the true energy factories: the cells supplied most generously with chloroplasts.

The lower leaf surface is different; its gray-green hue comes from a fine layer of wax over a much thinner waterproof covering. The covering itself is perforated at the breathing pores, each of which connects to a well-developed system of air spaces inside the leaf. This makes it possible for carbon dioxide and water vapor to diffuse rapidly among the cells that make up the leaf. These are the messy but critical conduits through which carbon dioxide is supplied for photosynthesis.

Mature ginkgo leaves are tough and resistant to decay, more so than the leaves of many other trees. The cuticle around the outside, together with strands of mucilage-like resin inside the leaf between the veins, makes them slow to decompose. Ginkgo leaves are among the last to succumb to the compost heap; their durability is one reason why they make such good fossils. Across Japan, the more than half a million ginkgos planted as street trees produce an overabundance of leaves in the autumn. In Tokyo, ginkgos make up about 12 percent of all street trees, and their leaves break down so slowly that for a time a research team for the city was working on how best to turn them into useful compost.[9]

As far as I know, no one has ever counted how many leaves there are on a large ginkgo tree, but even on a small tree it is a very big number. During his Ph.D. studies at Yale, Kirk Johnson, now director of the U.S. National Museum of Natural History, cut down a fifty-foot-tall red maple to count the leaves. He was interested in the total number of leaves produced in a typical patch of forest. Kirk's red maple was tall but slender, with a trunk just eleven inches in diameter at chest height. It was not what most people would call a very big tree, but it still took him and a friend eight hours to count all the leaves. The final total was 99,284. A much bigger ginkgo, perhaps twice as tall and more profusely branched, might have 300,000 to 500,000 leaves. An old and truly massive ginkgo, like some of those in China, Japan, or Korea, might bear close to a million.

Every leaf on a ginkgo tree is a triumph of modular design built through the unfailing translation of complicated instructions that are written in code in strands of ginkgo DNA. We have come a long way in learning how to read the language of DNA, but how it is translated to make a leaf through the carefully choreographed dance of a multitude of molecules still eludes us. Nevertheless, each leaf is built in a matter of months, according to complex specifications and with supremely high standards of quality control. Just as astonishing, after all that effort, is that every leaf is shed as winter approaches. The creation and disposal of hundreds of thousands of leaves every year is a testament to the power of photosynthesis, and to the productivity of a large tree.

In the spring, last year's hard-won energy is invested to grow and expand the new leaves. Through the summer each leaf will yield a good return. When light is plentiful, temperatures are warm, and water is in good supply, leaves produce much more energy than they consume. However, in the dark and cold of winter, when their delicate biochemical machinery may be damaged, when water may be hard to coax from frozen soil, and when energy is needed to keep them alive, leaves become a liability. In ginkgo and many other trees they are quickly and thoroughly discarded. This is the economics of growth in a deciduous tree. The energy and nutrients used to build the leaves are partly reclaimed, and the rest is written off.[10]

Every autumn, just at the moment before the leaves fall, ginkgo is at its most beautiful. By then the leaves are bright yellow; in the low-angled autumn sun they are "brilliant as a brimstone butterfly." Often the yellowing begins at the leaf edge and slowly expands across the whole blade. It reflects complicated, tightly controlled changes going on inside each leaf, all apparently triggered by the shorter days rather than the onset of cold weather. In the process, chloroplasts degenerate, chlorophyll is broken down, and photosynthesis slows. Valuable nutrients, especially nitrogen and phosphorous, are absorbed back into the tree; they are too precious to waste. And as the green

*In mid-November, bright yellow leaves cover the lawn beneath a ginkgo tree in the garden
of the President's House at Yale University in New Haven, Connecticut.*

of chlorophyll disappears, the yellow of other light-absorbing pigments, the carotenoids, come to the fore. Leaves of green turn to leaves of gold.[11]

For these few short weeks ginkgo is worth its place in any garden. A single ginkgo can be a dramatic horticultural focal point. In early November, around the time of the annual Yale-Harvard football game, when the tree peepers are enjoying the last of New England's autumn color, the single ginkgo that stands proudly in the garden of the President's House at Yale is at its best. Regrettably, the ginkgos at Harvard's Arnold Arboretum are still more spectacular.

A massed planting of ginkgos is enough to bring out the sightseers. In the heart of Tokyo the double allée of more than 140 large ginkgo trees that line the view to the Meiji Memorial Gallery in the Meiji-Jingu Park is a popular destination. When these trees were planted in the 1920s, they were about twenty feet tall; the largest now approach a hundred feet. They are meticulously maintained as one of the finest examples of Western-style landscape architecture in Japan. In the autumn they create a spectacle. Sightseers come to admire the view, eat roasted ginkgo nuts, and take tea as the late-afternoon light fades away and winter approaches.[12]

At the Meiji-Jingu Park, as everywhere else in the world, when the time comes for ginkgo leaves to be shed they go suddenly, almost all at once, seemingly without provocation. In Monroe, Michigan, there is a large ginkgo outside the Dorsch Memorial Library, planted from a seed reputedly given to Dr. Dorsch by the Chinese ambassador at Lincoln's inauguration. For many years there was a competition to guess the date on which the leaves would fall.[13]

Ginkgo has the most synchronized leaf drop of any tree I know. In a general way we understand how trees shed their leaves. It happens because of changes in a layer of cells right at the point where the stalk attaches to the branch. Ultimately, these cells die, the walls between them separate, and the leaf falls, but exactly how this happens and why, in ginkgo, it happens with eerie synchronicity, no one knows. The former U.S. poet laureate Howard Nemerov also wondered.[14]

Late in November, on a single night
Not even near to freezing, the ginkgo trees
That stand along the walk drop all their leaves
In one consent, and neither to rain nor to wind
But as though to time alone: the golden and the green
Leaves litter the lawn today, that yesterday
Had spread aloft their fluttering fans of light

What signal from the stars? What senses took it in?
What in those wooden motives so decided
To strike their leaves, to down their leaves,
Rebellion or surrender? and if this
Can happen thus, what race shall be exempt?
What use to learn the lessons taught by time,
If a star at any time may tell us: Now.

5
Growth

The wood will renew the foliage it sheds.

—Irish proverb

Matthew Hargraves, a curator at the Center for British Art at Yale, points out that in the late eighteenth to mid-nineteenth century, "the tree nearly supplanted the human figure as the best test of an artist's mettle." Any artist looking closely at ginkgo would quickly have seized upon its unusual form. Even when the leaves are gone, ginkgo is distinctive; for decades at the beginning of their long lives, ginkgo trees have a sparse crown, with widely spaced branches that poke out like long, thin, spiky fingers. Their characteristic silhouette differs from that of any other tree. Unlike the leaves, the architecture of a ginkgo tree is lacking in grace; it is no American elm.[1]

Branches of a mature ginkgo are also distinctive in another way. Twigs that are more than a few years old have sideways-pointing, spurlike, cylindrical pegs: so-called short shoots, each of which grows by only a fraction of an inch every year. In contrast, the long shoots on which they are borne may be very long. Each grows rapidly from a bud at its tip. Growth from a single bud in a single season may produce a foot or more of new twig. Similar differentiation into long and short shoots occurs in many other trees, such as apples. It reflects a division of labor between twigs with normal growth and widely spaced leaves, through which the tree gets bigger, and twigs that are repressed

Young leaves and pollen cones emerge from short shoots scattered along
the branches of an ancient ginkgo, Kyushu, Japan.

and bear their leaves in a tuft. In ginkgo the difference is especially pronounced. The long shoots grow straight and are few in number compared with the short shoots. The spikiness of the ginkgo habit in part reflects this special way of growing.

Short shoots add further character to the ginkgo silhouette. Each can be up to an inch or more long and bears the crowded scars left by the leaves of previous summers. On older branches a good-sized short shoot may bear the scars of a hundred or more leaves borne in a life that may last decades. Botanists in Europe and North America sometimes pride themselves on being able to identify trees in winter in the absence of leaves. It is a test of observation, deduction, and memory, which requires looking carefully at the bark, the scars of leaves that have been shed, and the pattern of branching. Only the most inexperienced student would have a problem in recognizing ginkgo. The characteristic cylindrical short shoots are unique.[2]

In spring, twigs on ginkgo trees that have been barren since autumn come back to life. Clusters of leaves, usually about four to six, emerge in miniature from the bud at the tip of each short shoot. They were formed in the previous growing season and survived the winter tiny and tightly packed inside their protective buds; each has a short stalk with a minute leaf blade that is rolled inward on either side. Their precocious development means that the leaves can be deployed quickly in the spring. With the coming warmth the bud scales fold back, the leaves are revealed, and the blades unroll. As the leaf stalk grows in length, the blade expands rapidly.

At bud burst, the leaves are small, lime green, delicate, and vulnerable to a late freeze. In April 2007, Chicago and the rest of the Midwest suffered a late cold snap as the wind chill dipped to minus 30° Fahrenheit soon after the leaves had emerged: most did not survive. Eventually, a few weeks later, the trees summoned the energy to replace them, but successive years of such natural abuse would quickly exhaust them.[3]

On a well-established ginkgo most of the leaves are borne on short shoots, and these are the leaves that bring in most of the tree's energy. Short shoots are a neat functional trick in the economics of tree growth. Each supports a cluster of leaves that yields a good return without the energetic expense of building a long branch on which to bear them. With their multiple leaves, short shoots produce a significant amount of energy with only minimal long-term investment in wood and other tissues. They also help make the crown open and light, rather than dense and compact. In summer, each long spiky branch of a young ginkgo is a feathery cylinder of leaf clusters: a single main long shoot bearing hundreds of short shoots and thousands of leaves.

Short shoots bring in energy but do little to increase the tree's stature or prevent it from being overgrown by its neighbors. It is the long shoots that are responsible for most of the physical growth of the tree. They elongate rapidly during the growing season and bear leaves alternately, one by one, along their length. In the spring, as the buds are breaking, new short shoots may start from each of the small buds left in the axils of last year's long-shoot leaves. In this way, long shoots extend the length of branches and expand the size of the crown, while short shoots fill in the crown with leaves.

The whole system is also beautifully flexible, because the difference between the long and short shoots is under simple control. The short shoots are held in check by chemical signals passing back from the dominant thrusting bud at the tip of the long shoot. With strict discipline controlled by simple chemistry, ginkgo short shoots defer to their leader. However, if signals from the tip of a long shoot are interrupted—for example, if the bud is damaged or lost—one or more short shoots soon break ranks, start to elongate, and quickly become new leaders.[4]

The leaves on long and short shoots are slightly different. On short shoots they have fan-shaped blades that are rarely deeply divided. The upper edge is usually more or less smooth, or at the most only shallowly notched. The same is true of the first few leaves that burst from the tip of a long shoot at the beginning of the season. However, leaves formed on long shoots later in the growing season have a deep central notch that may be up to two-thirds of the leaf blade. These are the leaves that served Goethe's poetic purpose and that Linnaeus had in mind when he coined the epithet *biloba*.[5]

The difference between leaves on long and short shoots is beautifully seen in a nineteenth-century illustration of ginkgo from Siebold and Zuccarini's *Flora Japonica*. The difference probably reflects the different conditions the leaves encounter as they develop. Leaves on short shoots, and the first-formed leaves on the long shoots, both begin their development inside tightly closed buds. They are well protected, develop slowly, and survive in near suspended animation over the winter. The long-shoot leaves are formed later, and they begin to develop only after growth has started in the spring; they develop and grow more quickly inside the bud at the tip of a long shoot that is itself elongating rapidly.

These same factors may also be responsible for the internal differences between the leaves on long and short shoots, and their different capacities to conduct water. Short-shoot leaves develop attached to a stubby branch that is already mature, is not growing rapidly, and that has well-developed tissues capable of providing an ample supply of water. Long-shoot leaves, on the other hand, are borne on shoots that are immature, are growing very actively, and have a greater need for water even as their internal water-conducting tissues are still being formed. Careful measurements show that long-shoot leaves, perhaps because they are more likely to be short of water as they grow, are more effective in water conduction.[6]

The need to keep the leaves fully supplied with water may be the crucial limitation that prevents ginkgo from growing really tall. All the water needed to keep the leaves fully hydrated comes from the soil, scavenged molecule by molecule by fine hairs on young roots. It is then moved to the bigger roots and upward through the outer parts of the woody column inside the trunk, out through the branches, and eventually to the leaves. The supply needs to be steady and consistent to replace the water lost through

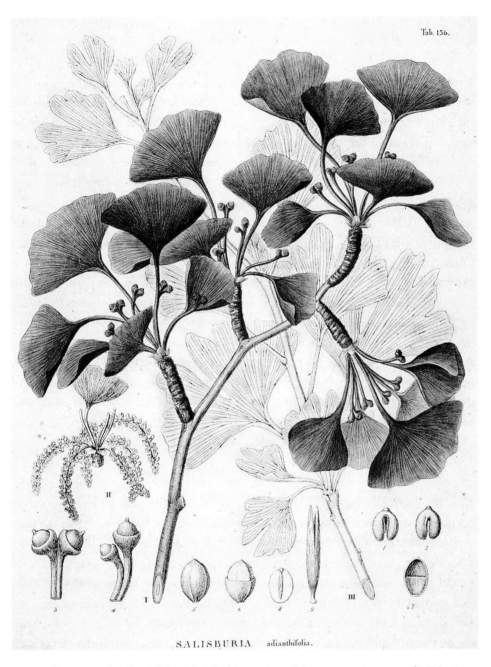

Tab. 156.

SALISBURIA   adianthifolia.

The illustration of Ginkgo biloba, *labeled with a variant of the once-popular name of* Salisburia adiantifolia, *from Philipp Franz von Siebold and Joseph Gerhard Zuccarini's* Flora Japonica.

the breathing pores, as well as smaller amounts lost by evaporation through the cuticle and used in photosynthesis.[7]

What is remarkable is that the entire process of water transport in plants is passive: there is no hidden microscopic pump and no expenditure of energy on the part of the tree. The tiny elongated cells through which the water passes are dead and empty, and except in the early spring, when stored sugars in the trunk and roots begin to be mobilized, and the sap starts to rise, there is little sustained pressure from below. In principle, the arrangement could hardly be more simple: as water is lost or used up in the leaves, it is replaced by water drawn up through the stem all the way from the roots.[8]

The amount of water needed to keep a big tree fully hydrated varies greatly, depending on the kind of tree and where it is growing, but it can be substantial. A thirty-year-old sessile oak growing in a plantation in eastern France might use only about a half-gallon of water a day, while the yevaro, a large tree that dominates certain kinds of tropical forests in the Venezuelan Amazon, might use five hundred times as much.[9]

All of this water is literally sucked up from the roots; the taller the tree, the greater the suction needed, and the greater the internal pressures that are generated. The negative pressure needed to lift the water against the force exerted by gravity increases by almost one and a half pounds per square inch for every three feet or so of height. The resulting tension in the water-filled cells of the wood of a tall tree is enormous. The elongated cells through which water is drawn have to withstand massive pressures; they have robust cell walls and specialized internal thickenings that help prevent them from imploding.

A further hazard created by such pressures is that air inside nearby empty cells may be sucked into the water column. The result is a kind of botanical embolism; a bubble of air that creates a blockage and prevents further conduction. In ginkgo, this danger is minimized by the minute pores that connect adjacent cells; they are just the right size to allow water through, but too small to let an air bubble pass. In this, as in many other details, every tree is an extraordinary feat of hydraulic engineering: all created by natural selection from natural variation over millions of generations.[10]

The wood that makes up the trunk of ginkgo and other trees not only conducts water but also provides support. It creates the scaffold on which the leaves are arranged, lifting them up toward the sun and away from the shade created by nearby plants. This competitive struggle for light was probably the ultimate driver in the evolution of trees. However, getting bigger has consequences. As trees become larger, water and other fluids become harder to move around: the distances involved, and the internal pressures on the water-conducting cells, increase. Bigger trees are also heavier, especially when laden with water after a downpour: the trunk has to be strong enough to carry the load.

There are also considerations of internal economy. Over the life of a single tree the more energy diverted to building the trunk and branches, the less energy is available for reproduction or other kinds of growth. The energetic cost of simple maintenance, sustaining basic life processes, is also considerable. It makes sense that truly massive trees live in places where growth never completely shuts down in the winter, and where water is readily available almost all of the time. For most trees, maintaining a gigantic branching scaffold and hundreds of thousands of leaves through tough times, when temperatures are low, the soil water is frozen, and no energy is coming in, may not be a winning strategy.

So like most things in nature, the height of trees is a trade-off, dictated in part by the ability of a big tree to conduct water, in part by the energetic costs and benefits of achieving and maintaining large size, and in part by the physical strength of the wood and how that relates to the form of the tree. In the case of ginkgo all of these factors may be at play, but with its broad leaves and their unusual architecture, maintaining a steady supply of water may be especially important. Ginkgo is perhaps not quite as efficient in holding onto water and supplying it to the leaves as some of the world's most imposing plant giants. Part of the explanation for the great size of the Yongmunsa Ginkgo, and other truly massive ginkgos, may be that their roots are tapped into a plentiful water supply.

6
Stature

... as carpenters carve wood, the wise shape their minds ...

—The Buddha, *The Dhammapada*

Puccini's opera *Madama Butterfly,* set in Japan in 1904, begins in the garden terrace of a small house overlooking Nagasaki Harbor. The tragic story of Cho-Cho San, which Puccini adapted from the novel by John Luther Long, was just one expression of increased fascination in the West with all things Japanese in the decades after Commodore Perry's Black Ships forced the reclusive nation into contact with the outside world. Much has been written about the possible models for Long's story: connections have been made to the book *Madame Chrysanthemum* by Pierre Loti, a French naval officer who was in Nagasaki during the summer of 1885, and to also to Thomas Glover, a Scottish merchant and key figure in the early industrialization of Japan who lived in Nagasaki in the late nineteenth century. Another connection is to Philipp Franz von Siebold, who was in Nagasaki in the 1820s and 1850s. Siebold was no Lieutenant Pinkerton, but some parts of his story will sound familiar to those who know Puccini's opera.[1]

From the mid-sixteenth century until the time of Commodore Perry, Nagasaki, with its fine natural harbor, was the only port that linked Japan to the West. Trade was controlled through the monopoly held by the Dutch East India Company. Company ships brought western luxuries from Holland to Japan and returned laden with fine ceramics and other goods that helped ignite Europe's craze for *japonism*. The elite of Amsterdam, London, or Paris could order customized sets of Japanese china made in the kilns of Kyushu. It was into this milieu that Siebold came when he arrived in Nagasaki in 1823 as the doctor to this far-flung outpost of the Dutch trading network. Following in the footsteps of Engelbert Kaempfer and Carl Peter Thunberg, two of his predecessors who were also employed by the Dutch East India Company, he became one of the three great early botanical explorers of Japan and a pioneer of Japanese studies in Europe.

Siebold studied medicine at the University of Würzburg, was a member of the Senckenberg Natural History Society in Frankfurt, and was much influenced by the adventures and writings of Alexander Von Humboldt. His dream was to emulate Humboldt's explorations in far-off countries, and when he was sent to Japan as surgeon major, Siebold seized the opportunity. He used his time there to gather detailed information about the country, its plants and animals, its culture, and the people that he met.

During the seven years that Siebold spent in Nagasaki in the 1820s, Japan was still almost entirely closed to the Western world. The Dutch were confined to the island of Deshima in Nagasaki Harbor, the import and export of goods was strictly monitored, and there was little contact with the local population. However, as a doctor, and particularly as an ophthalmologist and obstetrician with expertise in Western medicine, Siebold had unusual freedom to leave the island to treat Japanese patients and collect herbs. He also recruited students, who helped him purchase land and establish the first school in Japan to teach Western medicine.[2]

During house visits in Nagasaki, Siebold met and fell in love with a local woman, Kusumoto Taki, who was also given the name Sonogi. Siebold was twenty-seven and Sonogi sixteen when they met. He wrote to his uncle that he had "become quite attached to a sweet sixteen-year old Japanese girl, who I would not willingly exchange for a European one." Sonogi was one of the few Japanese allowed to stay on Deshima. Siebold called her Otaksa, and they had a daughter together, whom they named Oine.[3]

Siebold would perhaps have returned to Japan repeatedly except for an incident that took place in 1826 during the annual visit of the Dutch legation to Edo, modern Tokyo. Each year, with great ceremony, representatives of the Dutch traders traveled to Edo to offer gifts in tribute to the Shogun. Kaempfer and Thunberg had made the same journey and, like them, Siebold valued this rare opportunity to see other parts of the country and learn more about all aspects of Japanese life. In 1826, as on his earlier journeys

Statue of Philipp Franz von Siebold (1796–1866) as a young man, Siebold Museum, Nagasaki, Japan.

to Edo, Siebold collected many items that helped illuminate Japanese culture, but especially prized were several detailed maps of Japan and Korea given to him by his friend Takahashi Sakuzaemon, the court astronomer and supervisor of the Imperial Library. In return Sakuzaemon received a recently issued Russian map of the world. Siebold's motivation was simple curiosity, but for Japan, with its concern about outside influences, such maps were highly sensitive: possessing one was forbidden, and supplying one to a foreigner was a capital offense.[4]

Siebold must have understood the risk he was taking, but Kaempfer and Thunberg had returned with similar documents. According to some accounts, Siebold was already on a boat leaving Nagasaki Harbor when his good fortune ran out. The maps would probably have gone undetected had it not been for bad weather that brought his ship back to port. His belongings were unloaded and inspected, and the maps were discovered, along with another illegal item, a gown bearing the Shogun's crest. When this came to the attention of the Japanese authorities, Siebold was interrogated, as were about forty Japanese who were in contact with him during the court journey, along with fifty interpreters and many of his students. Takahashi Sakuzaemon fared the worst. He was arrested and died in prison soon after.[5]

The verdict on Siebold was reached more than a year later. The court found no evidence that he was a spy, and also took into account his service to Japan, as well as pleas from his employers in Europe, but the sentence handed down in October 1829 was nonetheless severe. Some of his Japanese colleagues were imprisoned or sent to remote islands. Siebold was banished from Japan for life and left Deshima on December 30, 1829. He said his final goodbyes to Sonogi, Oine, and two of his devoted students, Kō Ryōsai and Ninomiya Keisaku, on his ship, the *Cornelius Houtman*, as it left Nagasaki Harbor. He took with him two small lacquered boxes with portraits of Sonogi and Oine that contained locks of their hair. Oine was two years and eight months old when her father left; Siebold entrusted responsibility for her well-being and education to Kō Ryōsai and Ninomiya Keisaku.[6]

When Siebold left Nagasaki, Japan was still an isolated country, but more than to any other Westerner, history gave Siebold the opportunity to view at close quarters the massive changes wrought as the country opened its doors to the world. Two years after the arrival of Commodore Perry, Siebold returned to Japan in 1855 with the responsibility of delivering the Netherlands Trading Company's trade agreement for ratification. There was also an emotional reunion with Sonogi and Oine. Siebold had since married and had five children, Sonogi had married twice, and Oine was on her way to becoming Japan's first female doctor.[7]

Siebold accumulated a vast collection of specimens, books, and artifacts in Japan, some on his trips to Edo and others in return for medical services. He also preserved large collections of plants and animals. On his return to Europe he used these to document Japanese natural history and culture in a trilogy of important early works on different aspects of Japan. *Flora Japonica*, Siebold's work on Japanese plants, illustrated with lavish plates, remains an influential early work on the plants of Japan.[8]

The copy of *Flora Japonica* in the Kew Library was acquired by William Jackson Hooker, and after his death it was purchased for the nation, along with the rest of Hooker's library. As well as *Flora Japonica*, the books in Hooker's collection included superb art by Ehret, the Bauer brothers, and many others. The books became the nucleus of one of the greatest collections of botanical art anywhere. However, there is also interesting botanical art in the Kew Economic Botany Collection, and particularly unusual are a series of twenty-six boards, each with an illustration of a plant painted on its surface. Like *Flora Japonica,* these also have a link to Siebold.[9]

Each board is a little more than a foot long and nine inches wide, held in a frame made of pieces of a small branch. The frames, as well as the boards on which the paintings are made, seem to be made from the plants that are illustrated. One is labeled "*Ginkgo biloba,* Linn.," with the name of the plant written in Chinese Kanji and Japanese Katakana characters below; both read in the old-fashioned way, from right to left.

The records at Kew tell us nothing about when and where this strange xylotheque was made, or how it came to London, but similar collections of boards at the Koishikawa Botanical Garden in Tokyo, the Berlin-Dahlem Botanic Garden, and the Harvard University Herbaria, as well as a private collection in Britain, provide some clues. On the back surface many bear the same red seal of Chikusai Kato, the first plant illustrator employed in the Koishikawa Botanic Garden, which is now part of the University of Tokyo. The seals incorporate the line "Spring, New creative achievement," with the date "11 Meiji," 1878 in the Western calendar.[10]

Both the Kew and Berlin collections contain a board illustrating ginkgo. The Chinese characters on the Berlin board read: "Grandfather Grandson tree, more formally, Silver Apricot." The two illustrations of ginkgo twigs, with leaves and young seed stalks attached, are very similar and also closely resemble a painting of ginkgo produced by Chikusai Kato for the Koishikawa Garden. Parts of the twigs and short shoots in all three illustrations are almost identical, and the way in which the mature seeds on their seed stalks are drawn is exactly the same. Only a few details on the Kew board are not shown in the Berlin illustration, but all are in Kato's ginkgo painting.[11]

Chikusai Kato worked for Keisuke Ito, an early professor at Tokyo University Botanic Garden, who had met Siebold as a young man in 1826. Kato may also have been influenced by Keiga Kawahara, an artist who worked for Siebold during his time in Nagasaki. As a result the style of these three ginkgo portraits blends Japanese and European traditions and reflects the growing early influence of the West on Japan. The way that the twigs are represented has a clear Japanese influence but the separate illustration of botanical details is adopted from the Western scientific tradition that was almost certainly introduced by Siebold.

The ginkgo wood used to make the boards in the Kew and Berlin xylotheques has the same basic structure as that of most woody plants. Cut down a ginkgo and the

A portrait of ginkgo painted onto a board made from ginkgo wood and framed with young ginkgo branches. The portrait is by Chikusai Kato and was made for Tokyo University in 1878.

trunk looks exactly like that of a normal tree; only a specialist would be able to distinguish the wood from that of pine or cedar. Around the outside in a narrow band is the bark, and immediately inside is a zone of softer tissues; the rest of the trunk is made up of wood.

In its outer part, where water and nutrients are still being transported, the wood is light in color, but the center of the stem, the heartwood, is dark, and plays no part in water conduction. Its cells are often blocked by deposits of various kinds that build up over time. Nevertheless, the heartwood is vitally important: the strength of this dense central column is crucial for support of the tree.

Almost all of the tissues that make up the twigs, branches, or even the trunk, of a full-grown ginkgo tree are produced in the same way. Except for the bark, they can all be traced back to a delicate cylinder of living cells just a single cell thick. In most trees this cylinder, known technically as the cambium, occurs not far below the bark between the wood and the softer tissues toward the outside. As they grow, the elongated cells in this layer divide consistently and regularly at a tangent to the circular cross section of the trunk; in the process they form new cells toward both the outside and the inside.

As soon as the new cells produced by the cambium toward the inside are fully developed, they die. They become part of the central woody column and add to the mass of strengthened cells, xylem cells, that are conduits for water on its way from the roots to the leaves. Toward the outside, on the other side of the cylinder of dividing cells, the newly formed elongated cells remain alive and form the softer zone of fibrous tissue immediately inside the bark. It is these phloem cells, the functional counterpoint to the dead, water-conducting, xylem cells of the wood, that actively transport the sugars made in the leaves to the lower part of the stem and the roots.

Also produced by the actively dividing layer, but this time running horizontally rather than vertically through the wood, and scattered at different levels among the files of dead cells, are thinner-walled, elongated living cells, the so-called rays. Unlike the cells making up most of the rest of the wood, these cells are alive. They help transport sap horizontally through wood and are often the cells where starch is stored through the winter.[12]

A consequence of the way that the water-conducting cells in the trunk are formed is that they are arranged in radial files. These come about because each row of elongated cells, both toward the inside and toward the outside, traces its origin back to one of the

Close-up of the bark of a young ginkgo showing the successive layers produced by a zone of actively dividing cells near the outer surface of the trunk. In this tree, which was growing on a busy street in Seoul, South Korea, the bark has been partly smoothed by the hundreds of people who brush against it every year.

cells in the actively dividing layer. Each file preserves a record of the repeated divisions of that single cell through time. However, just occasionally one file can be seen to have split into two, and this reflects occasional radial divisions of one of the cells in the actively dividing cylinder. This is how trees increase in girth. Such divisions increase the diameter of the cylinder and allow it to expand to keep up with the increasing mass of wood in the center of the trunk or twig. Ultimately, the increased size of the cylinder results in an increased circumference of the trunk and, to the outside, the increase in girth is reflected in the fissures in the bark.

The bark is a layer of corky cells produced by a second and less-well-defined cylinder of actively dividing cells. The quantity and texture of bark produced by these cells varies greatly in different kinds of trees. Smooth bark, such as that of beech trees, forms where there is little cork and where its production is even around the stem. Rough and deeply textured bark like that of oak trees results from much more extensive production of corky cells. In ginkgo the bark is intermediate in texture, but a close

A cross section through the wood of living ginkgo, showing the large-diameter water-conducting cells produced in the spring and summer and the smaller-diameter water-conducting cells produced late in the growing season. Annual rings are the result of this rhythmic pattern of growth.

look clearly shows the rhythmic layers produced by the annual growth of an outer cylinder of actively dividing cells, the cork cambium.[13]

The individual cells that make up the wood of ginkgo are not much more than a few hundred thousandths of an inch across but may be up to half an inch long. For many years it was thought that these kinds of long narrow cells, which make up most of the trunk in ginkgo and conifers, were inefficient conductors of water; especially compared with what seemed to be more sophisticated systems in most other trees, where cells are arranged end to end and connected by holes in their end walls to form long pipes. However, this turns out not to be true; the conducting cells of ginkgo and the other conifers are equally sophisticated but employ a different system. Each is perforated with thousands of tiny valves that close when water is in short supply, but that still permit easy and efficient transport through the wood when water is plentiful and the valves are open. As often happens, over the eons evolution has invented more than one solution to the same problem.[14]

Ginkgo responds to the changing seasons on the outside by dropping its leaves. On the inside it responds by shutting down the cylinders of dividing cells over the winter. The effect is most obvious in the wood. New cells of the wood produced in the autumn have slightly smaller diameters than those produced in spring and summer; then cell division stops abruptly as the tree becomes dormant. Through the winter no new cells are produced toward either the inside or the outside. But when growth begins again in the spring, the new cells produced are of normal size. It is the difference between the small-diameter cells produced late in the growing season and the larger-diameter cells produced early in the next growing season that shows up, even to the unaided eye, as annual rings: a rhythmic record of the tree's annual growth. The rings on the young branches used to make the corners of the Kew ginkgo board show that each had grown for about eleven years.[15]

Ginkgo is not grown for timber, but its wood nevertheless has its uses. Technically, ginkgo is a soft wood like pine or spruce, but its more or less homogenous structure compared with the wood of most trees, especially hardwoods, makes it resilient. Even if it regularly gets wet and then dries out, ginkgo wood does not easily shrink, crack, or warp. As John Quin noted, it can be used as a base for lacquerware where shrinkage would cause cracking and peeling of the lacquer. In China, when the trees in a ginkgo orchard cease to be productive, they are cut down, chopped, and shredded to make particle board. In Japan, larger pieces of timber are used for chopping blocks, furniture, molds, and ornaments.[16]

The smooth, even texture of ginkgo wood, a result of its relatively simple structure, also makes it easy to work. Scott Strobel, a colleague at Yale with great skill on a wood lathe, says that ginkgo wood works more smoothly and easily than any other wood he has used. The chisel goes through it "like a knife through butter." Daoist shamans are said to have engraved their magical spells and seals on old ginkgo wood to communicate with the spirit world. The Vairocana, the Manjusuri, and the Samantabhadra statues at the Haeinsa Buddhist Temple in South Korea are also carved from ginkgo wood. Much less ornate, but no less impressive, are the remarkable sculptures housed in a small temple high above the small town of Ojiya in the mountains of central Honshu.[17]

The temple is a tiny jewel reached by a steep road that winds up through Japanese cedars and gives spectacular views to the valley below. Outside it is a modest, square, wooden building with a peaked roof in traditional Japanese style. In the winter the outside is protected by thick boards from the heavy snows. Inside it is also plain, but along the wall opposite the entrance are thirty-five sculptures, each a representation of the Buddhist Goddess of Mercy with the head centered in a stylized disk representing the sun. The largest figure, about five feet tall, has its right hand resting on its cheek. Around the head is an inscription in charcoal. To the right and left are two similar but smaller figures, and in alcoves on either side others are arranged in two groups of sixteen; each a little different from the next, and each with its own symbolic meaning.

The sculptures are all the work of Mokujiki Shonin, one of the most famous Buddhist monks of eighteenth-century Japan. Born in 1718 and graduating to the priest-hood in his early twenties, he practiced in the temple at Edo for more than twenty years before becoming attracted to a particular sect of Buddhism that eschews meat, fish, and rice for a diet of nuts, leaves, and fruits. From the age of sixty until his death in 1810 at the age of ninety-three, Mokujiki Shonin traveled from temple to temple in central Japan and carved more than one thousand wooden Buddha sculptures.[18]

He was already approaching his eighties when he first visited the temple in Ojiya, but on his return several years later the old temple was gone, burned down in an acci-dental fire. He set out to make the new temple unique. The townspeople brought him pieces of a large ginkgo tree that they had cut from farther down the valley. In a little over three weeks in August 1803, he crafted them into the remarkable figures that are the heart of the Ojiya Temple.

7
Sex

We are survival machines—robot vehicles blindly programmed

to preserve the selfish molecules known as genes.

—Richard Dawkins, *The Selfish Gene*

All living ginkgos are connected, part of an unbroken chain of genetic continuity that has survived through thousands of generations. With Darwinian hindsight we recognize now that this continuity links all living ginkgos to their ancestors that grew more than 200 million years ago, and that has been sustained, as in all living organisms, by an innate propensity for self-perpetuation through reproduction.[1]

Reproduction in plants, just like reproduction in humans, involves sex: the bringing together and fusion of reproductive cells, usually but not always from two parents, to create an embryo that develops into a new individual. Each sex cell—sperm from the male, egg from the female—carries just a single set of chromosomes, and after fusion the newly created embryo has two sets: one from the father, the other from the mother. In this respect ginkgo is not remarkable. But beyond these bare essentials, sex in ginkgo is a long and complex process.

In ginkgo, as in ourselves, there are separate male and female individuals. In the tree, pollination—the transfer of pollen from the male to the female, the essential precursor to sexual fusion—takes place in the spring. Sexual fusion, the initial step in the formation of an embryo, follows in the late summer or early autumn, and the result, a seed containing the embryo of a new young plant, is shed a month or two later. Generally, the time from pollination to the point at which the seed germinates is about a year.

A key feature of sexual reproduction, which comes about by combining the chromosomes from each sex cell, is that the resulting offspring are not exactly like either parent. In a complicated way, through mechanisms mediated at the level of DNA, the new young plant or animal combines features from both parents. In addition, during the development of the sex cells on both the maternal and paternal sides, DNA on the chromosomes is shuffled. As a result, the combination of features specified by the DNA in every sperm or egg cell ends up being different. These are the main reasons why brothers and sisters are seldom identical; for a mature male and female ginkgo, this means that the genetic makeup of every one of their offspring, the embryo inside every seed that develops on a female tree, is unique.

A further important part of the development of the sex cells, and a necessary prelude to sexual fusion, is that the chromosome number is reduced by half, which ensures that the resulting embryo does not end up with four sets of chromosomes, double the number of its parents. In ginkgo and in all other plants this halving of the chromosome number, which occurs at the same time as the shuffling of the DNA, takes place during the formation of the pollen grains, and also inside the developing seeds as the eggs are being formed.

The seventeenth-century German scientist Rudolf Jakob Camerarius was the first to begin to develop a scientific understanding of the crucial importance of the transfer of pollen for the production of seed. Earlier botanists, such as John Ray and Nehemiah Grew, had made similar suggestions, but while studying mulberries, Camerarius noticed that female plants growing far from males would produce fruits but no seeds. He also experimented on castor oil plants and maize by removing the pollen-producing parts, and showed that no seeds were formed.[2]

From the earliest human encounters with ginkgo it would have been clear that seeds are produced only on some trees and not on others. Realizing that trees of two different kinds are needed for seed to be produced would have required a further mental step, but was nevertheless recognized well before any scientific understanding of the process. What may be the earliest written reference to ginkgo in China, from the tenth century, includes the statement: "Let male and female ginkgo trees grow near one another, then fruit will form."[3]

Ginkgo is relatively unusual among seed-bearing plants in having male and female reproductive parts produced on separate trees. Pines, for example, have separate pollen and seed cones, but both are borne on the same plant and they are therefore hermaphrodites. It is even more common to have male and female parts closely associated in the same structure, as in most flowering trees, like a cherry or a magnolia, where the pollen-producing male parts surround the seed-producing female parts in the same flower.

Nevertheless, Darwin was the first to suggest that the complete separation of sexes in plants might, in the right circumstances, provide advantages through a division of labor in the production of sex cells. Since then, the conditions under which a switch from hermaphroditism to having separate males and females might be favored have been studied in detail, and among flowering plants it is now recognized that this must have occurred independently more than a hundred times. The basic idea is simple: if an advantage exists for a plant in terms of passing on more of its genes to the next generation by specializing as a male or a female, rather than by being both, then natural selection will tend to push successive populations in that direction.[4]

Simple differences in the arrangement of the sex organs in plants have important consequences. In plants like cherries and magnolias, where male and female parts occur in the same flower, Darwin recognized that self-pollination and self-fertilization may occur quite often even though many plants have evolved ingenious mechanisms to prevent it. In pines, which have separate male and female cones on the same plant, the chances of self-pollination may be reduced, but nevertheless there is still ample opportunity for self-pollination and potentially for self-fertilization. However, in ginkgo and plants like it, reproduction always requires a partner. Self-fertilization is a physical impossibility.

When male ginkgo trees are mature, the small conelike structures in which the pollen is formed are produced on normal short shoots at the same time as the young leaves begin to appear. The cones emerge from the bud at the shoot tip; each has a short stalk and is borne at the base of one of the tiny developing leaves. On a single short shoot up to half a dozen pollen cones may be produced among the cluster of young leaves. At first the pollen cones are slightly fleshy, and just two- or three-tenths of an inch long. They have a central stalk that is completely surrounded by small, tightly packed side branches, and each side branch has two small, downward-pointing yellow sacs at the tip in which the pollen is produced.[5]

On dry warm days in the spring, the central stalk grows quickly, the cones elongate, the side branches become widely separated, and each pollen sac splits open by a single lengthwise slit to release the pollen. A triangular bulge at the end of each side branch, right at the point where the two pollen sacs join, may be involved some way in the splitting process. For just a few days every year vast quantities of tiny pollen grains are released into the air and blow away on the wind. In the process the cones dry up, and soon after, they are shed; desiccated pollen catkins carpet the ground beneath the tree, their job done for another year. From then until new cones appear the next spring, the only way to spot a male ginkgo is through negative evidence: the lack of seeds on the tree or on the ground beneath it.

Each pollen grain is tiny, about a thousandth of an inch long, roughly the shape of a rugby ball, with a tough, flexible wall except for a single narrowly elliptical region where the pollen wall is thinner. The grains are not very distinctive and look much like the pollen of cycads or even some flowering plants. When they are released, the pollen grains dry out, the thin part of the wall folds inward, the grains become thinner and slightly longer, and further water loss is slowed.[6]

The output of pollen grains from a single ginkgo tree over just a few crucial days every year is extraordinary. Andrew Leslie, a former colleague at the University of Chicago, made a very rough estimate for the thirty-five- to forty-foot ginkgo trees growing on campus. He calculated that the pollen output from about seven cones on an individual short shoot might be as many as fifty-nine million individual pollen grains. With an estimated 17,500 short shoots on a tree of moderate size, this works out to an annual production of an incredible one trillion pollen grains. This is a vast number, but it needs to be. Imagine a single microscopic pollen grain blowing in the wind: what are the odds that it will find its way to exactly the right place on a developing seed on a female tree?[7]

At the same time that pollen cones begin to be produced on male trees, or perhaps just a little later, mature female trees produce ovules. These are the structures in which the egg cells develop; if pollinated, they will eventually develop into seeds. They are borne on specialized stalks, and like the pollen cones, each is attached to the short shoot at the base of one of the young developing leaves. The ovule-bearing structure of ginkgo is distinctive, quite unlike that of any other living plant; no one would mistake it for a flower, or for the woody seed cone of a conifer or cycad. Each has a single, simple stalk, usually with two young ovules borne at the tip. Each ovule is surrounded by a collarlike rim and its pointed tip projects upward and outward; if pollinated and fertilized, each has the potential to develop into a seed.[8]

At precisely the same time that ginkgo pollen is drifting on the breeze, each young ovule is tipped by a glistening, watery drop of fluid, the pollination drop. The drop helps trap passing pollen grains from the air, and any grains that are captured sink in the fluid. The pollination drop is repeatedly absorbed and restored each day until it has drawn grains inside the ovule.[9]

In plants, which are unable to move to find a mate, pollination is the riskiest part of reproduction, and this is especially the case for ginkgo and other species with separate male and female plants. The whole sequence of events needs to be beautifully coordinated, with pollen and pollination drop produced at the same time. Timing is everything; if pollination is successful, and if the male sex cells that develop within the pollen grain manage to fertilize the egg inside the ovule a few months later, then an embryo, a new young plant, will develop as the seed matures. However, any deviation from exact synchronization is penalized harshly. Trees with a poor sense of timing will produce no seeds and leave no offspring for the next generation. It is not hard to see how such exquisite timing in the reproduction of male and female trees has been coordinated by the evolutionary process of natural selection.

Normally, only one of the two ovules at the tip of a ginkgo seed stalk develops. It might be that the other was not pollinated, or loses out in the competition for resources from the mother tree, or aborts if there are insufficient resources for all the seeds that are pollinated to develop. Occasionally, however, both of the ovules on a single stalk mature, and even more rarely there may be three or four mature seeds on each stalk. In most seeds two or three egg cells are produced, but normally only one is fertilized or goes on to develop. In about two out of every hundred seeds more than one egg is fertilized, and more than one embryo develops. In these rare cases, when the seed germinates, two young plants may emerge.[10]

The mature ginkgo seed is a plumlike structure typically up to about an inch and a quarter long, and up to about an inch wide, with the embryo developing inside. When fully ripe, the pale yellow seeds have a distinctive silver sheen which gives the ginkgo its common name in Chinese, "silver apricot." Inside, a fleshy yellow pulp surrounds

Clusters of succulent apricot-like fruits on a female ginkgo in the late autumn.

a hard stone, and inside the stone is the developing embryo embedded in the nutritive tissue that at an earlier stage produced the eggs.

Ginkgo is easy to grow from seed; as long as the fleshy pulp is gone, there is no built-in dormancy that has to be overcome: during the winter development simply slows or stops, picking up again as temperatures rise in the spring. From almost the beginning of its development the embryo has a clear top and bottom. The top, which points toward the tip of the ovule, eventually develops into the root of the plant, while at the other end, toward the attachment to the stalk, growth from a zone of actively dividing cells eventually forms the aboveground parts of the new plant. First to be produced are a pair of seedling leaves that remain embedded in the nutritive tissue of the seed. Strangely, the seedling leaves are green, even though they are never released and never expand. The root grows through the apex of the seed, and as the seedling leaves elongate, the new shoot apex is pushed out of the seed in the same direction. The young shoot apex then becomes green and begins to grow upward, toward the light and against gravity, producing new leaves alternately, one after another on the tree's first long shoot.[11]

Alan Mitchell, one of the great British tree specialists of the late twentieth century, estimated that female trees first produce seeds when they are about twenty-five to thirty years old. Male trees probably reach maturity at about the same time. Before then it is impossible to tell whether a tree is male or female. But those two or three decades have not been wasted. The trees have grown in size and built up reserves of energy. Branching has also created opportunities to produce pollen cones and seeds in large numbers. When the embryo begins its growth, it has only one group of active cells that can produce a shoot and new leaves, but every time a new bud is formed a new group is added. By the time the tree is mature, there are tens of thousands of these meristems. Every one of them can potentially produce new shoots and leaves. Under the right circumstances they can also produce pollen grains and ovules with the sex cells inside. As those sex cells develop, the DNA of the new parent, with its unique set of genes, is once again prepared for its passage to the next generation.

8
Gender

We allow our ignorance to prevail upon us and make us think

we can survive alone . . .

—Maya Angelou, address to Centenary College of Louisiana

The most widely accepted explanation for why ginkgo and other trees have separate male and female plants is that it eliminates negative effects that might come from extreme inbreeding if self-pollination were followed by self-fertilization. This conjecture is supported by a great deal of evidence that plants produced by self-fertilization are less successful and leave fewer offspring in the next generation than those arising from cross-fertilization. Given this principle, it is not surprising, as observed by Darwin, that where male and female parts occur together in the same flower, there are often mechanisms or structures to prevent self-fertilization. However, having separate male and female plants, as in ginkgo, is a particularly blunt and unforgiving mechanism for ruling out self-fertilization. It makes absolutely certain that offspring are genetically different from both their parents.[1]

Ensuring that offspring differ from their parents is likely to be an advantage when a new generation of plants faces a range of different ecological conditions. Indirectly, it may also be beneficial over the long haul in a changing world. Variation, after all, as Darwin was the first to recognize, is the raw material of evolution. Together with the overproduction of offspring and the winnowing effect of selection, minute variations controlled by genetics create the opportunity for successive populations of ginkgo trees to respond to changing environmental circumstances. As different variants are favored by different local conditions, ginkgo populations have the potential to change their genetic makeup through time. This is the essence of natural selection.

However, even acknowledging the advantages in terms of producing new variation, an inflexible system with separate male and female plants has inherent disadvantages. In plants that cannot move to find a mate, there is no backup, no fail-safe mechanism that allows seed to be produced if there are no trees of the opposite sex close by. Many plants have mechanisms to help ensure cross-pollination, but many also have mechanisms that effectively keep their reproductive options open, just in case cross-pollination fails to work. Ginkgo does not have such an option. It is also more than a little ironic that while ginkgo seems to have a mechanism that forces the production of new genetic variation, it is also the botanical poster child for stasis, a striking lack of evolutionary change. Paradoxically, it has stayed much the same across vast spans of geologic time.[2]

The seeming simplicity of the male-female distinction in ginkgo made it an obvious focus of early research to try to understand how sex is controlled in plants. Just as Camerarius focused on maize and castor oil, which have separate male and female parts, Joseph von Jacquin, professor of botany at the University of Vienna in the early nineteenth century, turned to ginkgo to carry out an interesting early experiment into what determines sex in plants.[3]

Ginkgo was introduced into cultivation in Europe in the eighteenth century, and, according to Jacquin, who carefully reviewed these early introductions, the first record of one of these plants reaching sexual maturity is from 1795, when two trees at Kew, one of them the Old Lion, produced pollen cones. Cuttings from the ginkgo trees in England had been sent to the Imperial Palace of the Habsburg Empire at Schönbrunn in 1781 and were the first ginkgo trees to be grown in Vienna. Perhaps from the same stock, Jacquin's father, Nikolaus Joseph von Jacquin, planted the large male ginkgo that still grows in the botanical garden of the University of Vienna.[4]

Movement of water through the stems of plants had been well known since the work of the English clergyman Stephen Hales in the early eighteenth century. Jacquin wondered therefore whether a female branch grafted onto a male tree might change sex. He obtained a cutting from a female tree, probably from the first female ginkgo to be recognized in Europe, and made the graft onto the male tree at the University of Vienna botanical garden.[5]

The graft was successful; the branch flourished, but it remained female. It continued to produce seeds, even though it was now attached to a male plant. It also maintained its independence in another way; every spring its foliage emerged two weeks later than that of the rest of the tree, and in the autumn its leaves were still green when the rest of the tree's leaves had turned yellow.[6]

Jacquin's observation was simple, and not unexpected, but it was also profound. It showed that plants are made of individual parts that are to some extent autonomous, and that sex and other characteristics are determined not at the level of a whole organism, as in most animals, but potentially in individual tissues in different parts of the plant. It highlighted the tension between thinking about a tree as a whole organism, like a bird or a person, and thinking about a tree as a collection of potentially independent organisms, something more like a coral colony. It is a distinction that has practical consequences. "Dolly the sheep," the first cloned mammal, was a major scientific accomplishment not achieved until the mid-1990s. In contrast, any gardener can establish a new plant from a cutting; people have been cloning plants for thousands of years.

Not until genetics started to be better understood in the early twentieth century were the next steps taken toward deeper knowledge of how sex in plants is controlled. Stimulated by the rediscovery of Gregor Mendel's work on pea plants, as well as new research by the Dutch scientist Hugo de Vries in the 1890s, it became clear that the characteristics of organisms, including "maleness" and "femaleness," were inherited in a "particulate" manner. Before then it had been assumed, by Darwin and others, that inheritance operated by some kind of "blending," based on the observation that offspring were often somewhat intermediate in their features compared with their parents. Writing thirty years after Darwin's *Origin of Species,* de Vries had no idea what the nature of these particles might be. He could not see them and could only infer their existence from his and Mendel's experiments. He called them pangenes. We now refer to them simply as genes.[7]

De Vries's breakthrough laid the foundation for the rapid development of the science of genetics in the early twentieth century, including the recognition, achieved independently in 1902 by Theodor Boveri and Walter Sutton, that the "particulate" hereditary factors were borne on chromosomes. This was later confirmed by Thomas Hunt Morgan and his team at Columbia University in New York, based in large part on discoveries made in 1905 by Nettie Stevens and Edmund Beecher Wilson that the two sexes in insects like the common mealworm also differ in chromosomes in their cells. This was the first time that physical differences among organisms could be linked to observable differences in their chromosomes, and it also explained Jacquin's observations on ginkgo. The chromosomes in the cells of his graft bore the genes for "femaleness," while the chromosomes in the cells on the rest of the tree bore the genes for "maleness."[8]

Nettie Stevens observed that in the female larvae of her mealworms there were twenty large chromosomes, while the male larvae possessed nineteen large chromosomes of equal size, with one of the tenth pair reduced in size. She called the short chromosome of the unequal pair the Y chromosome and the longer of the two the X chromosome. Both she and Edmund Beecher Wilson recognized that the XY configuration results in males, while the XX configuration results in females, and we now know that this same mechanism operates in humans.[9]

The rediscovery of Mendel's laws, the recognition of the importance of chromosomes, the discovery of sex chromosomes, and Morgan's work on fruit flies all came at about the same time as an upsurge in interest in ginkgo. As a result, ginkgo was among the first plants to be examined for its chromosomes, and it is now well established that every living cell in a ginkgo plant has twenty-four chromosomes, twelve pairs: a dozen brought by the male reproductive cell, which develops in pollen grain, the other twelve contributed by the egg that develops inside the ovule.[10]

When I was a student, I was taught that just as in humans and mealworms, differences among one of the pairs of chromosomes in male and female ginkgo trees indicated a system of sex chromosomes, which results in males and females in approximately equal proportions. However, careful observations over the past few decades have shown that in ginkgo the supposed X and Y chromosomes are not consistently different. Furthermore, even though both of these chromosomes have small, so-called satellite fragments of DNA attached to them, which may be important for sex determination in other plants, in ginkgo these fragments vary in ways that make them unlikely determinants of the sex of a mature tree.[11]

Nevertheless, whether visible or not, there must be real differences in the chromosomes of male and female ginkgo trees at the level of genes. This implies that the sex of a ginkgo embryo is determined at the time that the egg is fertilized, just as it is in humans. However, there is also evidence that sex determination in ginkgo may not be quite as strongly fixed as it is in many animals, and even that it may not be completely stable through the life of every tree.[12]

In the summer of 2006 a small twig on the Old Lion at Kew, a male tree, spontaneously produced three seeds. I published a note on this in the *Kew Magazine* later that year and had letters and emails from several colleagues suggesting that this was probably the reemergence of an old graft rather than a sexual switch in part of the tree. This certainly could have been possible because the practice of grafting female branches onto male ginkgo trees became widespread in the late nineteenth and early twentieth centuries, both for reasons of curiosity and to produce viable seeds. We know that such a graft was made on the Kew tree in 1911, and it apparently fruited copiously before it was pruned off accidentally by an overenthusiastic arborist. Similar uncertainties about whether or not a graft was involved also apply to the large, nearly 200-year-old male tree in the botanical garden in Jena, Germany, that began to produce seeds on a single branch in the early 1990s.[13]

However, in other cases there is clear evidence of male ginkgo trees spontaneously producing seed without human intervention. Martin Hamilton, a native of Kentucky and one of the many wonderful students trained in horticulture during my time at Kew, drew my attention to an example that piqued his interest as a young botanist. In the eastern and southeastern United States some of the best places for marvelous specimen trees are historic cemeteries, and the Cave Hill Cemetery in Louisville, Kentucky, is an excellent example. It boasts a wonderful collection of trees dating from the mid-1800s, including several massive ginkgos. Among them is a large male tree, one of the largest and most spectacular in North America. High in the canopy it has a witches' broom, a dense mass of shoots that have somehow become feminized and that produce large quantities of seeds every year. These seeds contain perfectly good embryos and grow normally. Similar spontaneous partial sexual switches have been recorded several times on old male ginkgo trees in Japan.[14]

Equally definitive evidence of the same phenomenon comes from the Ginkgo Plantation at the Blandy Experimental Farm in Boyce, Virginia, where more than six hundred ginkgo trees were planted between 1929 and 1947, all from seeds produced by a large female ginkgo on the University of Virginia campus in Charlottesville. About half the trees survived and began producing either pollen cones or seeds after twenty or thirty years. Surveys to determine the sex of each tree were carried out in the late 1970s and early 1980s, and in May 1982 an especially careful survey was made using a mechanized lift so that the tree canopy could be carefully inspected.[15]

For all of those trees for which sex could be determined, the ratio was roughly one to one: 157 female trees to 140 male trees. But the 1982 survey turned up a surprise. Four trees recorded as females in previous surveys, based on the presence of seeds, turned out to be predominantly male. These trees were watched especially closely in 1982, and three of them produced from one to seven seeds each.

Disentangling the factors responsible for such leakiness in sexual expression is not so easy, but it is perhaps not surprising that a male tree might occasionally produce a few ovules. Over the long haul, this might be a useful trick if you are a lonely male without a mate. It also does not completely undermine the normally strict outbreeding enforced by the rigid separate sex system. At worst, a few ovules on a "leaky male" are likely to use only a tiny fraction of the pollen grains that would otherwise have gone to pollinate females, and at best such spontaneous production of ovules, if they mature into seeds, might provide an intergenerational fail-safe when no mature females are available.

From an evolutionary point of view, "leaky females"—female trees that occasionally produce pollen catkins—would be a much less advantageous form of gender modification. Pollen is produced in such large quantities that even a few rogue pollen cones could result in massive self-pollination of the ovules on a mainly female tree and a large-scale override of the normal system that ensures that sex is possible only between different individuals. "Leaky females" would also be more difficult to detect than "leaky males" because pollen catkins are on the tree for only a short time early in the season, but for the moment, the evidence suggests that while most ginkgo trees are strictly male or female, when deviations occur, they usually take the form of males producing seeds rather than the other way round. If this is true, ginkgo would fit the pattern seen in other plants with separate male and female plants, as well as the theoretical predictions: "leaky males" are much more common than "leaky females."[16]

9
Seeding

If you can look into the seeds of time,

And say which grain will grow and which will not,

Speak then to me . . .

—William Shakespeare, *Macbeth*

I like to think of the old male ginkgo that grows at Kew as one of the celebrity ginkgo trees of the world: a tree worth making a special effort to meet. As a living connection to King George III and Sir Joseph Banks, as well as the early introduction of ginkgo into the West, it would have a rich story to tell if you could engage it in conversation. However, an equally special celebrity ginkgo grows on the other side of the world in the Koishikawa Botanical Garden at the University of Tokyo. It is a living link to pre–Meiji Era Japan. Keisuke Ito, who had been inspired by Siebold, would have known this tree when he was appointed professor there. Planted more than three hundred years ago, today it stands over eighty feet tall and has survived many vicissitudes, from the Great Kanto Earthquake of 1923 to the firebombs that consumed much of Tokyo toward the end of World War II. It is a precious old female tree, and it has a special connection to early development of modern science in Japan.[1]

The Koishikawa Garden was founded on the site of what was once the medicinal plant garden of the Tokugawa Shogun. In 1868, at a time of great upheaval in Japan, the land passed to the new imperial Meiji government, and the land became the botanical garden of the newly founded Tokyo University. The transfer, however, was deeply unpopular with the samurai, the traditional military class, who opposed the new regime. They showed their displeasure at the changes that swept Japan in the 1860s in many ways, but at the Koishikawa Garden one manifestation was the felling of some of the large trees just before the time of transfer. The Koishikawa Ginkgo was very nearly among those that were lost.[2]

Although a time of rapid and difficult social change, the late nineteenth century was crucially important for the future development of science in Japan. After the installation of the Meiji government and the creation of a modern system of education, Japanese scientists had their first opportunity to collaborate with colleagues in the West. In plant science, strong early connections developed with leading botanists working in Germany. Both Ryokichi Yatabe and Jinzo Matsumura, the first two professors of botany at Tokyo University, studied in Germany in the late 1880s. A few decades later, so did Manabu Miyoshi, who returned to found the discipline of plant physiology in Japan. Also from this group came Kenjiro Fujii, who studied in Munich under Professor Goebel, did important work on ginkgo, and founded the Japanese scientific journal *Cytologia*.[3]

Since the Meiji Restoration, the ginkgo at the Koishikawa Garden has been visited by many scientific luminaries, but few have had a life quite as varied and colorful as the precocious young Englishwoman Marie Stopes, who studied there in 1907 and 1908. Stopes came to Tokyo through her connection to Kenjiro Fujii. They met in 1903, when both were working in Munich, and they did important research together on fossil plants from Hokkaido. Both were familiar with the large female ginkgo in the Koishikawa Garden.[4]

Marie Stopes is best known for her controversial pioneering work promoting family planning in Britain, and her influential book *Married Love*, described by some as a sex manual, was banned as obscene in the United States until 1931. In one survey of the most significant books of the twentieth century *Married Love* scored only a little behind *Das Kapital* and ahead of *The Meaning of Relativity*. Through her contributions to women's rights and her shaping of women's expectations within marriage, Marie Stopes is credited with almost single-handedly leading women in Britain out of the repression of the Victorian Age into a more enlightened age of sexual awareness.[5]

However, Marie Stopes began her career focused on a more mundane pursuit: studying plants. At the age of eighteen, she was granted a science scholarship at University College, London, and she finished a double degree in botany and geology in only two years. After completing her Ph.D. on reproduction in cycads, she became Britain's youngest doctor of science at the age of twenty-five. The University of Manchester hired her immediately as a lecturer. She did important work on several different kinds of extinct plants and also wrote a classic paper on different kinds of coals, which is still widely used. In an era when women faced enormous obstacles in pursuing a career in science, her geological and paleontological contributions were extraordinary.[6]

Stopes and Fujii worked under Goebel in Munich, but another important group of botanists studying plant biology and evolution at that time was in Bonn under the leadership of the great German botanist Eduard Strasburger, who became interested in the details of reproduction in ginkgo in the 1890s. Richard von Wettstein, professor of botany at the University of Vienna, sent him seeds from the graft that Joseph von Jacquin had made many years earlier. They were dispatched every two weeks from June through to early September, and based on this material, Strasburger described many aspects of sexual reproduction in ginkgo. However, he missed a crucial detail; a detail filled in only later by Sakugoro Hirase, working at the Koishikawa Garden.[7]

Hirase was an accomplished technician and illustrator. In 1895 he published a meticulous account of how the embryo developed into a new young plant inside a ginkgo seed. Through the summer of 1896 Hirase kept up an especially detailed study of the seeds collected from the Koishikawa tree, and on September 9, 1896, he was the first to observe the previously overlooked final stage of sexual reproduction in ginkgo.

What Hirase saw was a botanical sensation: pollen grains drawn into the ovule by the pollination drop produced a branching tube that had grown into the tissues at the top of the ovule. The bulging base of the tube, hanging down on the inside of the ovule, contained a pair of large sperm cells. Hirase saw that these two sperm were released into a cavity in the top of the ovule. Propelled by the synchronized movements of spiral bands of thousands of tiny flexuous hairs, they swam the short distance to fertilize the egg. He recognized immediately that this curious means of fertilization was unlike that of any other plant known at that time. Swimming sperm were well known in ferns, mosses, and similar plants but had never been observed in any seed-producing plant.[8]

Hirase's discovery was followed just a few months later by the report of exactly the same phenomenon in the common Sago Cycad by Seiichiro Ikeno, who was work-

*Sakugoro Hirase, the Japanese scientist who was the first to
observe the swimming sperm of ginkgo in 1896.*

*Close-up photograph of two swimming sperm just before their
release. Each sperm cell is a little less than four thousandths of
an inch long and is propelled by the synchronized movements
of thousands of tiny hairs arranged in a spiral.*

ing in the College of Agriculture of the Imperial University, Tokyo, and who had been guided by Hirase's work. Both discoveries were remarkable. They were also two of the first research contributions of Japanese scientists that made an impact on the international stage. In 1897 Ikeno and Hirase summarized their results in a joint paper in English, and in 1912 they were the first biologists to receive the Imperial Prize of the Japan Academy.[9]

From the standpoint of understanding plant evolution, the discoveries of Ikeno and Hirase were an unanticipated breakthrough. Until their work it had been assumed that fertilization in ginkgo and cycads was like that in conifers, the male sex cell being delivered to the egg through a tube growing from the pollen grain. The discovery of swimming sperm showed that reproduction in ginkgo and cycads was very different and more like that of ferns: it harked back to an earlier time in plant evolution when plant reproduction was more reliant on water.

Marie Stopes, who was at Tokyo Imperial University just a few years after Hirase made his discovery, and who published a diary of her experiences in Japan, refers several times to the excitement of repeating Hirase's observations. Her entry for September 17, 1907, records: "This is just the last day or so for the swimming out of the spermatozoids of Ginkgo, and I spent a few delightful hours in the laboratory watching their infusorian-like movements and the quick vibrations of their spiral crowns of cilia. Fancy cutting up dozens of juicy Ginkgo seeds!"

A year later, her entry for September 9, 1908, alludes to the effort needed to catch fertilization in ginkgo in the act: "I went early to the Institute, where there is grand excitement over Ginkgo; the sperms are just swimming out, and they only do it for a day or two each year. It is no such easy business to catch them, in 100 seeds you can only get five with sperms at the best of times, and may get one and be thankful. I spent pretty well the whole day over them, and got three, and several in the pollen tube not yet quite ripe. It is most entertaining to watch them swimming, their spiral of cilia wave energetically." She spent the next two days "hunting Ginkgo sperms nearly all the time."[10]

The significance of Hirase and Ikeno's work was further underlined by a paleobotanical discovery made just a few years later in Britain. For many years botanists had puzzled about why many of the large fossil fern leaves discovered in the coal measures of Europe and North America never had the spore-producing clusters on the underside of the leaf that are typical of most modern ferns. In 1903, F. W. Oliver, working at University College, London, and D. H. Scott, working at Kew, solved the problem by

The young Marie Stopes working at her desk in 1904, a few years
before she traveled to Japan for her botanical work at Tokyo Imperial University.
Her pioneering work on family planning in Britain came later in her career.

convincingly linking one such leaf with a fossil seed, thus demonstrating that many so-called fossil fern leaves were not produced by ferns at all. Their work revealed the existence of several different kinds of extinct plants, that were, in a way, intermediate between ferns and living seed plants. Such fossil plants came to be called Cycadofilicales, or seed ferns. Together, the discoveries made by Hirase and Ikeno on ginkgo and cycads, and by Oliver and Scott on their fossils, went some way to narrowing the evolutionary gap between living ferns and living seed plants.[11]

This line of thinking was also reinforced by reports of aberrant ginkgo trees that occasionally produced seeds and pollen sacs on the margins of otherwise normal-looking leaves. In 1896 Kenjiro Fujii described three such trees from Yamanashi Prefecture in Honshu, Japan, and since then others have noted the same phenomenon. Fujii suggested that the formation of seed or pollen sacs on the margins of leaves, together with Hirase's discovery about the fertilization process, might suggest that ginkgo evolved from fernlike ancestors.[12]

An interesting question that has loomed large in the thinking of some botanists is whether fertilization in ginkgo occurs while the seeds are still attached to the tree, as is normal in living seed plants, or whether it occurs after the seeds have been shed, as suggested by the French paleobotanist Louis Emberger. In some ways this would be more like what happens in ferns, but in ginkgo this is clearly not normal. Fertilization usually occurs in late summer or early autumn, and most seeds are not shed until a month or two later, by which time the embryo is already well developed. That said, it is not impossible that fertilization does occasionally occur on the ground. After pollination, the developing seeds grow almost to their mature size before fertilization. At the later stages of their development—for example, in August—young seeds do occasionally fall from the tree, and if that happens just prior to the release of sperm, then viable embryos might still be produced.[13]

In Chicago, most ginkgo seeds are on the ground just as the weather is turning cold in mid- and late November. On a few trees the seeds fall more gradually through the winter. However, whether it is in the late autumn or after a winter storm or cold snap, when there is a layer of ginkgo seeds on the ground, no one could mistake them for apricots, plums, or anything else. The smell is distinctive and powerful. A plant chemist would describe it as rich in butyric acid. Others would recognize a strong similarity to the unmistakable stench of human vomit. A friend who is a councilor in southwest London had to convince local residents that the smell about which they were complaining was not the result of binge drinking by hooligans outside her local train station but the equally malodorous output of a large female ginkgo. And if the smell isn't enough incentive to avoid the fallen fruits, the pulp contains ginkgolic acid, an allergy-causing chemical. Like poison oak or poison ivy, it can produce a nasty rash. If you get it in your eye it can put you in the hospital.[14]

10

Resilience

Strength . . . comes from an indomitable will.

—Mahatma Gandhi, *Non-Violent Resistance (Satyagraha)*

Brian Mathew is one of those botanical enthusiasts for whom a lifetime at Kew was an irresistible attraction. People who somehow have plants in their blood often end up working at Kew for their entire careers, and their connections to the place, through their friends and through their plants, persist long after retirement. In Brian's case, most of his working life was spent in the vast collection of preserved plants at Kew. Begun by William Jackson Hooker, this massive herbarium now contains about eight million specimens of dried plants. It may be the largest of its kind in the world, and Brian's role was to use that massive reference collection to accurately name the plants being grown in the gardens by Kew's horticulturalists.[1]

For Brian, as for many people at Kew, plants are not just a job but also an avocation, and his botanical passion finds expression in many different ways. He became an expert on bulbs, like crocus and snowdrops, and this led him to work on their conservation in countries like Turkey and Georgia, where bulbs are collected from the wild for export into the horticultural trade. He also wrote several books, becoming a successful author; like many at Kew, of course, he also became a passionate gardener.

In 1999, Brian was asked by friends in his village to design and plant a garden on the land around his local church. Rarely able to say "no" and always willing to help, he set to work. No money was available, but some volunteers gave a hand with the planting, others pitched in with donations of plants. One such donation was a young ginkgo, growing in a nearby garden that was becoming too big for the space available. It was a small tree, healthy, well established, and about ten feet tall. Moving a tree, even a relatively small one, is always harder than it looks, but Brian took up the challenge. He thought that if the root system was reasonably well developed, the tree would stand a good chance of survival if he dug it up while it was dormant in the winter. Brian's idea was to move it first into a pot where it could be cosseted for a season before being replanted.

Unfortunately, what seemed straightforward in theory turned out to be harder in practice. The first problem was that over the years the soil had been mounded up around the trunk. As Brian dug down, he had difficulty finding any roots at all, let alone getting to the bottom of the main root mass. Eventually, with the hole around the tree a couple of feet deep, and deeper digging difficult without uprooting nearby plants, Brian came across just a few small side roots coming off the main trunk. Below that the trunk just continued down. At this point, Brian felt that he could go no farther without impinging too much on the generosity of the donors and disrupting much of their garden. So, warning that the tree probably would not survive the move, he cut the trunk below the small side roots and planted it in a large pot. As an experienced gardener Brian knew that in order to maximize the young tree's slim chances of survival, he needed to reduce its crown dramatically. If the tree leafed out the next spring, one or two roots were not going to be enough to keep all the leaves supplied with water. So Brian pruned the ginkgo severely, placed the pot out of direct sun, and waited.

To his surprise the young tree burst into leaf the following spring. Healthy leaves appeared on the unpruned branches, and the plant looked fine through the summer and into the autumn. The next winter he found a good place to plant it, in an area used by children of a preschool. He thought that the ginkgo's distinctive leaves would make it an entertaining curiosity for the youngsters. In the pot the roots had developed well, there was now a nicely developed root mass, and the plant was flourishing despite its earlier trauma.

The next summer the tree again leafed out beautifully. Unfortunately, however, sometimes young people and young plants don't mix. The tree got snapped off, right in the middle of the trunk, about four feet above the ground. The plant was now just a bare trunk, broken at one end with no branches and no leaves. There was nothing for it; it had to go back into intensive care. Brian dug it up for the second time, planted it back in its pot, and waited to see whether it would recover. Again it came back from the dead; this time sprouting at soil level from around the broken stem. Soon there were several new trunks, and its recovery was rapid and long-lasting.

Brian's experience with the indomitable resilience of ginkgo is not unusual. Its tenacity in the face of abuse is one of the reasons that it does well as a street tree. It is also easy to propagate from small cuttings, which readily take root and establish new plants. This is probably how ginkgo was spread from garden to garden when it was first introduced into Europe. And as Brian found out, ginkgos have a peculiarity of their anatomy that creates new opportunities for growth if the main part of the plant is damaged. Ginkgo, in other words, has an alternative way of reproducing itself that does not require completing the cycle of sexual reproduction, with all its complexities, risks, and potential for failure.

In a few trees, such as willows, such mechanisms of self-propagation are especially well developed. Willows, often streamside plants, perpetuate themselves easily in the wild. Twigs break off, get washed downstream, become lodged on a bank, and then take root. A relative of the willow, the aspen, is equally vigorous. It sends up new branches from its roots that look like separate trees, but they are all connected belowground and are genetically identical. The establishment of new plants in this way is not common in ginkgo, but it does have a similar inbuilt means of sustaining itself. This allows it to occupy a single place for a long period, and as in the case of Brian's sapling, if the main branch is damaged, new branches soon form and can quickly take over.[2]

A characteristic feature of some ginkgo trees, and especially of very old specimens, is the production of peculiar downward growths from otherwise normal branches. These develop like stalactites from the roof of a cave. Several may develop along a single branch, giving the Daliesque impression that the branch is made of wax and has somehow begun to melt. Some of these downward growths, or burls, may be six feet or more from top to bottom, and as each reaches the ground, it can resprout and send up new shoots. Each may eventually develop into a substantial new trunk that can become separated from the parent tree. In very old trees, where these growths are extensive, the original main trunk can sometimes be hard to find.

In Japan these downward-growing branches are called chi-chi, literally "breasts," from the folklore that mothers who pray to them can increase their breast milk. In

China they are called zhōng rǔ, "stalactites." It is not known what stimulates their growth, but careful studies show that they form from small buds that become embedded in the wood of a branch as it grows and that somehow are later reactivated. Every leaf of a ginkgo tree has a small bud, a potential new shoot, in its axil, and as a branch increases in girth, buds that were once at the surface sometimes become buried by the tissues around them. While ginkgo trees rarely develop chi-chi until they are very old, the potential of regrowth from embedded buds seems always to be present in latent form.[3]

Peter Del Tredici, one of the world's experts on ginkgo, who works at the Arnold Arboretum of Harvard University, has shown that such embedded buds are established from the seedling stage onward, right at the beginning of the life of a new plant, and are a normal feature of growth in ginkgo. In all ginkgo seedlings the tiny buds that occur in the axils of the seedling leaves, the first leaves produced by the embryo, remain inactive and become embedded in the base of the tree as it grows. If the sapling is then damaged in some way, as happened to the tree nursed along by Brian Mathew, one of these leftover buds grows downward from the base of the trunk to produce a woody basal chi-chi, a so-called lignotuber, which can produce new trunks and branches, as well as new roots. Del Tredici suggests that these lignotubers are also helpful anchors on unstable soils.[4]

The combined effect of extensively developed chi-chi and the associated capacity for rampant self-propagation is seen most spectacularly in some of the remarkable giant ginkgos that grow in Aomori Prefecture in northern Honshu, Japan. One of the most impressive is the Kitakanegasawa Ginkgo, a real giant that grows squeezed between the crowded houses of Kitakanegasawa village and the main coastal road. It flourishes right at the point where the steep mountain slopes join the narrow coastal plain. It is only a few hundred yards from the sea, but its roots draw water from natural freshwater springs, and it has a sheltered spot tucked into the lee of the Cape of Odosesaki, where it is protected from the full force of the winter storms that roll in from the west.

From the small parking lot built for visitors by the local Department of Tourism, the Kitakanegasawa Ginkgo is a picture of health. The crown is huge, vigorous, and flourishing, with few dead branches or yellow leaves, but inside its personality changes. The inner crown contains lots of dead wood, and the internal scaffold of the tree reflects its irrepressible growth. Multiple trunks and branches, many of them snapped off, show the battering that the tree has taken over centuries from heavy snow, from typhoons,

The solid trunk of a large ginkgo, showing a mass of thin sucker shoots growing up from the base.
This is one of several large ginkgo trees in and around Shanghai, China.

perhaps even from earthquakes. A tangle of roots shows through where the thin soil and leaf duff have been worn away by the feet of tourists.

Even the largest ginkgos do not come close to the vast size of a coast redwood, or the great age of a bristlecone pine, but a truly massive ginkgo, like that at Kitakanegasawa, is nevertheless a great natural wonder. This is a tree that you walk into, rather than beneath, and as you enter you are engulfed by a complex tangle of large upward and horizontal branches, mixed with chi-chi growing downward from the larger limbs. Everywhere there are slender, vigorous, upward-pointing sucker shoots sprouting from the trunks and larger branches. The Kitakanegasawa Ginkgo is less like a tree and more like a thicket; the combined circumference of its trunks is more than seventy-two feet.

Among them are three substantial new trunks, already clothed by sucker shoots that are starting life as opportunists from their mother. With their potential for independence, they provide good insurance for the future.[5]

So ginkgo, through evolution, has hedged its bets. It can survive and propagate itself in different ways. The cycle of pollination, fertilization, dispersal, germination, and establishment is a seemingly fragile process. It relies on precise synchronization in the development of pollen and ovules borne on different trees, and benign conditions for seedling development and growth, but evidence from prehistory shows that it works well. These processes have gone on continuously, and probably in ways not much different from those of the living species, for more than 200 million years. Female ginkgo trees readily produce large quantities of seed, even with just a single male in the vicinity. In ginkgo, as in some other species, males are almost, but not quite, expendable. Ginkgo also has the fail-safe that at least on the male side it might very occasionally produce a few seeds. It is not beyond the bounds of possibility that very, very occasionally a single seed might produce two new plants, one male and one female.

Ginkgo is also resistant to physical abuse, which means that time is on its side. Potentially it can wait until its mate arrives. It resprouts vigorously from buds buried in its underground parts. It is irrepressible; its capacity for self-preservation has helped it survive through millions of generations.

The Kitakanegasawa Ginkgo exemplifies the resilience of its species. Local people think that it is the biggest in the world. It was designated as a national treasure only in September 2004. Somehow, it was overlooked in the initial national surveys of Japan's great ginkgos. It is now recognized as the fourth-largest tree in Japan, and it is certainly the most massive ginkgo. One enthusiast is gathering the evidence that he hopes will give it a place in the Guinness book of world records. But in a world where so much of the landscape has been altered by human activity, more important is that the connection of local people with this marvelous tree is deep. To them, and indeed to anyone with empathy for nature, this great plant is a testament both to its tenacity and also to its lasting power.

Origin and Prehistory

11
Origins

To every thing there is a season, and a time to every purpose under the heaven:

a time to be born, and a time to die . . .

—Ecclesiastes 3:1–2

Carl Linnaeus, the eighteenth-century naturalist who coined the name *Ginkgo biloba*, stands with Arrhenius and Celsius as one of the foremost of all Swedish scientists. When you arrive at Arlanda Airport in Stockholm, Linnaeus is among the collage of well-known Swedish citizens who welcome you to their "hometown," and his image appears on the one hundred–kronor banknote. Linnaeus spent his entire life in Europe, but his name is known worldwide, in large part because of his intrepid students who journeyed all over the globe to collect plants and bring them back for study. Linnaeus was the leader in efforts to make sense of the rapidly increasing knowledge of plant and animal life that came from eighteenth-century European exploration in far-flung parts of the world.

Linnaeus was also an omnipresent force in Sweden. He held sway at the University of Uppsala for decades and, as physician to the king, had the ear of the royal family. He was a charismatic teacher who inspired his students. His natural history excursions into the countryside around Uppsala were legendary. There would be music and lavish picnics as Linnaeus held forth about the trees, the flowers, the birds, and the insects.

Linnaeus could be vain and self-important, but there is no doubting his intellect or his extraordinary energy.

Linnaeus was among the founding members of the Royal Swedish Academy of Sciences, which today is best known for awarding the Nobel Prizes in physics, chemistry, and medicine. However, biology remains an important part of the academy's work, and its interest in the diversity of plants continues through its support of the nearby Bergianska Botanical Garden and its links, through history and present-day members, with the Swedish Museum of Natural History. Each of these three great Swedish institutions—the academy, the Bergianska Garden, and the Museum of Natural History, clustered together in Frescati on the northern edge of Stockholm—has played an important role in expanding our understanding of ginkgo and its place in plant evolution.[1]

The Swedish Museum of Natural History houses one of the world's great collections of fossil plants. Successive scientists who have worked there have been leaders in paleobotany since the renowned Arctic explorer Alfred Nathorst was appointed as professor in the newly founded Department of Archegoniates and Fossil Plants in 1884. Much of what we now know about plants of the past can be traced to the discoveries of Nathorst and the techniques for studying fossils that he developed.[2]

My longtime colleague Professor Else Marie Friis now occupies Nathorst's former position at the Swedish Museum of Natural History and oversees the massive collections of fossil plants accumulated by him and his successors. Today a quarter of a million specimens from all over the world are housed in more than seven thousand drawers and hundreds of cabinets on three floors of a building that Nathorst helped design. The collections are matched in size, scope, and quality only by those in the Natural History Museum in London and the U.S. National Museum of Natural History in the Smithsonian Institution in Washington, D.C. Together with collections in other museums and universities, these great storehouses of plant fossils are used by scientists from all over the world and house the physical samples on which our knowledge of the history of ginkgo and other plants is based.

In the Stockholm collections are ginkgo leaves collected by Nathorst on expeditions to Greenland and elsewhere. There are also beautifully preserved specimens collected by him in the late eighteenth century during the heyday of coal mining in southern Sweden, but among all the ginkgo material in the Stockholm Museum, the most beautiful specimen is a slab of gray siltstone collected in the 1970s from Ishpushta in central Afghanistan. Bequeathed by the German paleobotanist Hans Joachim Schweitzer,

Fossil leaves of an ancient ginkgo, Ginkgo cordilobata, *from the Early Jurassic of Ishpushta, Afghanistan, about 190 million years before the present.*

it is covered with the beautiful shiny black imprints of more than eight almost complete ginkgo leaves, which Schweitzer and his colleague Martin Kirchner named *Ginkgo cordilobata.*[3]

The great ginkgo slab from Afghanistan comes from rocks of Early Jurassic age that were laid down near the shores of an ancient sea around 190 million years ago. Each leaf is divided into six segments, and each segment is itself deeply bilobed. They look more like the leaves of ginkgo seedlings than like the leaves of mature trees, but there is no dispute about their identity; it does not take a specialist to immediately see the link to the familiar tree of our streets and gardens. This slab alone is enough to suggest the great antiquity of the ginkgo lineage, but it raises a still more fundamental question: where did ginkgo come from?

Through studies of living and fossil plants we now have a reasonably clear outline of the main events in the history of plant life and a rough idea of when they took place. In a few cases, fossil plants have also turned out to be key missing links that help connect seemingly separate pieces of the botanical evolutionary puzzle. For example, in the same way that *Archaeopteryx* from the Jurassic helps link modern birds to carnivorous dinosaurs, *Archaeopteris,* a fossil plant with a confusingly similar name, helps link living seed plants to strange extinct trees that reproduced by spores, like modern ferns, rather than by seeds. And while our understanding of the early evolution of animals has been informed by research on exquisitely preserved fossils from the Burgess Shale in British Columbia, studies of fossil plants from the Rhynie Chert, an equally remarkable example of exceptional preservation in Scotland, have added new depth to what we know about the early evolution of plants.

Discovered by the geologist William Mackie in 1912, the Rhynie Chert lies beneath a grassy slope just outside the small Scottish village of Rhynie, not far from Aberdeen. The chert is not even visible at the surface, but fortunately Mackie was an alert field geologist who noticed strange-looking rocks scattered in the field and built into nearby walls. Their source was later confirmed by digging trenches. More recently a team from the University of Aberdeen has drilled a series of holes to understand how this unique deposit was formed.[4]

The Rhynie Chert preserves the fossilized remains of a series of ancient peat bogs, one on top of the other, from about 400 million years ago. Different layers preserve entire ecological communities, with plants, algae, and fungi alongside a variety of spider-like creatures, as well as the oldest known insect. Everything is completely embedded in hard, glasslike silica exactly where it once lived, and many of the fossils are preserved with exquisite fidelity.[5]

The paleontologist Stephen Jay Gould devoted a whole book to the significance of the Burgess Shale for understanding the early history of animal life. The Rhynie Chert has had no such booster, but it is equally important for understanding the early history of plants. The five papers written on the fossil plants of the Rhynie Chert between 1917 and 1921 by Robert Kidston of the British Geological Survey and William H. Lang of the University of Manchester are classics of plant science. Together with superb work done toward the end of the twentieth century by others, especially Winfried Remy at the University of Münster, they revealed peculiar plants and important clues about the early development of life on land.

The Rhynie Chert provides gloriously detailed evidence that by around 400 million years ago plants had already begun to move out of the ponds, puddles, and tide pools in which they had evolved and established a significant foothold—roothold—on the land. All these plants are small, none more than a foot or so tall, and, compared with most plants of today, all are very simple in structure. Most consist of slender branches, sometimes with a capsule at the tip of each branch in which the spores were produced. A few show features reminiscent of living mosses or hornworts, while others, such as *Asteroxylon,* one of the most distinctive of the Rhynie Chert plants, are very like modern clubmosses. However, there is no hint that any of these small, simple plants was able to form wood, and none have large, complicated, leaves such as we find today in ferns, cycads, or ginkgo. The Rhynie Chert captures an early moment in plant evolution before the origin of trees, large leaves, and many other features that we take for granted in our modern world.[6]

Before the Rhynie Chert, even earlier evidence that plants were beginning to make the transition from living in water to living on land comes from dispersed plant spores with tough resistant walls. These spores are first recognized in the Middle Ordovician, about 450 million years ago, and are similar to those of some living liverworts. A little later, toward the end of the Silurian, about 420 million years ago, minute and very fragmentary plant fossils, often no bigger than the head of a pin, provide direct evidence of the tiny simple plants that produced some of these spores and that were the probable progenitors of the Rhynie Chert plants.[7]

The Rhynie Chert seems to have been formed near hot springs like those occurring today in Yellowstone National Park in the United States, or at Rotorua in New Zealand. Plants growing in the peat were intermittently flooded with hot water and rapidly impregnated with silica. In places the plants were dead and had already decayed before they were fossilized. Occasionally, however, living plants must have been overwhelmed, and fossilization evidently occurred within a few days or even a few hours. The sudden engulfment preserved delicate structures and ephemeral stages of ancient life cycles that can be hard to observe even in living plants.

All of the plants from the Rhynie Chert show the basic structural specializations that help make plant life possible on land. These include a water-resistant, waxy outer covering, the cuticle, which helps reduce water loss, and breathing pores, stomata, which regulate the exchange of carbon dioxide, water vapor, and oxygen between the environment and air spaces inside the plant tissues. Most of the Rhynie Chert plants also have elongated cells in the center of their stems: by analogy with living plants, the innermost of these helped transport water from the soil to the aerial parts, while the outer ones carried the sugars produced by photosynthesis to both the aboveground and belowground parts of the plant. So by 400 million years ago, many of the fundamental structures and processes seen today in ginkgo and other living land plants were already in place, and that these characteristics are shared by almost all land plants strongly suggests that plants, unlike animals, colonized the land only once.

The evolution of trees from small, simple plants like those preserved in the Rhynie Chert was probably driven mainly by competition for light. Tall plants have several potential advantages that may give them an edge in the struggle for survival, such as improved opportunities for spore dispersal, but light is indispensable for plant growth, and plants that are taller than their neighbors have access to more of it. During the Devonian, several different groups of plants show a tendency toward increased size, aided by different innovations that provided different kinds of structural support. Several groups also developed large complex leaves, not too different from those of ginkgo, cycads, and ferns, probably stimulated by the advantages that came from greater efficiency in harvesting the sun's energy. Such leaves apparently developed from elaborations of the simple branching seen in the Rhynie Chert plants.

In hindsight, major changes, such as the origin of leaves and the origin of trees, seem like dramatic evolutionary breakthroughs, but it is not hard to imagine how they could have come about rather gradually through the process of natural selection envisaged by Darwin. For example, in *Asteroxylon* from the Rhynie Chert, the elongated cells in the center of the stem have characteristic, often more or less spiral, internal thickenings, just like the water-conducting cells of most living plants. Initially, in the lives of individual plants growing more than 400 million years ago, water may occasionally have been in short supply; this seemingly minor difference may have been enough to give plants with weakly developed internal thickenings in their conducting cells a slight ecological edge over their neighbors. It might have been just sufficient to help the cells withstand pressures that would otherwise have caused them to collapse.

Over time, providing that the slight difference improved survival and could be inherited and passed on to the offspring of the survivors, the internal support of these cells may have been further strengthened in successive generations. Eventually, the same specialized cells may have been co-opted by natural selection to support increasingly large stems. The thickenings that initially helped prevent the cells from collapsing now provided an advantage in structural support. The elaboration of such cells, their production in large numbers, and their eventual modification for both water conduction and support allowed plants to grow into trees and also develop large leaves.

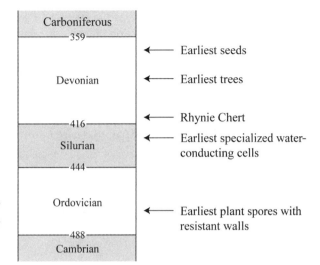

Key dates in the early history of land plants between about 500 million and 300 million years ago.

Similar kinds of contingent processes, building one upon the other, would also have been at work as plant reproduction became ever more attuned to life on land. Plants able to withstand desiccation would have survived preferentially when small freshwater ponds and damp places dried up, so it is no surprise that all of the land plants found in the Rhynie Chert, as well as their precursors from the Ordovician and Silurian, produced spores with tough walls. Resistant to drying out, dispersed by wind through the air and by water through the soil, these spores would have been the vehicle by which early plants survived tough times, and they would also have provided the means by which successive generations moved from place to place.[8]

When conditions were right, the spores would have germinated to produce a new free-living plant, but this time one that produced sex organs rather than spores. Remarkably, a few of these gamete-producing phases in the life cycle have been discovered in the Rhynie Chert, and in some it is just possible to make out the young sperm developing inside the male sex organs. The sperm cells would have swum through soil moisture to fertilize the eggs, exactly as happens today in mosses, liverworts, hornworts, clubmosses, and ferns. The motile sperm of the Rhynie Chert plants would have been a holdover from more fully aquatic ancestors, just as they are in ginkgo and in humans.

The evolution of seeds introduced new levels of complexity to the rather simple kind of sexual reproduction seen in the Rhynie Chert plants. It enabled further independence from water in reproduction, allowed for subtle modification of spores into pollen grains, and ultimately facilitated the de facto transfer of male gametes through the air. It also meant that the spore, from which the egg cell would ultimately be produced, could be retained and protected on the parent plant. This opened up new possibilities to support the growth of the spore, as well as the subsequent growth of the embryo, from the resources of the parent. These innovations effectively eliminated the free-living, gamete-producing phase of the life cycle and meant that male gametes no longer had to swim through soil moisture for fertilization to occur. They also created a new organ of dispersal, the seed, containing a prepackaged young plant.

Seed plants with this new way of reproducing themselves are first known from about 360 million years ago in the Late Devonian. They foreshadow the kind of sexual reproduction seen in ginkgo. Taken at face value, this sets a rough lower limit on the possible age of the ginkgo lineage. The leaves of *Ginkgo cordilobata* from Afghanistan set an upper limit of around 190 million years. One challenge for understanding the origin of ginkgo is to pinpoint the age of the lineage more precisely. Another is to place the living tree in its proper evolutionary position in relation to other plants.

12

Ancestry

Is it on his grandfather's or his grandmother's side

that the ape ancestry comes in?

—Samuel Wilberforce, Wilberforce-Huxley debate, June 30, 1860

To connect living ginkgo to the extinct seed plants that were alive between about 190 million and 360 million years ago, the traditional approach is to look for ancestors, beginning with those fossils that seem securely related to modern ginkgo, then working outward and backward to consider other fossils that might provide a connection to other kinds of seed plants. *Ginkgo cordilobata* from Afghanistan shows that plants with leaves like those of living ginkgo had already evolved by the Early Jurassic, between about 175 million and 200 million years ago. Similar leaves are also known, slightly earlier in the fossil record, from the Late Triassic, and especially interesting are rich and beautiful collections of fossil plants from rocks of the Molteno Formation in the Karoo Basin of South Africa.[1]

The Molteno fossils were collected and studied over many years by John and Heidi Anderson at the South African National Biodiversity Institute in Pretoria. Their massive collections, more than twenty-seven thousand specimens, come from almost seventy different localities that stretch in an ellipse from Little Switzerland and Golden Gate in the north of the Karoo Basin, to Askeaten, Aasvoëlberg, and Bamboesberg in the south. The fossils are preserved as beautiful impressions in dark to yellowish gray clays and silts. The quality of their preservation is much less impressive than the much older Rhynie Chert, but what these fossils lack in details of structure they make up for in the numbers of specimens available and the care with which huge collections have been assembled.

Collecting the fossil plants of the Molteno Formation was the work of a lifetime for John and Heidi Anderson. Both grew up in South Africa and despaired of the apartheid system, but in their professional lives, through decades of near scientific isolation, they spent many summers and countless weekends befriending farmers and ranchers and exploring their lands for new localities. The specimens were brought back to their laboratory in Pretoria, where they were carefully catalogued, described, and photographed. Their first major report focused on the most common fossils in the Molteno Formation. The second placed the Molteno in the context of all other fossil plant occurrences from South Africa. Most recently, they published two magnificently illustrated catalogues of the rich variety of extinct Molteno seed plants and ferns that lived in the lush forests and floodplains of what is now the dry southern tip of the African continent.[2]

Fossil ginkgo leaves are present at about one of every five localities in the Molteno Formation. The Andersons recognize six different species, which they cautiously assign not to *Ginkgo* itself but to a very similar genus, *Ginkgoites,* that paleobotanists often use for ginkgolike leaf fossils. Some of these leaves, such as *Ginkgoites koningensis* and *Ginkgoites matatiensis,* are similar to Schweitzer and Kirchner's *Ginkgo cordilobata* from Afghanistan. Others, such as *Ginkgoites muriselmata,* have leaf lobes that are more sharply pointed. *Ginkgoites telemachus* is a variation on the same theme, but with leaf lobes that are irregularly toothed. All these species, at least in terms of their leaves, are unmistakably related to living ginkgo, and they occur repeatedly with seed-bearing structures, which John and Heidi named *Avatia,* and probable pollen catkins, which they called *Eoasteria.* As far as we can tell these seeds and catkins are not very different from those of living ginkgo, and their relationship to the living tree seems pretty secure.

The Molteno fossils date from about 220 million years ago; they extend the fossil record of ginkgo back about 30 million years before *Ginkgo cordilobata.* Fossil ginkgo leaves of roughly similar age are known from many other parts of the world, including Arctic Canada, eastern North America, the southwestern United States and northwestern Mexico. Ginkgolike wood has also been described from just outside the Petrified

Forest National Park in Arizona. However, farther back, obvious ginkgo fossils quickly disappear. According to the Andersons' compilations, the oldest occurrence of ginkgo leaves in South Africa is in the early part of the Middle Triassic, about 240 million years ago. In the Southern Hemisphere as a whole early ginkgo is widespread and the earliest record is from the Sydney Basin of Australia, dating from near the end of the Early Triassic at about 245 million years ago. There are no older records in the Northern Hemisphere, and before this the obvious fossil history of ginkgo peters out.[3]

Tracing the ancestry of living ginkgo still farther back is difficult. The fossils become less like the living tree, and their relationships become less secure. Rudolph Florin, one of the great paleobotanists of the twentieth century, a member of the Royal Swedish Academy and director of its Bergianska Garden, was among the first who were bold enough to make a strong link between ginkgo and fossil plants from the Permian, the geologic era before the Triassic. Florin focused on a kind of fossil plant that had first been described more than seventy years earlier by the brilliant French paleobotanist Gaston de Saporta.

Rudolph Florin is best known for his contributions to understanding the history and evolutionary position of living conifers, but his most important work on ginkgo was published in 1949, based on an examination of three fossils similar to those that Saporta had named *Trichopitys heteromorpha*. All of the original material of *Trichopitys*, including that examined by Florin, is from old coal mines in the region around Lodève, southern France, which are Early Permian in age and produce fossils that date from about 275 million to 290 million years ago. Saporta had already suggested that *Trichopitys* might be related to ginkgo based on the similarity of the ribbonlike leaves to some deeply divided but unmistakably ginkgolike leaves from much younger rocks. Florin supported this idea and added the point that some of the leaves seem to have a small branch attached near their base, and that some of these branches have seeds at their tips. He compared these branches with the seed-bearing shoots of living ginkgo, and also pointed out that living ginkgo sometimes produces abnormal, branched seed-bearing shoots that may bear as many as ten small seeds.[4]

On a warm summer day in 1986 I had the chance to examine the original material of *Trichopitys* described by Saporta from Lodève in the collections of the Natural History Museum in Paris. With me was Sergei Meyen, one of the foremost Russian experts on plant paleontology and a specialist in Permian plants. We were both traveling back from a conference in Montpellier; Sergei asked whether I would like to see the specimens, and I followed along. They are preserved alongside ancient conifers and other plants as black fossils on a hard gray slate. The preservation is not very good; it is not possible to make out microscopic details, and the fossils are not easy to understand. We puzzled over them together for some time. Eventually, though, with deference to Florin's many other paleobotanical accomplishments, I had to agree with Sergei that the structure of these fossils may not have been quite as Florin described and that the link to ginkgo was not completely convincing.

One problem is that without more detailed information, the relationship between deeply divided ginkgolike leaves from the Triassic and those of *Trichopitys* from the Permian is problematic, especially given the presence of similar leaves in other groups of extinct plants. Second, as Sergei pointed out, what Florin interpreted as a branched seed-bearing axis attached at the base of a highly divided leaf might equally have been, at least in the specimens that we examined, just another deeply divided but somewhat flattened leaf segment. Florin's linking of *Trichopitys* to ginkgo depends on the seed-bearing structures' being borne at the base of a leaf, just like those of ginkgo, which is far from certain.[5]

Given these unresolved issues, *Trichopitys,* long held up as a likely ginkgo ancestor, the link between unequivocal ginkgo from the Triassic and earlier seed plants, for the moment is best treated as of uncertain relationship. Indeed, as more and more fossils have come to light from Permian rocks, *Trichopitys* increasingly seems to be part of a much more complicated picture, just one among many interesting but enigmatic fossil seed plants from around that time. In some general way, these enigmatic Permian fossils seem to provide a bridge between the ancient seed plants of the Paleozoic and the more modern seed plants of the Mesozoic, but exactly how they relate to one another and to older and younger plants is still uncertain.[6]

Sergei Meyen struggled with these kinds of questions, using his unrivaled knowledge of Permian fossil plants from Russia and elsewhere. Toward the end of his life he tried hard to develop a comprehensive overview of seed plant evolution that would account not just for living plants but also for the vast diversity of Permian and other extinct forms. He saw ginkgo as of particular importance and placed it at the center of what he called Class Ginkgoopsida, which he assembled by "congregational analysis taking into account as much information as possible."[7]

Sergei put a huge amount of care and thought into the development of his ideas, but his concept of Ginkgoopsida is extraordinarily broad and includes plants that display huge differences in many of their features. Many of the key fossil plants are also not very well understood. Most are known only from leaves or leaves with associated seeds. Frustratingly few can be compared point for point with living ginkgo, and while this is a normal problem with fossil plants, it is a huge impediment to understanding how they might relate to one another and to living plants.

In the fossil record of the past 350 million years it is rare that anything approaching a complete plant is preserved. In the early phases of land plant evolution, at the time of the Rhynie Chert, all of the plants were small and relatively simple, and often they were preserved more or less intact. As a result they are relatively easy to understand. However, as plants became larger and more complicated, the fossil record increasingly becomes one of isolated fragments: a leaf here, a seed there, with stems, pollen cones, and pollen grains somewhere else. Imagine the fallen leaves, twigs, fruits, and seeds from a rich patch of forest being washed out into an ancient lake, settling to the mud on the bottom and then being chipped out of the rock a few million years later. How would you know which piece goes with which?

Needless to say, this problem greatly complicates efforts to use fossil plants to understand plant evolution. For example, between about 270 million and 300 million years ago, around the time of *Trichopitys,* several other fossils might be important for a deeper understanding of the ginkgo life story. The Russian paleobotanist Serge Naugolnykh has described seed clusters from the Ural Mountains similar to those linked with ginkgolike leaves in the Triassic. From other sites in the same area he has also collected intriguing fossil leaves called *Kerpia,* which look a lot like those of ginkgo. The problem is that in both cases they are just isolated pieces. It is hard to make sense of them, still less to understand their significance, without more information about the rest of these ancient plants.[8]

Paleobotanists with an interest in plant evolution therefore spend a good deal of time trying to reconstruct something approaching a whole fossil plant from its separated pieces. A colleague once called it the Humpty Dumpty game: the aim is to put the pieces back together again. When the conditions of preservation are favorable, or where we are lucky enough to have fossils that are attached to each other, then we may be able to understand which fossil leaf goes with which seed or which pollen cone. In these fortunate situations it is sometimes possible to link the stems, leaves, seeds, pollen-producing structures, and pollen that were all produced by the same extinct species. Such reconstructions provide the key anchor points for understanding the evolutionary history of plants in what can otherwise be an extraordinarily confusing imbroglio of isolated bits and pieces of ancient plants.

One group of Permian plants that is relatively well understood and that may be relevant to the evolution of ginkgo are the so-called glossopterids. Sergei Meyen also placed glossopterids within his Ginkgoopsida. Glossopterid leaves, generally assigned to the genus *Glossopteris,* are relatively simple, with many fine veins that are all of about the same thickness. In some ways they are comparable to the leaves of ginkgo, but they differ in one important respect: the leaf veins form extensive elongated reticulations.[9]

Glossopterid leaves have been well known to paleontologists for more than a century and were among the specimens found with the bodies of Captain Robert Falcon Scott and his party, who perished as they returned from the South Pole in 1912. The occurrence of glossopterid leaves in all the Southern Hemisphere continents was important early evidence for the now widely accepted theory of continental drift.[10]

Botanically, glossopterids are much better understood than *Trichopitys.* Their seed-bearing and pollen-producing organs are borne very clearly at the base of the leaf, and in many cases they are directly fused to it. We also know that glossopterids were large trees with woody trunks, and sometimes had distinct long and short shoots not dramatically different from those of living ginkgo. There are also beautifully preserved specimens from Australia that provide a hint that glossopterids produced sperm cells like those observed by Hirase in ginkgo.[11]

A further contender for an early ginkgo relative is one of the plants described by John and Heidi Anderson from the Molteno flora of South Africa. The leaves, which they called *Kannaskoppifolia,* are much more ginkgolike than those of *Glossopteris.* They are wedge-shaped and deeply lobed, and while all the specimens that the Andersons assign to this genus have some reticulations among the leaf veins, this varies among the ten species that they recognize; in some, as in living ginkgo, reticulations are relatively uncommon. There are also many specimens showing that the seed-bearing structures, which they named *Kannaskoppia,* were borne at the base of the leaves. However, in this case the reproductive structures look quite different from those of living ginkgo. The seed-bearing structures are recurved and cuplike, and borne in large numbers on complex branches. Each recurved structure presumably contained one or more seeds, but

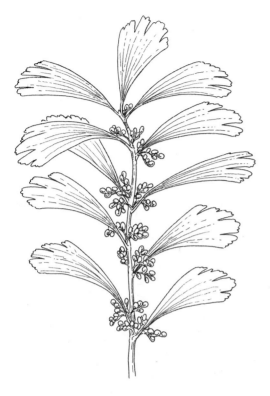

Reconstruction of the possible ancient ginkgo relative Kannaskoppia, *with its partly divided leaves (*Kannaskoppifolia*) and branches bearing numerous small, recurved seedlike structures (*Kannaskoppia*) from the Late Triassic of South Africa, about 220 million years before the present.*

no details are known. The possible pollen-producing organs, which were almost certainly produced by the same plant, were given the name *Kannaskoppianthus*. They also look strange. What seem to be pollen sacs are borne on a peculiar curled-over structure that resembles fingers pressed onto the palm of a hand.[12]

The main practical problem in evaluating the relationships of plants like glossopterids and *Kannaskoppifolia-Kannaskoppia* to each other and to living ginkgo is that even in these relatively well-known cases, just as with *Trichopitys*, we are missing important information. For example, other than in glossopterids, there are no details of the internal structure of the seeds of these plants, no information about pollen, and no knowledge of the internal tissues of the stem. Similarly, in cases like *Kannaskoppianthus*, we are also uncertain about how the basic structure of the fossils should be interpreted. With so much information missing and so much uncertainty about key features, it is hard to compare these plants with each other, still less with a living plant like ginkgo.

However, there is also a theoretical problem: along with the similarities, there are many differences. For example, in glossopterids the details of the pollen sacs and pollen grains are quite different from those of living ginkgo. So in trying to track down the origin of ginkgo, we need not only more information on the relevant fossil plants but also a method that helps us know what to make of the various patterns of similarity and difference.

13
Relationships

Maternity is a matter of fact, paternity is a matter of opinion.

— Proverb

There are many fossil plants, especially from the Permian and Triassic, which may be important for understanding the origin of living ginkgo, but, as in the case of glossopterids, *Kannaskoppifolia-Kannaskoppia,* and *Trichopitys,* it is hard to decide which are the most important. In some cases this is because the fossils are not known in sufficient detail to allow useful comparison, but there is also the problem of how to account for and understand the similarities and differences we observe. For example, should we focus more on the leaves, or is it similarities of the seeds that should be given greater weight? And how should we choose among different competing ideas that might link ginkgo to glossopterids on the one hand, or *Kannaskoppifolia-Kannaskoppia* on the other? This uncertainty suggests that we might be in need of a different way to think about this problem, rather than simply trying to trace the ancestry of ginkgo backward in the fossil record.

Instead of looking for ancestors, an alternative approach is to ask instead: to which group of living or fossil plants is ginkgo most closely related? For example, is ginkgo more closely related to cycads than it is to conifers, or vice versa? Or is ginkgo more closely related to glossopterids or *Kannaskoppifolia-Kannaskoppia* than it is to any living plant? Framing the question in this way forces evaluation of the evidence for alternative ideas about the relationships among the different groups of seed plants. What we know and what we don't know needs to be clearly spelled out, and in such a way that we can choose among different competing ideas. This approach, based on relative degrees of relatedness, is the primary way that these kinds of "origins" questions are addressed today across the whole of biology, whether the focus is the origin of mammals or the origin of the HIV virus.

The modern methods used to assess the evolutionary interrelationships of organisms developed from the work of the German entomologist Willi Hennig. Hennig, a specialist on living and fossil flies, began his research in the 1930s, but his most influential work, *Grundzüge einer Theorie der phylogenetischen Systematik,* was begun at the end of the Second World War during his time as a prisoner of the British. The book appeared in 1950, and an English translation, undertaken at the Field Museum in Chicago, was published in 1966.

Subsequently, during the 1970s and 1980s, Hennig's ideas were developed and elaborated through lively, and often acrimonious, debates mainly centered around key participants at the American Museum of Natural History in New York and the Natural History Museum in London. For a young scientist working and teaching in this area on both sides of the Atlantic in the late 1970s and early 1980s, it was fascinating, although not always comfortable, to observe this scientific revolution at close quarters. It was a true paradigm shift in modern evolutionary biology, and even on the fringes of the intellectual action it was heady stuff. The outcome was a new, theoretical basis through which questions about evolutionary relationships could be approached and answered.[1]

Hennig's breakthrough was to recognize there is a hierarchy among the different features that organisms share; and that this hierarchy makes sense in terms of a simple model of the evolutionary process. The features might be characteristic structures, or they might be components of DNA sequences, but Hennig's point was that they can be used to define successively less and less inclusive groups nested one inside the other like a matryoshka doll. Membership within those groups defines successive levels of relationship: successive degrees of relatedness.

What this means in practice is that relationships are defined in relative terms. For example, ginkgo and conifers are regarded as more closely related to each other than either is to a moss, because both belong to a group of plants, vascular plants, defined by the presence of water-conducting cells with specially reinforced walls. In the same

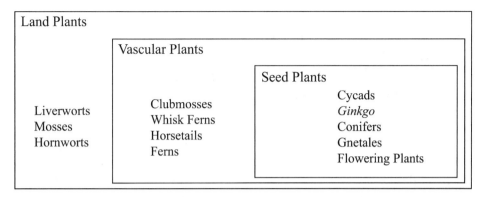

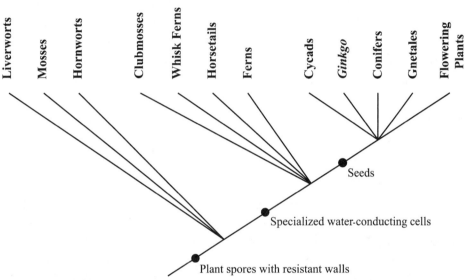

The twelve major groups of living plants, classified into three larger groups, above, that are nested like matryoshka dolls of different sizes. These nested relationships can also be expressed in a treelike diagram, below, that approximates evolutionary relationships.

way, ginkgo and conifers are regarded as more closely related to each other than either is to a fern because both belong to a group of plants defined by the ability to produce the massive woody tissues seen in the trunk of a normal tree. No living fern produces wood or becomes a tree in the same way. At the next level in the hierarchy, ginkgo and conifers also belong to a more inclusive group, seed plants, that reproduce by seeds rather than by spores, and so on.[2]

Obviously, in many cases the hierarchy of features is not quite so clear-cut, making it more difficult to understand relative degrees of relatedness. Sometimes, different features suggest different hierarchies, and therefore imply conflicting patterns of relationship, and when fossils are included for which key information is lacking, the situation gets still more complicated. Often there are different alternative hierarchies of relationships that could potentially explain the same data, and the number of potential alternative patterns of relationship increases sharply as the number of different groups under consideration increases. For four different groups there are only fifteen possible arrangements, but to assess the potential relationships of ginkgo to nine other plants the number of possible alternative arrangements among the ten groups under consideration is a staggering 34,459,425.[3]

In such cases the method used to choose among competing patterns of relationship is to invoke the standard scientific principle that the simplest explanation is preferred. What this means from an evolutionary point of view is that, for a particular set of plants and a given set of features, the preferred explanation minimizes the number of separate evolutionary events needed to explain the origin of these features. For example, it makes more sense to support an explanation in which the production of woody tissues and seeds each evolved only once rather than an alternative explanation that requires multiple origins of each feature.

Obviously, with many plants and many different features, and therefore a very large number of possible explanations, it may be hard to find the one that really is the simplest. There may be multiple patterns that are equally, or almost equally, simple. For this reason the application of Hennig's ideas drew huge new impetus from the development of computers and appropriate software through which the large datasets could be analyzed. Even with the most sophisticated approaches, finding the simplest explanation requires either complex calculations that approximate an answer or exhaustive mapping of many different features onto an almost overwhelming number of different potential patterns of relationships, or "trees," as they are sometimes called. For any complex analysis involving more than a few plants and more than a few features it is impossible to do this by hand.[4]

A further obvious problem is that it is not easy to compare organisms that differ greatly in their structure and biology. For example, how do you compare ginkgo with a moss, or a sea urchin with a shark? Point-by-point comparisons are often hard to make. For this reason the study of evolutionary relationships received a further boost from the recognition that relatively short DNA sequences extracted from very different kinds of organisms could be obtained rather easily and, with the right analytical software, could be compared very straightforwardly. In plants, the first large-scale applications that used DNA data and computer analyses to apply Hennig's ideas were done in the 1990s, and since then there has been rapid progress with understanding how different groups of plants are interrelated. We now have strong evidence, for example, that magnolias are more closely related to bay laurels than to water lilies, and that the sacred lotus is more closely related to poppies and plane trees than to water lilies, grasses, or palms. For the more than 350,000 species of flowering plants, there is now a well-corroborated framework of relationships in which other aspects of plant evolution can be studied and against which the fossil record can be compared.[5]

At first glance, especially given the very rapid progress that has been made in other areas where a much larger number of plants is involved and where broad consensus on the pattern of relationships has been achieved, figuring out where ginkgo fits among living seed plants ought not to be too difficult. After all, beyond ginkgo itself, there are only four other living groups: conifers, cycads, flowering plants, and a peculiar and rather obscure group known as Gnetales. We also have a huge amount of information about the similarities and differences among their DNA sequences. The cost of obtaining longer and longer DNA sequences, or even sequencing the entire DNA of an organism, also continues to get cheaper and cheaper all the time. It ought to be easy to determine which one of the 105 potential patterns of relationships for the five groups is most strongly supported by all the available evidence.[6]

The question of how the five groups of living seed plants are interrelated has received an enormous amount of attention, beginning with papers published thirty years ago, long before the application of DNA techniques to these kinds of problems. The focus has mainly been on trying to understand the relationships of flowering plants, and hence to learn something about their origin. However, there is still no consensus on which of the many different patterns of relationship that have been revealed by slightly different kinds of analyses, on slightly different kinds of data, more accurately reflect the actual pattern of evolution. When the many extinct groups of seed plants are also included—and the data are thus restricted to the limited information that can be gleaned from fossils—a still different set of possible answers emerges. In many analyses based on DNA sequences, and others that include fossils, ginkgo seems to be more closely related to conifers than to any other group of living seed plants. However, other

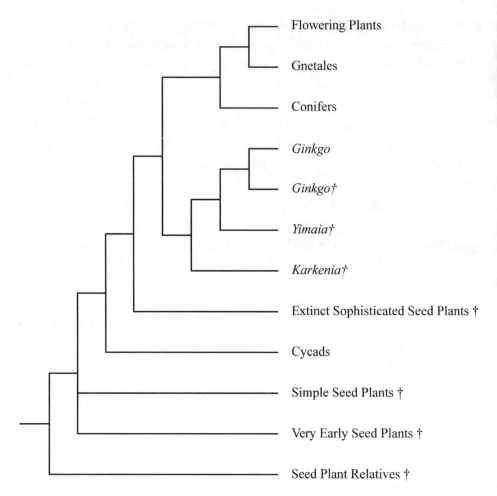

One plausible treelike diagram that summarizes the possible relationships among selected groups of living and extinct seed plants. Dagger (†) indicates extinct groups. For more on Yimaia *and* Karkenia *see Chapter 15.*

analyses come to different conclusions and find instead that conifers are closely related to Gnetales, with ginkgo equally closely related to both.

Given the effort already invested to try to solve this problem, it may be that a truly definitive answer to the question of whether ginkgo is more closely related to conifers, cycads, flowering plants, or Gnetales will never emerge from the brute-force application of more and more molecular data. But more important, even if the relationship of ginkgo to other living seed plants could be solved, the answer is unlikely to be helpful with regard to what we really want to know, which is how ginkgo fits into the broader constellation of living and extinct plant diversity and how its characteristic features evolved. Solving these problems will require probing more deeply, more thoughtfully, and in an integrated way, the similarities and differences between ginkgo and other seed plants, especially those seed plants that are known only as fossils, about which we still have much to learn. We will also need to see through apparently conflicting signals from different lines of evidence and reconcile them into a pattern of relationships that most closely approximates the actual course of evolution.[7]

At the moment we can only fall back on the lame generalization that ginkgo probably had its origin among those groups of seed plants with relatively simple leaves and flattened seeds that diversified during the latter part of the Paleozoic. These groups, which included fossil plants like *Kannaskoppifolia-Kannaskoppia*, glossopterids, and perhaps *Trichopitys*, represent a second wave of seed plant evolution beginning about 300 million years ago, which succeeded the more ancient seed plants of the Devonian and Carboniferous. It is all frustratingly vague, but beyond that it is impossible to go at the moment. Despite our best efforts, exactly how ginkgo fits into the grand scheme of plant evolution remains elusive.

14
Recognition

Too much light often blinds gentlemen of this sort.

They cannot see the forest for the trees.

—Martin Wieland, *Musarion oder die Philosophie der Grazien*

Many of the most famous figures in the history of plant paleontology have written about ginkgo, but none has done more to illuminate its long evolutionary history than the Chinese paleobotanist Zhou Zhiyan. Beginning with a series of influential studies in the 1980s, Zhou discovered several different kinds of early ginkgolike plants and came to understand them not just from their leaves but also from their seeds and other parts. He also placed his new discoveries in the broader context of what earlier researchers had learned. As a result, the history of the ginkgo lineage is now better understood, and also much more complicated and interesting, than it was thirty years ago. In 1994, Zhou's work was recognized by one of the most prestigious awards given by the Chinese Academy of Sciences, and the following year Zhou was elected to the academy, a high honor in a culture where science is held in great esteem. Zhou is the most respected of all Chinese scholars of the plant fossil record, and his work is central to a full understanding of living ginkgo.[1]

Zhou Zhiyan spent the first half of his career in China, but in 1980 he traveled to the United Kingdom to study with Professor Tom Harris at the University of Reading. It

Tom Harris (1903–1983), the British paleobotanist who contributed much to our understanding of ancient ginkgo and other fossil plants. Photographed at the University of Reading in summer 1980.

was one of the pivotal experiences of Zhou's life. His priority was to study a collection of fossils that he brought with him from China, but had he known that ginkgo would figure so prominently in his later career, he might have spent his time differently. Harris had also made important contributions to knowledge of ginkgo fossils. They would have had much to speak about.[2]

Harris arrived at Reading in 1934 as one of the first professors to be appointed after the university was founded, and he worked there until his death in 1983. By the early 1980s Harris was in his late seventies, but he was still a force to be reckoned with: every day he came to the Department of Geology to work on the latest plant fossils to command his attention. Tall and gangly, with a strong personality, Harris had a steel-trap mind that saw the world in black and white.

Harris was generally warm to those around him, and he was nothing but kind and generous to me in my years at Reading, but he did not suffer fools. He had little patience with those who saw the world in shades of gray. He built his reputation in the middle of the twentieth century based on clear thinking combined with energetic efforts to collect new fossils. The resulting stream of published research spanned almost six decades. His magnum opus, at least in the second half of his career, was his comprehensive five-volume treatment of *The Yorkshire Jurassic Flora*. He became the dominant force among British paleobotanists in the decades after the Second World War, and the cumulative impact of his work was enormous.[3]

Harris went to Cambridge intending to become a doctor, but he was something of a child prodigy. At the age of eighteen, traveling every day from his home in Leicester to the University of Nottingham, he had already received his bachelor of science degree. It was through the influence of H. S. Holden, whom Harris met at Nottingham, that he first became interested in fossil plants. Holden had been trained at the University of Manchester, which had long been a stronghold of scientific paleobotany in Britain, and his enthusiasm for such work evidently rubbed off. Harris quickly discarded medicine in favor of botany.[4]

At Cambridge, Harris became part of the circle around A. C. Seward, probably the foremost paleobotanist of his time. Twenty years after Charles Darwin's death, Seward had edited a volume of his letters with Darwin's son Francis, but his main interest was fossil plants, and he wrote prolifically on all aspects of the subject. Seward was also a skilled administrator, and by the 1920s, when Harris arrived, Seward was well established, well respected, and well connected.[5]

Seward made important contributions in many different areas of paleobotany and worked on a range of fossils from many different parts of the world, but he was particularly intrigued by fossils of flowering plants. It was Seward who was largely responsible for popularizing a phase used by Darwin, who once remarked that the origin of flowering plants was an "abominable mystery." With a gentle push from Seward, that phrase has been quoted more than Darwin could ever have imagined, and perhaps more than he ever would have wished.[6]

In the 1920s Seward was interested in fossil flowering plants from about 100 million years ago and especially in material that was then becoming available from the work of Danish geologists in West Greenland. He traveled there himself in the summer of 1921 and, soon after, was sent a collection of fifteen packing cases of fossil plants from much older rocks collected on the other side of the icecap, from the fjords about halfway up Greenland's east coast. As Harris told it, the crates were sent to Seward by mistake. The material had been collected in 1900, on an early geological expedition by the Danish geologist Nikolaj Hartz. Harris seized the opportunity, made the collection the focus of his research at Cambridge, and never looked back.[7]

In 1925 Harris visited Stockholm and Professor Thor Halle, Nathorst's successor at the Swedish Museum of Natural History, to learn the techniques needed to study the Greenland material. Harris used those same techniques, with little modification, for the rest of his career. The following year, over dinner at Seward's home in Cambridge, he met the Danish geologist Lauge Koch. In Harris's words, Koch was a "huge, rather fierce looking man in charge of the Greenland Geological Survey. He asked me — could I come with him on an expedition to East Greenland for a year, starting next month. It appeared to me instantly that it was one of those situations where thought does not lead to a wiser decision, so I said 'yes.'"[8]

In the summer of 1926, after a brief stop in Copenhagen and a three-week passage to the east coast of Greenland, the small party lead by Koch landed at Scoresby Sound. Along with Koch and Harris, the group consisted of Alfred Rosenkrantz, a Danish geologist and engineer, two Eskimo hunters, and about fifty sledge dogs. Their aim was to help understand the geology of that part of East Greenland by making extensive collections of fossils, and especially to expand the collections of fossil plants made by Hartz.

The expedition had enough supplies to last the winter, but their lives were not easy in their isolation. The climate was harsh, the terrain vertiginous, and the slopes slippery on top of the permafrost. Living conditions were basic, but in the year they spent there, much of it during the long winter and using primitive equipment, Harris and his colleagues amassed a large collection of fossil plants from a geologic section about three hundred feet thick. Harris remarked, "Each bed had one or a dozen species often all different from those of a neighboring bed," and in some beds there were abundant fossil leaves unquestionably similar to those of modern ginkgo. The party was picked up again the following summer. Tons of rocks with plant fossils, including many ancient ginkgo leaves, were shipped back for study.

In all, Harris worked on the fossils from East Greenland for about a decade, and these were the most paleobotanically productive years of his life. He described a rich variety of ferns, horsetails, and clubmosses, as well as conifers, cycads, and a wide range of peculiar extinct seed plants. He established the East Greenland flora as one of the best-known fossil floras in the world, and it quickly secured his reputation; he moved to Reading as the first professor of botany at the age of thirty-one.

Harris carefully described the ginkgolike leaves from East Greenland and recognized that they were more deeply and more regularly divided than the leaves of modern *Ginkgo biloba*. He called them *Ginkgoites taeniata* and noted that they were more like the leaves produced by ginkgo seedlings, or the early leaves produced from regen-

erating shoots, than leaves from a mature tree. Also, using the classic but rather drastic paleobotanical technique that he had learned from Halle, and later taught to Zhou,
Harris employed strong acids, followed by a strong alkali, to dissolve away the coal-like
material of the fossil leaves and leave just the resistant, waxy outer cuticular covering
of the upper and lower leaf surfaces. He then used similarities in their cellular details
to show that although variable in shape and size, all the ginkgolike leaves were most
probably from a single species. Harris also suggested that some of the seeds he found
associated with *Ginkgoites taeniata* leaves were probably produced by the same trees.
He could not prove it, but he was convinced that they were part of the same plant based
on their consistent association at different localities and similarities in their cuticular
coverings.[9]

When the Greenland work was completed, Harris, by that time at Reading, needed
to decide what to work on next. Rather than continuing with the Greenland material,
he turned his attention to something new and embarked on a comprehensive revision
of the beautifully preserved Jurassic plants from Yorkshire in the United Kingdom.
These fossils had been collected and studied since the earliest days of scientific paleontology, and by the early 1930s the Yorkshire material had already been worked on
extensively, including by Nathorst in Stockholm and Hugh Hamshaw-Thomas, one of
Harris's senior colleagues at Cambridge. Thomas especially had done extensive fieldwork and made major new collections. Harris, however, was undeterred; in his view
there was plenty still to be done, and Yorkshire was easier to get to than East Greenland.

Harris never worked again on the material from Greenland, although many years
later he did allow that perhaps this was a pity. Harris once said to me that he would
have made more exciting discoveries had he focused on the Greenland fossils; but he
then added, with his characteristic grin and bob of the head, that he also would never
have enjoyed so many holidays in Yorkshire. Many of those holidays, with his family
in tow, were apparently spent collecting at the classic localities, and walking the North
Yorkshire Moors looking for new sites that might give new information or produce
new kinds of fossil plants.

Through *The Yorkshire Jurassic Flora*, Tom Harris made major contributions to
understanding the plants that grew in the estuaries, back swamps, and floodplains of
the ancient coastline that existed about 150 million years ago in what is now northeastern England. That coastline, which today is often swept by biting winds off the North

Fossil leaves of an ancient ginkgo, Ginkgo huttoni, *from Scalby Ness, Yorkshire, collected from the sands of an ancient meandering river that flowed out toward what is now the North Sea about 170 million years ago.*

Sea, was then home to diverse luxuriant, more or less tropical vegetation of conifers, cycads, ferns, and many kinds of plants now extinct. It was entirely devoid of modern kinds of mammals, as well as birds, butterflies, bees, and many other animals that today we take for granted. Instead, it was home to dinosaurs and pterosaurs; beetles and early flies would have been among the common insects. Harris devoted the bulk of his career to bringing the plants of these ancient ecosystems to life. Building handsomely on what had gone before, his work established the fossil flora from Yorkshire as the benchmark against which all other fossil floras of this age are compared.[10]

Ginkgo leaves occur at several of the classic Yorkshire localities but are especially common in the Scalby Ness plant bed, which is exposed on the coast just north of Scarborough. I collected them there for the first time in the autumn of 1974 with Harris, his former student Joan Watson, and her students from the University of Manchester. I have returned many times since and have never been disappointed.

The cliffs at Scalby Ness offer a crude cross section through the sands of an ancient river, now hardened by time, that once meandered out to sea. The fossil plants occur tumbled together exactly where they settled among the river sands millions of years ago. By far the most common fossil plants are the deeply divided leaves of *Ginkgo huttoni*. Over the course of his career, Harris also found seeds very like those of modern ginkgo at Scalby Ness. They were similar to those that he had collected in Greenland, and Harris thought that they were probably produced by the plant that bore *Ginkgo huttoni* leaves. He also described a single pollen-producing structure similar to that of modern ginkgo.[11]

The species of ginkgo that Tom Harris described at different times in his career, first from East Greenland and then Yorkshire, are unquestionably part of the lineage leading to modern *Ginkgo biloba;* they are certainly more closely related to ginkgo than any other living plant. However, our knowledge of these plants is incomplete in several crucial ways, especially in showing how the seeds were borne. Harris's painstaking work did much to clarify the exact similarities between the leaves of fossil and living ginkgo, but he more than anyone else recognized that not much more could be done until some fossil ginkgolike plants were known from more complete information. Harris always emphasized that when the opportunity arose, assembling whole fossil plants should be the top priority. Without such reconstructions Harris knew that it would be impossible to make more useful comparisons between the living and fossil species. It was a central theme in his life's work and a philosophy he passed on to all of his students, including Zhou Zhiyan.

15
Proliferation

Be fruitful, and multiply, and replenish the earth.

—Genesis 1:28

Zhou Zhiyan arrived at the University of Reading in September of 1980, and for almost a year we worked side by side in the same laboratory. I was in my mid-twenties and coming to the end of my first job in the Department of Botany. Zhou was in his late forties and a senior scientist at the Institute of Geology and Palaeontology of the Chinese Academy of Sciences in Nanjing. He was on his first trip outside China. At that time, the competition for such opportunities was fierce, and Zhou was among the first Chinese scientists allowed to travel overseas in the years following the death of Mao. He brought with him interesting fossil plants to study, but his broader objective was simple. He intended to learn all he could and make up for lost time: the years spent on what he calls "unprofessional activities"; the political meetings, manual labor, and other activities unrelated to science that he endured during the Cultural Revolution.[1]

Zhou was born in Shanghai in 1933. In the early 1950s he studied at Nanjing University and gained a position at what was then the Nanjing Institute of Geology and Palaeontology. Like Tom Harris, he specialized in fossil plants of the Mesozoic, but while Harris had been a botanist almost from the beginning, Zhou's training was in geology. Most of his early work used fossil plants to support geological exploration; his fossils

Zhou Zhiyan, the Chinese paleobotanist whose work revolutionized our understanding of ancient ginkgo based on studies of fossils from China and elsewhere. Photographed in early 2009 outside the Museum at the Nanjing Institute of Geology and Palaeontology of the Chinese Academy of Sciences.

were useful indicators of geologic age, and his broader goal was to better understand the commercially important coals that would be crucial for China's economic development. Only later, and especially after his stay in Britain, did Zhou turn his attention to botanical questions, including the area in which he is now preeminent: elucidating the history of ginkgo and its relatives. More than anyone else, in his own gentle and understated way, Zhou Zhiyan has helped make sense of the fossil ginkgolike leaves from between about 60 million and 225 million years ago and has illuminated what they tell us about the evolution of the single living species.

Unless they devote themselves entirely to theory, or to working on collections made by others, all paleontologists need a bit of luck. One way or another they need to have good specimens that are sufficiently well preserved to give new information and useful new insights. However, in paleontology as in other areas, fortune favors the prepared mind. Good paleontologists are always on the lookout for interesting new material, and when serendipity brings it their way, they know what to do with it. Zhou Zhiyan's studies of early ginkgo fit this profile: the right material came along at just the right time, he recognized its significance, and he did what needed to be done.

In the mid-1980s, not long after returning from Britain, Zhou was contacted by Zhang Bole, a mining engineer working in the Yima region of Henan Province, in northern China. Zhang was a professional geologist, and he knew a good deal about the coal mines in that area, but he was also a dedicated fossil hunter with a particular interest in plant fossils. He and his family devoted much of their spare time to collecting fossil plants from the spoil heaps of the large open-cast coal mine at Yima, and the hours that they had invested had yielded spectacular fossil ginkgo leaves, many of them beautifully preserved in a soft gray siltstone.

Zhou recognized the potential of Zhang's material immediately. It was of Jurassic age, dating from about 170 million years before the present, and was much better preserved than the specimens he had studied earlier in his career, including the ones that he had taken with him to Reading. Most Jurassic plant fossils from elsewhere in China—for example, from the coal mines around Beijing—occur in rocks that have been subject to great temperatures and pressures after they were formed; as a result, the fossil plant material is badly squashed, chemically altered, and almost hopeless for providing detailed information. In contrast, the Henan material was preserved in softer rocks that had not been deeply buried. The fossils looked more like some of the well-preserved material from the Jurassic of Yorkshire that Zhou had seen in Britain. He quickly realized that the waxy cuticular coverings of the leaves and other plant parts were still preserved and that, with the right techniques, these fossils would yield useful information.

An immediate priority was to visit the site and collect more fossils, so in 1986 Zhou traveled to Henan and spent several days at the Yima coal mine collecting specimens with his student Xuanli Yao. He also pored over the large collections that Zhang and his family had made and recognized that among the beautifully preserved ginkgolike leaves were not just seeds but also the structures on which the seeds were borne. The leaves, seeds, and seed-bearing structures were so common that they were almost certainly produced by the same plants.

The first report that Zhou Zhiyan and Zhang Bole published on fossil plants from the Yima mine was a preliminary announcement of two new types of ginkgolike seed-bearing structure. They noted that "as a detailed study will take quite a period of time, we are disposed to announce the important features and to present briefly a preliminary assessment here for reference to colleagues who are interested in the past history of *Ginkgo.*" They recognized that both these seed-bearing structures were entirely new discoveries, and they drew the obvious conclusion; there was more than one kind of ginkgolike plant preserved at the Yima locality.[2]

The first of the two fossil plants from the Yima coal mine to be worked out in detail by Zhou and Zhang was a very ginkgolike fossil plant known from leaves and seed-bearing structures. They named the leaves *Ginkgo yimaensis.* Zhou and Zhang strengthened their earlier argument that the two different organs were produced by the same species, based not only on their consistent association and abundance at one particular level in the Yima coal mine but also on the structure of their waxy outer covering, which was similar, but not identical, to that in the modern tree. Zhou and Zhang also noted that the leaves were "more deeply lobed than those of *G. biloba*" and that the seed-bearing stalk was branched with each of the five or six seeds borne on a long stalk.[3]

Zhou and Zhang then turned their attention to the second plant that they had recognized in their preliminary account of the Yima fossils. They named the seed-bearing structures *Yimaia recurva.* Each had a simple stalk with a cluster of about eight to nine seeds at the tip. They were associated with quite different, and much more finely divided, leaves than had been described by H. C. Sze, a distinguished Chinese paleobotanist of the previous generation. Sze had named these leaves *Baiera hallei* after Thor Halle, the Swedish paleobotanist who had helped Tom Harris and done much early work on fossil plants from China. The leaves of *Baiera hallei* were so highly divided that the individual leaf segments were almost grasslike.

Zhou and Zhang named the reproductive structures associated with *Baiera hallei* after the locality at which they were collected, *Yimaia.* In this case the fossil was known not just from leaves and isolated seeds but also from distinctive twigs with obvious long and short shoots. The different plant parts occurred massed together in the Yima coal mine a little below the level from which *Ginkgo yimaensis* was described. At the time they wrote the paper, Zhou and Zhang had about fifty specimens of *Yimaia recurva* and hundreds of *Baiera hallei* leaves.[4]

After the original description of *Ginkgo yimaensis* and *Yimaia recurva* from the Middle Jurassic of Henan, Zhou went on to discover similar fossil plants of about the same age from elsewhere in China. In the space of eighteen years Zhou and others have described several different species of *Yimaia* and clearly established it as one of the best-known ancient early relatives of living ginkgo. The picture that emerges is that *Yimaia*-like plants were reasonably common in the vegetation of the Northern Hemisphere all the way from Europe to China during the middle of the Mesozoic, between

Leaf of an ancient ginkgo relative, Ginkgo australis, *from the Early Cretaceous Koonwarra fish beds, Victoria, Australia, about 130 million years before the present.*

160 million and 200 million years ago, and that, at the same time, these same ancient landscapes were also home to other plants, such as *Ginkgo yimaensis,* that were even more like modern ginkgo.[5]

In their initial studies of fossil plants from the Yima coal mine in the late 1980s, Zhou and Zhang recognized two different ginkgolike plants, but surprisingly, almost fifteen years later, they were also able to recognize a third. Zhou assigned this third ginkgo-like plant to the genus *Karkenia,* a group of fossils first described in the mid-1960s from the Early Cretaceous of Tico, Santa Cruz Province, Argentina, by the pioneering Argentinean paleobotanist Sergio Archangelsky. Archangelsky had also worked with Tom Harris earlier in his career, and he too had been thoroughly inculcated with the importance of reconstructing ancient plants from their different, dispersed parts.[6]

The Tico flora is beautifully preserved and dates from about 130 million years ago. It contains common ginkgolike leaves that Archangelsky named *Ginkgoites tigrensis.* Associated with them are seed-bearing organs for which Archangelsky created a new name, *Karkenia incurva. Karkenia* presents an interesting puzzle. On the one hand the leaves are very like those of a deeply divided modern ginkgo leaf; it is easy to see the relationship. However, the associated seed-bearing structures are very different, so different that when I first read about these fossils in the mid-1980s, I was skeptical whether they had anything whatever to do with living ginkgo. The seeds are tightly packed in masses of more than a hundred, almost making a cone. They are quite different from the simple seed stalks of living ginkgo that bear just one or two seeds. In addition, in *Karkenia,* each seed has its own stalk and is curved back on itself, so the tip of the seed faces inward, back toward the cone axis. However, there is no disputing the ginkgolike leaves, and the evidence for linking the isolated fossil leaves from Argentina, China, and elsewhere with the same kind of seed-bearing structures is strong. Sergio Archangelsky's initial deduction of the link between the leaves and seeds has been completely borne out by later work and in hindsight is now seen as a key step in expanding our knowledge of the biography of living ginkgo.

The *Karkenia* that Zhou and his colleagues recognized from China is much less common in the fossil flora from Yima than either *Ginkgo yimaensis* or *Yimaia recurva.* It comes from a level in the coal mine a little below that from which *Yimaia* was described. Only about five seed-bearing structures are known, but these discoveries add to evidence of diverse ginkgolike plants during the Mesozoic. In a single coal mine in northeastern China, Zhou had now discovered three quite different fossil plants that were all more closely related to ginkgo than to any other living plant.[7]

Given the presence of *Karkenia* in both the Northern and Southern Hemispheres, an important question is whether plants more like ginkgo or extinct *Yimaia* ever existed in the Southern Hemisphere. Undeniably ginkgolike leaves have been recorded from many places in Africa, Australia, and South America, for example, and also from India, which at that time was also part of the mass of continents aggregated together in the Southern Hemisphere. However, in no case have these leaves been associated with typical ginkgo or *Yimaia* reproductive structures. They may all be *Karkenia*-like plants, or perhaps further kinds of ancient ginkgo that we do not yet understand.[8]

Zhou Zhiyan made his key discoveries of fossil ginkgolike plants decades ago, but he has continued to add meticulous descriptions of new fossil material from other localities. These new fossils tended to confirm, rather than expand, the picture already developed from studies at the Yima coal mine, but in 2003 Zhou made another breakthrough in understanding the fossil history of ginkgo.

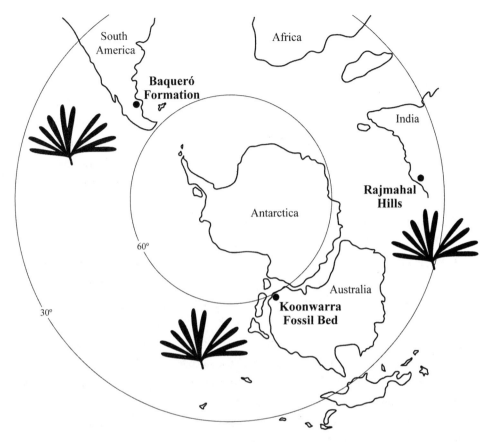

*The relative position of the southern continents around 100 million years before the present
as they separated from Antarctica, showing the presence of similar ginkgolike leaves
in South America, India, and Australia.*

During the 1990s some of the most sensational paleontological discoveries of all time began to emerge from a remarkable fossil deposit in northeast China. This so-called Jehol Biota, collected from the Yixian Formation in Liaoning Province, is of Early Cretaceous age and dates from about 120 million to 125 million years ago. It has continued to make headline after headline as it has yielded one spectacular animal fossil after another. Especially prominent has been a remarkable series of early birds and related dinosaurs, some of which show clear evidence of feathers or the downy feather-like structures that were probable feather precursors. There are also early mammals and amphibians, a great variety of insects, and a wide range of fossil plants, including ferns and even mosses. The seed plants include conifers and some of the most informative and earliest fossils of flowering plants.[9]

It is surprising, given the widespread occurrence of ginkgolike plants through the Jurassic and Cretaceous, that few fossils relevant to ginkgo's life story are found in the Yixian Formation. However, in 2003, Zhou Zhiyan and his colleague Shaolin Zheng described unequivocal ginkgo fossils from these beds for the first time, and these fossils help fill the gap between the truly ancient *Ginkgo yimaensis* and essentially modern ginkgo. These new ginkgo fossils were not from the classic Jianshangou Bed of the Yixian Formation near Jinzhou, which has yielded the most spectacular animal fossils, but from the Zhuanchengzi Bed, which outcrops to the southeast on the southern slope of Yinwoshan Mountain near Toudaohezi Village. The two beds are thought to be roughly the same age, but they contain a slightly different assortment of fossil plants.

Fossil collecting at Yinwoshan Mountain has turned up about a dozen indisputable ginkgo leaves, along with seven seed-bearing structures that seem to have been preserved at slightly different stages of maturity. The leaves are unusually small compared with the leaves of modern ginkgo, none of them much more than about an inch long, and they are deeply divided. The seed-bearing structures are also small, and in all cases they were more or less unbranched. At their tips they bore up to about three to six tiny seeds. Among the specimens were a few seed stalks with immature seeds still attached, but there were also some with a large seed attached and large scars from which others had been shed. Zhou and Zheng named the leaves and the associated seed-bearing structures *Ginkgo apodes.*

The importance of *Ginkgo apodes* is that it fits nicely in the age gap between *Ginkgo yimaensis,* at about 170 million years, and fossils that are essentially the same as modern ginkgo, at about 65 million years. It is also conveniently intermediate in structure. *Ginkgo yimaensis* has a seed-bearing stalk that divided into three or four branches, each bearing a single seed at its tip. However, in *Ginkgo apodes* the branches of the seed stalks are almost nonexistent and the seeds are attached close together at its tip. This is much more like modern ginkgo, although in the living species there are normally just two seeds on each seed-bearing structure rather than three to six. Zhou Zhiyan concluded that the pervasive trend in the evolution of ginkgo, at least in the seed-bearing structures, has been one of reduction from about six seeds on separate stalks in *Ginkgo yimaensis* to about two seeds on an unbranched seed stalk in the living plant.[10]

16
Winnowing

People from a planet without flowers would think we must be mad
with joy the whole time to have such things about us
—Iris Murdoch, *A Fairly Honourable Defeat*

Since Zhou Zhiyan began his work in the Yima coal mine a quarter of a century ago, what we know about the fossil record of ginkgo and its relatives has expanded dramatically. New information and new discoveries continue to accumulate and have revealed an astonishing variety of ancient ginkgolike plants. This unexpected diversity changes the way we think about the evolution of the single living species. Studies of fossil leaves had hinted at the existence of such diversity, but until more was known about their corresponding seeds and other parts, the real plants lurking behind the isolated leaf fossils remained enigmatic and difficult to compare with the living tree. Work by Zhou Zhiyan and his colleagues has changed all that.[1]

Given what we know now, it is possible to begin to consider the major patterns in the evolution of living ginkgo and its extinct relatives. Again, it is Zhou Zhiyan who has led the way based on some simple analyses made possible by his unrivaled knowledge of the relevant fossils.

As an initial attempt to gain a sense of the changing diversity and abundance of ginkgolike plants through time, Zhou made a graph showing the number of different kinds of ginkgolike leaves that have been distinguished as different genera through the long fossil history of the group. Setting aside *Trichopitys* and other Permian plants, which may or may not be related to the ginkgo lineage, Zhou showed that from four different kinds of ginkgolike leaves in the Early Triassic, the number increases over about 50 million years to six in the Middle Triassic, and twelve in the Upper Triassic. The number remains high through the Jurassic and into the Early Cretaceous, varying from seven to eleven between about 100 million and 200 million years ago, before declining to four in the Late Cretaceous and just one and two in the Paleogene and Neogene. There are fewer ginkgo-related reproductive structures, but the pattern is the same: numbers for the Late Triassic through Early Cretaceous range from five to three, whereas only a single kind of reproductive structure is known after about 100 million years ago. In both cases the patterns are crude but nonetheless revealing.[2]

Because of potential confusion about which fossil leaves can be securely related to the ginkgo lineage, Zhou also took a more conservative approach to see whether the pattern would still hold up. He focused on the fossils he knew best and traced through time the number of different species of the three main leaf types that he had worked on from the Yima coal mine: *Baiera,* the leaf of *Yimaia; Ginkgoites,* the leaf of ginkgo; and *Sphenobaiera,* the leaf of *Karkenia.* Again, the bulk of the diversity is right in the middle of the Mesozoic, from the Upper Triassic to Early Cretaceous, between about 100 million and 225 million years ago, but drops off rapidly thereafter. In the Early Cretaceous of China, Zhou recognized twenty-two different kinds of ginkgo leaves, ten different kinds of *Baiera* leaf, and ten different kinds of *Sphenobaiera* leaf, but in the Late Cretaceous all had disappeared except for a single kind of *Ginkgoites* leaf. However you look at these simple statistics they show an astonishing and consistent pattern of decline in the number of ginkgolike plants. From some point in the middle of the Cretaceous, around 100 million years ago, the world of ginkgo began to change.[3]

In parallel with the decline in the number of different kinds of ginkgolike plants, the importance of ginkgo and its relatives in ancient vegetation seems likely to have declined substantially as well. We might expect that these plants steadily became less common on Mesozoic landscapes. This is harder to assess from what has been published on the fossil record, but Zhou was able to gain a rough idea of how widespread these different kinds of ginkgolike plants were by looking at the number of counties in China from which they had been recorded. In the Early Cretaceous the number of counties with records of ginkgo was eight, compared with thirty-one for *Ginkgoites,*

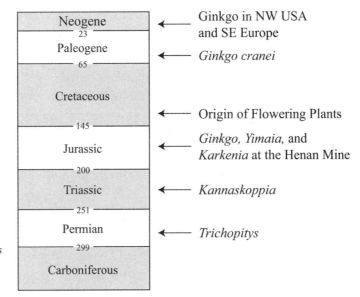

Key dates in the history of ginkgo and its relatives over the past 300 million years.

eighteen for *Baiera,* and eleven for *Sphenobaiera.* Thereafter, from the Late Cretaceous onward, there are no records of *Baiera* or *Sphenobaiera* at all, just one for *Ginkgoites,* and five for ginkgo itself. Again, the pattern is crude but the conclusion is clear. For ginkgo and its relatives, the middle of the Cretaceous period was a time of transition.

It is possible only to speculate about what might be behind the apparently pervasive decline of ginkgolike plants about 100 million years ago, but one obvious potential cause is competition with flowering plants, a new group of highly successful plants that rose rapidly to dominance during the middle part of the Cretaceous. Of course, there were many other environmental changes during the mid-Cretaceous; for example, accelerated rates of continental drift created new configurations of land and sea, making possible new kinds of currents in the oceans and atmosphere, which must have resulted in new kinds of climates. However, it is also hard to believe that the explosive increase in the diversity and abundance of flowering plants in the Cretaceous did not have some impact on the ginkgolike plants that had previously been so prominent. By the Late Cretaceous *Yimaia, Karkenia,* and similar plants seem to have disappeared. The ginkgo group as a whole was a shadow of its former self, and the survivors were those ginkgolike plants most like the single living species.[4]

The period between about 65 million and 100 million years ago, immediately after the initial rise of flowering plants, was an interesting time in the history of life on land: a time when familiar plants grew alongside unfamiliar animals. It was a moment when magnolias were food for *Triceratops,* and hadrosaurs made their nests among groves of ancient plane trees. During this time, about three-quarters of the known mammals belonged to an extinct group called the multituberculates, small marsupial-like animals that are often compared with rodents. It has been speculated that these small mammals may have fed upon and dispersed ginkgo nuts.[5]

One of the best insights into the strange ecosystems during this last gasp of the age of dinosaurs comes from fossils preserved in the Horseshoe Canyon Formation of central and southern Alberta. These rocks have provided us with some of the best known of all dinosaurs — carnivores, such as *Tyrannosaurus rex* and *Albertosaurus,* as well as herbivores, such as *Triceratops* and *Maiasaura.* Fossil plants from these deposits have received less attention but are important because they were the ultimate source of energy on which the dinosaurs and other animals living on these ancient landscapes depended. The plants of the Horseshoe Canyon Formation are also important because of the ways in which they are preserved. Sometimes they are squashed and flattened, but other times they are petrified by various minerals including calcite, phosphate, and silica.

A detailed study by Kevin Aulenback shows that the plants of the Horseshoe Canyon Formation included mosses, clubmosses, horsetails, and various ferns, as well as several different kinds of seed plants. Flowering plants included aroids and gingers, as well as trees similar to living swamp gum, dove tree, and hornbeam. Conifers similar to living dawn redwood, cypress, and China fir are also present alongside leaves and seeds that are very like those of living ginkgo. Many of the living relatives of these conifers are still found in eastern and southwestern China, not far from those areas that are home to living populations of ginkgo.[6]

Farther south and east of Alberta, in Montana and North Dakota, ginkgo is also widespread in the sands and silts laid down on the ancient floodplains of the Hell Creek Formation. Fossils from the Hell Creek Formation provide some of the last known occurrences of dinosaurs before they disappeared during the mass extinction at the end of the Cretaceous. From the standpoint of animal evolution, the impact of the end-Cretaceous mass extinction was devastating. The losses were selective but widespread, and many groups of animals, both in the ocean and on land, were eliminated. Their extinction left an indelible mark on the trajectory of animal evolution.

Interestingly, the long-term effects of the end-Cretaceous extinction on the history of plant life were seemingly much less profound. Over the short term, many of the previously dominant plants were lost, and a careful study by Peter Wilf and Kirk Johnson in southwestern North Dakota showed that almost two of every three plant species that were present below the boundary became extinct. However, while there is strong evidence of regional extinction, it is less clear that major groups of plants were lost at a global scale.

In southeastern North Dakota only a few of the plants known from below the boundary reappear above it: one of them is ginkgo. Ginkgo leaves are present at many fossil localities in the Hell Creek Formation and have been collected just a few feet below the boundary, but they are back again, not long after the apparent cataclysm, in the overlying Paleocene rocks of the Fort Union Formation. This pattern speaks well to the capacity of ginkgo for survival, but it makes the great ginkgo decline in the middle part of the Cretaceous, which does not coincide with a massive extinction in the world of animals, still more striking. It underlines the point that in some ways plant and animal evolution is decoupled. Somehow, it seems, plants and animals dance to a different evolutionary beat.[7]

Taken as a whole, the large-scale patterns in the evolution of the ginkgo group expand the way that we should think about the evolution of the living tree. We might have thought that we could trace ginkgo back through a series of ancestors to ancient plants, perhaps like *Trichopitys*, that lived in Permian or older times. Today, however, as a result of the work of Zhou Zhiyan and others, we have a very different picture, a picture that is complicated by a great deal of previously unrecognized diversity. We now know that many species of ginkgolike plants lived at the same time in the past and perhaps even alongside one another in the same kinds of plant communities. This may have been the case with *Ginkgo yimaensis*, *Yimaia recurva*, and *Karkenia incurva*, which grew in similar kinds of habitats 170 million years ago in the coal swamps of northern China, but it was probably also true of ginkgo and its relatives in many other ancient plant communities between about 100 million and 225 million years ago.

When we delve deeply into the fossil record of many groups of plants or animals, the situation that has come to light for ginkgo turns out to be quite common. For example, in the case of the modern horse, we have not just one ancient lineage leading in ladderlike progression to the single living species; rather, we have fossils of many kinds of extinct horses, some of which browsed together in the same ancient landscapes. The pattern in the fossil record of horses is a bush of many closely related species that was winnowed by extinction over time to a single lucky winner.

Based on Zhou Zhiyan's work on different kinds of ancient ginkgolike fossils, and his painstaking analyses of the ginkgo fossil record worldwide, a similar picture emerges of once-great diversity among ginkgo and its relatives, followed over time by the gradual emergence of a single winner and many losers. From the Jurassic onward, and perhaps from the latest Triassic, there is strong evidence of plants similar to the living species. However, at first, this lineage was just one among many, just as the lineage leading to ourselves was one among many hominids on the savannah of East Africa between a half-million and a million years ago. There were once other lineages of ginkgolike plants, and each of those also had multiple species. Whether it is the evolutionary history of horses, of ginkgo, or of ourselves, the pattern is the same: sudden appearance and proliferation, followed inevitably and inexorably by many losses, in which most species fall by the wayside. In the cases of ginkgo, horses, and ourselves, only one species survived.[8]

17

Persistence

The "tendency to persevere," . . . it is this that in all
things distinguishes the strong soul from the weak.
—Thomas Carlyle, *Chartism: Past and Present*

My own small contribution to what we know about the prehistory of ginkgo came
from work on fossils much younger than those from the Triassic and Jurassic studied
by Tom Harris and Zhou Zhiyan. It began in the late summer of 1982 as I was driving
back to Indiana University after a long trip collecting fossils with my colleague David
Dilcher. I was just wrapping up a year of work in David's laboratory before moving to
the Field Museum in Chicago. The trip had started with fossil hunting in the drylands
of eastern Oregon, followed by a long swing through Idaho, Wyoming, and Colorado.
Now in North Dakota, with the end of summer approaching, I was getting ready to
move on. Ginkgo could not have been farther from my mind. We had seen no trace of
it at any of the fossil sites that we had visited that summer.[1]

As the last stop on our trip, we were trying to track down the source of an interesting collection of fossil plants that David had been given by Rudi Turner, a colleague at
Indiana. Rudi, an enthusiastic fossil collector in his spare time, had come across these
fossils at various rock and fossil shows organized by amateur and commercial collec-

The leaf of an ancient ginkgo, Ginkgo cranei, *from the Late Paleocene of Almont, North Dakota, about 57 million years before the present.*

tors. He had traced the material to a newly discovered site a little west of Bismarck, North Dakota.

As we pulled the car off I-94, the main east-west highway, the signs were not encouraging. The road headed north through flat agricultural land. There were no road cuts or badlands to explore, mostly just fields of sunflowers, until eventually, at the designated spot, we came to a single figure bent over by the side of the road among low piles of sharp, yellow-brown shale. As we pulled up, he greeted us warmly, and the first fossil he showed us was an exquisitely preserved ginkgo leaf, laid out complete and absolutely perfectly on the hard shale.

Ginkgo leaves were the main target for our new friend and other local fossil collectors who were working at the site. The fossils were attractive, reasonably common, and easily recognized, often with all the fine veins nicely picked out in red and brown against the ochre background. It was easy to see why these distinctive fossils, dating from about 57 million years ago, were a prized curiosity in the now nearly treeless landscape of this part of North Dakota.

On this first visit to what we later called the Almont locality, the focus of local collectors on ginkgo leaves was a boon to David and me. We had broader interests, and the heaps of discarded material were packed with other fascinating plant fossils, many of them beautifully preserved. Especially common were the flying saucer–shaped fruits of the Asian wheel wingnut. Circular in outline, and about an inch across, these were the fruits that had first attracted David's interest. Also prominent in the Almont collections were the leaves, pollen catkins, and characteristic spiny fruits of an early extinct relative of hazels and hornbeams, which turned out to be similar to fossils I had described a year or two earlier from a locality of about the same age in southern England. The collecting that afternoon was some of the most exhilarating I have ever experienced; there was something new on almost every piece of shale. It quickly became obvious that there were many interesting fossils at Almont, especially a rich assortment of fossil fruits and seeds, which would be worthy of detailed study.[2]

Along with the ginkgo leaves, it was also easy to recognize ginkgo seeds among the fossils in the Almont shale. With the leaves so common, and with so many other kinds of fruits and seeds present, it would have been strange if they had not been there as well. Often the leaves and seeds occurred together on the same piece of shale. Usually, the only part of the seeds preserved was the hard inner shell, about half an inch long, with a distinctive raised longitudinal seam running round the outside and a small point at the top. However, in a few cases the hard layer was enclosed in a shiny covering almost certainly formed from the remains of a silvery, waxy coat like that of modern ginkgo seeds.

It was also striking that both the leaves and the seeds had the same characteristic red-brown appearance and shiny, leathery texture, probably a result of their thick waterproof covering and the resin in the original tissues. It was obvious that they "be-

The seed of an ancient ginkgo, Ginkgo cranei, *from the Late Paleocene of Almont, North Dakota, approximately 57 million years before the present.*

longed together" and had been shed from the same kind of tree. Somehow they had been washed into a small pond on the ancient floodplain, buried in mud, and preserved together until we collected them on that August afternoon 57 million years later.

After several hours sifting through piles of shale discarded by previous collectors, and with the sun going down, we wrapped the last fossil and loaded the last box. Almont was a paleobotanical treasure trove that was hard to leave. The fossils were superbly preserved and partly impregnated with silica. It was clear that Almont would provide more detailed information on ancient plants than any of the other fossil localities that we had seen that summer. The quality of preservation was better than that of any fossil locality of similar age anywhere in the world. We could not resist coming back the next morning for a few more hours, and when we eventually eased back onto the interstate heading for Chicago and Indiana, David's aging Oldsmobile was well down on its springs.

Over the next few years I returned to Almont several times, and on two or three occasions I sent a small team from the Field Museum on the long drive from Chicago to collect there. Other paleobotanists found their way there, too. The collecting was easy, the landowner was more than tolerant, and the fossils turned out to be just as informative as we had anticipated back in the early 1980s. Eventually, David Dilcher, Steve Manchester, and I published an overview of the Almont fossil flora, detailing not only the ginkgo leaves and seeds but also — importantly for ginkgo's biography — some new information on how the mature seeds were borne on this ancient fossil plant.[3]

On that first and many subsequent visits to Almont, we collected many plant fragments that were puzzling. We knew that there was plenty of work to be done, and just because we didn't immediately understand what a particular fossil was, that didn't mean that we would not collect it. We brought all the oddities back and puzzled over them in the laboratory. Among them were some especially enigmatic fossils that looked as if they ought to be somehow connected to the ginkgo leaves and seeds. They had the same texture and color, and were superficially like the stalk of a ginkgo leaf, but they didn't look quite right. For one thing they never seemed to be attached to a leaf blade, and for another they had a peculiar knobbly structure at one end.

Back in Chicago, under a microscope these strange fossils provided a bit more information, and eventually I decided to rummage around under a few living ginkgo trees to see what was left in the leaf litter beneath them. This unconventional approach has served me well over the years. Textbooks often provide a beautiful portrayal of a plant in all its structural details, but for a paleontologist such images are excessively sanitized. When you are trying to interpret fossils, what you really want to know is what a tree looks like when it falls apart: which pieces normally become detached, and what do they look like as they start to decay? This is the best way to find out what pieces of a plant have a chance of ending up as a fossil.

A couple of minutes under an old female ginkgo tree were enough to get me what I needed. Among the rotten fruits were exactly the same kinds of knobbly stalks that we had collected from the Almont fossil site, and among the other ginkgo debris it was easy enough to figure out what they were. The unusual pieces of the Almont Ginkgo were the shed stalks on which the mature seeds had been borne. In living ginkgo the seeds and the stalks fall from the tree at about the same time, becoming separated from each other in the process. In the fossil ginkgo from Almont, as in the living species, there was usually only one large seed scar on each stalk, indicating that only one of the two seeds on each seed stalk had matured. However, at the tip of both the fossil seed stalks and those from living ginkgo, the remains of the other young seed that had failed to develop were often easy to make out. It was obvious that the fossil, like the living species, had two seeds on each seed stalk but that normally only one of those seeds developed.

Apart from this minor triumph, which solved one of the small problems raised by the Almont fossils, the recognition that the ginkgo growing in North Dakota around 57 million years ago bore its seeds in exactly the same way as the living species was useful new information. It showed that the similarities between this ancient ginkgo and its living relative extended beyond the form of the leaf and the shape and size of the seeds to other aspects of the plant. This was one more piece of evidence to justify using the Latin name *Ginkgo* for these leaves, with the important implication that the rest of this ancient plant, if we understood it in all of its details, would not differ significantly from the living tree.[4]

One of the interests of the late Tom Schopf, a former colleague at the University of Chicago, was "living fossils," and as one of his last projects Tom wrote a commentary about the supposed living-fossil status of a wide variety of plants and animals. Classic examples from the world of animals include the clamlike creature *Lingula,* which still lives in the cold shallow waters off northwestern Europe and appears to have survived more or less unmodified from the Silurian period, more than 450 million years ago. Among animals with backbones, the coelacanth first occurs in rocks about 350 million years old but survives in deep ocean waters off the Comoros in the Indian Ocean, and Malaysia in the South China Sea. Other well-known living fossils include the horseshoe crab, the paddlefish, and the platypus.[5]

From this compilation Tom pondered some of the questions that living fossils raise for our ideas of plant and animal evolution, especially whether the apparent absence of change over vast periods of geologic time reflects reality or illusion. He wanted to know whether the absence of information misleads us into thinking that living and fossil organisms are the same, when we would clearly recognize them as different from each other if we knew more about them. Tom raised the possibility that significant changes could be occurring in these organisms through time but were going undetected because they are invisible in those parts of organisms that are preserved as fossils.

Dianne Edwards, a paleobotanical colleague at the University of Wales, uses the analogy of the Mini, a car first introduced in the United Kingdom in 1959 and now manufactured by BMW. From the outside the 1960s version and the twenty-first-century version are unmistakably similar, but inside they could scarcely be more different. From the computer chips in the engine to the liquid crystal displays on the dashboard, the technology of the modern Mini is rather different from that of its progenitor. But if you had only a photograph of the outside, how would you know?

Tom asked, just because the shells of *Lingula* from the Silurian and today look the same, does that mean that the animals inside were identical? If we could go back in a time machine and take our microscopes and laboratory with us, would we find that the ancient Almont Ginkgo of 57 million years ago was more or less the same as or quite different from that of today? I suspect that we would find the former. There is much else in the landscape of 57 million years ago that we would find strange, but the evidence we have so far suggests that we would find the ancient ginkgo comfortingly familiar. Even though the paleontological lens through which we view these ancient plants is obviously imperfect, it seems that the more we find out about the Almont Ginkgo, the more it looks like its modern counterpart. The similarity of the fossil seed-bearing stalks from Almont to those of living ginkgo was one more piece of evidence pointing in that same direction.

Of course, it is possible that future research will discover significant differences—for example, in the structure of the pollen cones, pollen, or wood—but I suspect that the reverse is much more likely to be true: as we find out more about the Almont Ginkgo, it will come to look more and more like the living tree. It will be interesting to see whether the next generation of paleobotanists to turn their attention to Almont will take up this question. For example, among the small catkins that Steve Manchester, David Dilcher, and I described is one that we perhaps passed over too quickly; now, with the benefit of hindsight, it looks very like a shed ginkgo pollen catkin. The fossil pollen that we extracted from a similar specimen doesn't look right, but fossils like these certainly deserve a more careful look. The prediction is that if it is a ginkgo pollen catkin, its structure and its pollen grains will turn out to be more or less identical to that of the living species.[6]

Just as the structure of ginkgo seems to have remained virtually unchanged for tens of millions of years, it seems likely that aspects of its ecology have also remained constant. Dana Royer from Wesleyan University, Leo Hickey from Yale, and Scott Wing from the U.S. National Museum of Natural History analyzed forty-eight separate fossil occurrences of ginkgo leaves from western North America, ranging from about 55 million to 65 million years ago, all roughly similar in age to the Almont fossils. In all but two cases the geological situation in which the fossils occurred suggested that the leaves were preserved in ancient muds and sands and deposited in, or close to, river channels. Occurrences of ginkgo in lake deposits, in deltas, or in other situations were either nonexistent or very rare. Dana Royer and his colleagues concluded that these ancient ginkgos consistently favored streamside habitats, and they pointed out that these are also the kinds of environments where ginkgo seems to flourish today.

In general, modern ginkgo dislikes being deeply shaded by other plants. It grows best on sites that are partly or fully exposed. Living ginkgo trees in China show that they favor steep rocky slopes, cliff edges, and—most significant for comparison with the fossils—stream banks. They like to be partly in the open and partly in the shade. In addition, as many of the great ginkgo trees of Asia, such as the giant Yongmunsa Ginkgo, show us, ginkgo does especially well where its roots have good access to water. The same was probably true for ancient ginkgo that grew 60 million years ago in the American West. They thrived in partly open habitats and especially along rivers. By a happy accident, this also happens to be the right place for their leaves, seeds, and other parts to be incorporated into the river mud and sand accumulating nearby. Long-term constancy in its ecology, combined with its tough and easily recognized leaves, is one of the reasons why ginkgo has such an excellent fossil record.[7]

18
Prosperity

Everything in the world may be endured except continued prosperity.

—Johann Wolfgang von Goethe

John Starkie Gardner is not among the truly great paleontologists of the later nineteenth century, but he was nevertheless energetic and talented. Between about 1879 and 1887 he produced several publications on fossil plants, including the two-volume *British Eocene Flora*, before abruptly bringing all his paleontological work to a close and devoting himself instead to a second career as an expert on decorative ironworks. He created Victoria Gate in London's Hyde Park and the iron gates and screens at Edinburgh's Holyrood Palace. As such a radical and sudden change of direction might suggest, Gardner was a forceful character who held strong views.[1]

In the 1880s Gardner clashed spectacularly in the pages of the scientific journal *Nature* with one of his German contemporaries, Baron Constantin von Ettingshausen, over the scientific reliability of some of the latter's work. The two had previously collaborated on a study of ancient ferns, conifers, and ginkgo from southern England, but they parted company on how to deal scientifically with the massive collections of fossil leaves that Gardner had accumulated from his excavations along the cliffs at Bournemouth in Hampshire. These leaves, from localities now long obliterated or covered by development in this popular seaside town, date from the Middle Eocene, about 40 million to 50 million years before the present.[2]

What bothered Gardner was Ettingshausen's tendency to rush into print with new identifications and formal Latin names for fossils that had not yet been studied thoroughly. He disapproved of Ettingshausen's cavalier attitude toward linking these leaves with specific living plants. It was true that Ettingshausen had studied the patterns of veins in the leaves of different kinds of broadleaved plants more carefully than anyone else, and that he had also published a beautiful atlas of the leaves of living plants, which he had carefully "skeletonized" to reveal their delicate vein patterns. The illustrations in his book were striking white-on-black silhouettes. However, Gardner was right: Ettingshausen's way of working was to match fossil and modern leaves without careful comparison. If they looked broadly similar in shape and vein pattern, he was quick to say the fossil and living plants were the same. Gardner was exasperated. He knew that the numbers of plants to which the fossils had to be compared was overwhelming, and he knew also that the leaves of even distantly related plants often look confusingly similar. He was acutely aware of the possibility of making massive, misleading mistakes. Gardner could not sign on to an approach that he regarded as patently unscientific.

Fortunately, identifying ginkgo in the fossil record is much less problematic than identifying the often rather nondescript leaves of flowering plants that make up most fossil floras from the middle of the Cretaceous onward. One of the wonderful things about ginkgo, from a paleobotanical point of view, is that the leaves are so distinctive: they are unlikely to be overlooked or confused with anything else. So given how widespread ginkgo had been during the Jurassic and Cretaceous, it is perhaps significant that in the whole of Gardner's massive collections of fossil leaves from Bournemouth, there is no example of anything like a ginkgo leaf. In their early work together, before their falling out, Gardner and Ettingshausen had described fossil ginkgo leaves from Scotland, but there is not a single ginkgo leaf among the many hundreds of specimens collected from Bournemouth, which today occupy many tens of drawers in the Natural History Museum in London.

Ginkgo is also strangely absent from another massive and important collection of fossil plants from southern England, which also dates from the Eocene. Fossils collected from the beaches of the Isle of Sheppey in the Thames Estuary, just forty miles east of central London, provide an unusually detailed window into the plant life of between about 45 million and 55 million years ago. This so-called London Clay flora has been collected since the earliest days of scientific paleontology. Largely through the efforts of Eleanor Reid and Marjorie Chandler, two pioneering women paleontologists, what has been learned from the London Clay fossils exceeds that from any other collection of fossil plants of similar age. Reid and Chandler's classic work, *The London Clay Flora,* published by the Natural History Museum in 1933, is a hefty tome. It described and illustrated more than four hundred species and set new standards of accuracy in the comparison of living and fossil plants. Its importance rests in part on its being based on studies of beautifully preserved fossil fruits and seeds rather than of fossil leaves.[3]

Fossils from the London Clay are preserved in three dimensions in iron pyrite, a mineral sometimes called fool's gold. They weather out from the soft, chocolate-brown cliffs along the northern shore of the Isle of Sheppey, and are concentrated on the muddy foreshore by the tides and the waves. Collections from Sheppey are especially extensive, but similar fossils have been collected from other places around the coast of southeastern England, particularly from Herne Bay in Kent, Bognor Regis in West Sussex, and Walton-on-the-Naze in Essex. However, at all localities the fossils are unusual because they occur in clays that were clearly deposited in the sea. Fruits and seeds are found alongside the teeth of sharks and rays, the shells of marine snails, and the fossilized carcasses of crabs and shrimps. This is a little strange; it is much more common for plant fossils to be preserved in muds and sands laid down in ancient freshwater lakes, ponds, and river systems on the land. In the case of the London Clay, the fruits and seeds were washed out to sea, eventually sank, and were then literally petrified in the fetid mud at the bottom of the ocean.

These peculiarities of the London Clay flora bring with them several advantages over fossil plants collected from ancient floodplains. First, because the fossil fruits and seeds had drifted out to sea, they provide an unusually broad sample of the plants that grew along the ancient shoreline of the London Clay Sea and the banks of rivers that emptied into it. Second, the preservation in iron pyrite means that the tissues of the fruits and seeds are well preserved, not just on the outside but also on the inside. Often exquisite details of internal structure are visible, enabling these ancient fruits and seeds to be compared carefully with their living counterparts.

The London Clay flora also has a third advantage; it has been collected intensively

386

NIPA FRUTICANS:—Wurmb.—Blanco.— Miq.

A nineteenth-century botanical print of the living stemless nypa palm, which grows today in tropical mangrove habitats around the coast of southeast Asia. Abundant fossil fruits of nypa have been recovered from the London Clay, on the Isle of Sheppey in the Thames Estuary. The presence of nypa indicates the warmth of the climate about 50 million years ago.

for more than 150 years. Hundreds of people have spent untold hours on the beach at Sheppey and elsewhere looking for these fossils. Many thousands of specimens have been collected. The London Clay flora is by far the richest and most informative Eocene flora currently known. As at Bournemouth, though, ginkgo has so far failed to put in an appearance in the London Clay. The hard inner shell of a ginkgo seed is distinctive, would be readily preserved, and would quickly have been recognized by Eleanor Reid, Marjorie Chandler, and the many other specialists who have studied the Sheppey fossils over the years.[4]

An important clue as to why ginkgo might be missing from Gardner's Bournemouth collections, and also from the London Clay, comes from the most conspicuous plant fossils collected from the beach at Sheppey; the distinctive fruits of the stemless palm nypa. Today nypa is native to tropical southeast Asia and entirely restricted to brackish, often partly tidal, mangrove habitats that are protected from full exposure to the ocean. Not only does good representation of such a coastal plant in the London Clay make ecological sense, but extrapolation from its living relative strongly suggests that the vegetation of London Clay times grew in a tropical or near tropical climate.[5]

Other plants that are common in the London Clay flora, such as palms and members of the sweetsop family, support this idea. They point to dramatically warmer temperatures than those in the Thames Estuary of today. For example, there are abundant climbers, from the grape and moonseed families, as well as true mangroves. In the flora as a whole, plants that are now characteristic of tropical southeast Asia are especially well represented. There are also bay trees, magnolias, and a variety of other plants, which while not truly tropical, prefer warmer climates. The strong overall impression is of an ancient coastal plain clothed with lush subtropical-to-tropical vegetation. The London Clay flora, together with other fossil floras from elsewhere in Europe, tell us that climates during the Eocene, even at midlatitudes, were very warm.

To the north, however, in slightly older rocks that seem to have deposited in slightly cooler climates, the situation is different. The columnar basalts exposed along the ruggedly beautiful southwest coast of the Isle of Mull in Scotland rival those at the Giant's Causeway on the other side of the Irish Sea in County Antrim, Northern Ireland. They date from around the Early Paleocene, about 60 million to 65 million years before the present, and were formed by successive outpourings of molten basalt from a deep geological rupture in the Earth's crust that ultimately helped create the North Atlantic. On the Giant's Causeway, on Mull, and also on the Isle of Staffa, where they create Fingal's Cave, the hot basalts cooled into layers of tall hexagonal columns.[6]

Ardtun Head, near the southwestern tip of Mull, is formed by several layers of massive black columnar basalt, but as early as 1851 the landowner, the Duke of Argyll, who also had a keen interest in the emerging science of geology, described the details of the rock section exposed in the cliffs. He was also among the first to draw attention to the softer rocks between the basalts and the fossil plants that they contained. The main features of the geological section that the Duke of Argyll described are still clearly visible on Ardtun Head today. They show three main layers of basalt that range from ten to forty-eight feet thick, between which are siltstones laid down not by volcanic activity but by water. As successive layers of basalt cooled, they were colonized by vegetation, just as happens today, for example, on Iceland or on the Big Island of Hawai'i. In the muds created by the streams, ponds, and lakes on this new landscape, leaves and other plant parts became entombed as fossils before they were eventually buried more deeply, and for the long term, by a new outpouring of lava.[7]

It was to these fossils, preserved in the soft layers between the basalts on the Isle of Mull, that John Starkie Gardner brought the same enthusiasm for collecting that he brought to the cliffs at Bournemouth. When challenged by the massive hard black basalts on the Isle of Mull, he resorted to dynamite, and with great effect. Spectacular plant fossils from Mull are now housed along with those from Bournemouth in the collections of the Royal Scottish Museum in Edinburgh, the Hunterian Museum in Glasgow, the Natural History Museum in London, and elsewhere. Together with ginkgo, the fossils that Gardner collected include the leaves of ancient hazel, oak, laurel, and katsura trees. Taken together, they point to cooler climates than those of the London Clay.[8]

This pattern seen in Britain, where ginkgo appears to be missing from more tropical floras but is present in fossil floras from cooler climates, seems to hold across Europe from about 40 million to about 65 million years ago. For example, ginkgo is absent from the remarkable fossil assemblage collected in an old oil shale mine at Messel not far from Frankfurt, Germany. The truly exceptional preservation of ancient animal life at Messel includes mammals with the remains of skin, hair, and gut contents, as well as a hummingbird with feathers. The abundance of bats and crocodiles is one of several indications of tropical conditions, which is also confirmed by the plants, which include palms, ferns, and aroids. In the past few years, a superb study by the young paleobotanist Selena Smith showed that one especially puzzling plant fossil, which had remained enigmatic for many years, is actually the fruiting stalk of the Panama hat palm, one of a group of plants that today grows only in tropical Latin America.[9]

In North America, the picture is the same. The Eocene fossil floras from Kentucky and Tennessee have been collected extensively for more than a hundred years, especially in the early twentieth century by E. W. Berry and those who worked with him. More recently, during the 1960s, 1970s, and 1980s, enormous collections were made by David Dilcher and his students. There are tens of thousands of specimens from the Eocene fossil floras of Kentucky and Tennessee in the collections of many museums across the United States. The collections of the National Museum of Natural History in

Washington, D.C., and the Field Museum in Chicago, as well as the Dilcher collections now at the Florida Museum of Natural History, are especially rich. Among all this material, however, there is not a single ginkgo leaf.

Ginkgo is also missing among the rich collections of Eocene fossil leaves made from Castle Rock near Denver by Kirk Johnson and his team from the Denver Museum of Natural History. This locality came to light in the early 1990s, when excavations were made to broaden the interstate highway that climbs up from Denver through the Front Range of the Rocky Mountains. The Castle Rock plants have a distinctly tropical look to them, and many species have the large leaves characteristic of tropical plants, but again there is no ginkgo. Ginkgo has also never been found among the large collections of fossil plants made from the Green River Fossil Basin of Colorado, Utah, and Wyoming. Palms and many other warmth-loving plants are well represented, and there are crocodiles and turtles among the animals of "Fossil Lake," but among the collections of fossil leaves from these deposits there is no evidence of ginkgo.[10]

Ginkgo is present, however, in one of the most intensively studied and most informative Eocene fossil floras from North America: that from the ancient mudflow of Clarno Formation, not far from the town of Fossil in eastern Oregon. The fossil plants at Clarno are preserved as jumbled fruits, seeds, and leaves in an ancient mudflow created by nearby volcanic activity. Like the fruits and seeds preserved in the London Clay, the fossils have been literally turned to stone, in this case not by iron pyrite at the bottom of a stagnant ocean but by silica dissolved in the warm waters produced by nearby volcanoes.

Fossil fruits and seeds from the Clarno fossil assemblage have been painstakingly collected over several decades, especially by the late Tom Bones, with the help of generations of high school students through programs run by the Oregon Museum of Science and Industry. Steve Manchester, one of the graduates of that program and now a professor at the Florida Museum of Natural History, has provided the definitive description of the Clarno flora. Thousands of fossil fruits and seeds have been collected, as well as hundreds of specimens of leaves, but so far the only clear evidence of ginkgo is a piece of petrified ginkgo wood and a single unmistakable ginkgo leaf. Again the flora is strongly indicative of warm conditions. There are fruits of palms and the sweetsop family, and as in the London Clay there is an abundance of climbers. But in this case ginkgo was present, although apparently rather rare.[11]

Similarly, not too far away from the Clarno Formation, and in rocks of about the same age, ginkgo puts in an appearance in the rich fossil floras preserved in ancient lakes during early evolution of the northern Cascade Mountain Range. These lakes, which were gradually filled with ash and other fine volcanic debris, preserved fish, insects, and occasional mammals, as well as the leaves and other parts of plants that grew nearby. Several of these fossil-producing ancient lakes straddle the U.S.-Canadian border in the Okanagan Highlands of northern Washington State and adjacent British Columbia, but by far the most thoroughly studied is that around the small town of Republic in northeastern Washington, not far from where the Columbia River passes into Canada.

Like the fossil plants at Clarno, those from Republic have been extensively collected, not only by teams of professionals but also by many schoolchildren led by the late Wes Wehr, working from a base at the Burke Museum of the University of Washington in Seattle. The dedication of this multifaceted and truly lovely man resulted in the collection of tens of thousands of specimens from the Republic site, among which were many spectacular fossils. The flora looks much more temperate than that from Clarno and the Green River; in place of palms and tropical climbers there are currants and witch hazels, along with ancient representatives of the birch, elm, oak, rose, and walnut families. Ginkgo is also present, but extremely rare; it is represented by just one or two fragmentary but unmistakable leaves. Floras of similar age from China that include ginkgo are also known from Liaoning Province.[12]

One explanation for the presence of ginkgo in the Republic flora and perhaps at Clarno, in contrast to its absence among the fossil plants from the Green River Formation and from Kentucky and Tennessee, may be slightly increased elevation, as well as higher latitude and hence slightly cooler temperatures. At the time that the leaves were washed into the Republic Lake, they may have been about 2,300–3,000 feet above sea level. Ginkgo seems to prefer these cooler temperatures, and this preference is dramatically confirmed by the fact that at roughly the same time ginkgo was growing at very high latitudes within the Arctic Circle, not too far from the North Pole.[13]

On Ellesmere Island, Canada, high in the Canadian Arctic, Jim Basinger and the late Elizabeth McIver of the University of Saskatchewan described ginkgo among the ancient plants growing in and around high-latitude peat bogs. The flora, which is not rich compared to those farther south, includes a few different kinds of broadleaved trees alongside early firs, pines, and the unmistakable leaves of the dawn redwood. The presence of forests with ginkgo in the Arctic around 55 million years ago is a vivid reminder that the familiar configuration of our planet, with permanent ice at the poles and year-round warmth only in the tropics, is the exception rather than the norm in the past 200 million years of life on Earth.[14]

PART IV

Decline and Survival

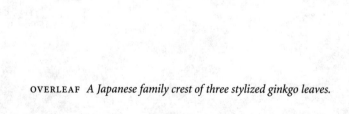

OVERLEAF *A Japanese family crest of three stylized ginkgo leaves.*

19
Constraint

What does not kill me, strengthens me.

—Friedrich Nietzsche, *Nietzsche contra Wagner*

Today, ginkgo can be grown easily in many parts of the world, including over most of Europe and the United States, as well as much of eastern Asia, but it is confined mostly to what might loosely be called the temperate regions. For example, in Europe, ginkgo flourishes from Paris in northern France to Montpellier in the south, and the connection to Goethe has made ginkgo popular throughout Germany. It is not a tree that survives across most of Finland, though, or that thrives at the other end of Europe's climatic spectrum—for example, in Sicily or the Greek Isles, where both high temperatures and scarce water are problematic. Similarly, in Australia, ginkgo grows happily outdoors in Melbourne, and also in Sydney, but you will not find it on the streets of Cairns or Darwin. There seems to be a reasonably clear band of latitude within which ginkgo flourishes but outside of which it struggles and eventually dies. There are constraints of some kind, probably a combination of constraints, which limit the survival of ginkgo both toward the poles and toward the tropics.[1]

The different aspects of climate that govern the distribution of ginkgo and other tree species are complicated, and made more so by the fact that local conditions, microclimates, and the availability of nutrients and water in the soil can vary dramatically over short distances depending on aspect, elevation, or proximity to fresh and salt water. All gardeners know that whether a plant flourishes, struggles, or dies in a particular place depends on many different things. However, a key constraint on the growth of ginkgo toward the poles and at high elevations is obviously cold temperature. In the plant hardiness zones recognized by the United States Department of Agriculture, which are widely used by gardeners in North America as a guide to the likely tolerances of particular plants, the northern limit of ginkgo is zone 5, defined by average annual minimum temperatures down to minus 20 degrees Fahrenheit.

In China, ginkgo grows well in climates with an average annual temperature in the fairly broad range of 50–65 degrees Fahrenheit and an annual rainfall in the range of twenty-four to forty inches a year, but from the North American hardiness zones it is also clear that ginkgo's tolerances are broad. The tree can tolerate very low winter temperatures and very high summer temperatures, at least in short bursts. This makes it possible for it to thrive in places like Chicago, where winter temperatures can go down to minus 27 degrees Fahrenheit and summer temperatures can top out at 108. Similarly, ginkgo can get by in Minneapolis–Saint Paul, where winter temperatures are even more extreme and the winters even longer. Nevertheless, there are obviously limits. Ginkgo will not survive in North American hardiness zones 1 and 2, or colder places where winter temperatures may plummet to minus 45 or 50 degrees or lower.[2]

Ginkgo can survive, although not always happily, in all fifty U.S. states, including Alaska. Along the East Coast of North America, ginkgo grows well from Charleston, South Carolina, to Montreal in Quebec, but there are no ginkgo trees in Labrador or the southernmost parts of Florida. In Hawai'i ginkgo grows well only in the mountains. Along the West Coast, ginkgo grows from San Diego to Vancouver, but farther north in Alaska, or farther south in Baja, it is a different story. Nursery owners in Juneau, southern Alaska, report that the tree can hang on for a number of years but it does not thrive, and it will not survive farther north in Fairbanks.

The situation is similar in Europe. For example, in Copenhagen, Denmark, or even Lund, southern Sweden, ginkgo trees grow outside without special protection, but a little farther north—for example, around Stockholm or Uppsala in central Sweden— ginkgo can normally be grown only inside. Ginkgo trees in the garden of the Komarov Botanical Institute in Saint Petersburg, at nearly 60 degrees north latitude, are among the most northerly in Europe. However, ginkgo cannot be grown outside in Moscow, away from the ameliorating influence of the Baltic.

Part of ginkgo's success in tolerating extreme cold is its deciduousness. As in most

In early spring clusters of tiny ginkgo leaves emerge from the buds in which they have been protected all winter. The scars of last year's leaves show the traces of the two strands of conducting tissue that supplied the leaves with water.

temperate broadleaved trees, shedding the leaves is one way to avoid the damage that would be caused by ice crystals forming in the living tissues or by water loss through the leaves, during the time when the groundwater is frozen. In effect, deciduousness shifts the problem of winter survival to the easier task of protecting the young leaves inside the overwintering buds. In ginkgo the buds are small, rounded, and well protected by a mass of tightly packed, overlapping, light brown, papery bud scales. Some temperate trees, and perhaps also ginkgo, also have an ingenious mechanism by which the buds are actively supercooled so that chances of ice crystal formation are further minimized. However, this mechanism appears to fail at temperatures of about minus 40 degrees Fahrenheit: at colder temperatures ice crystals form spontaneously and damage the living tissues. It seems likely that poleward limit of where ginkgo can grow outside in part reflects the extent to which it can protect the next season's leaves inside their buds.[3]

While absolute winter temperatures are one factor that helps set a poleward limit for the growth of ginkgo, the length of the growing season is probably just as important. If the leaves can't harvest enough energy in a short summer to cover what we might think of as "the costs of maintenance," then growing in those places is probably not a viable proposition. In most cases where there is a tree line, a clear observable limit to tree growth, either with increasing latitude or with increasing altitude, these seem to reflect not winter temperatures that are too cold for trees to survive but rather some critical limit in the length and warmth of the growing season. This supposition also fits with the observation that in many tree species photosynthesis still works at low temperatures, but if levels of activity are insufficient, then the energy is stored and saved for later rather than used directly for growth.[4]

The susceptibility of ginkgo and other trees to extreme cold also varies considerably depending on the stage of life at which a particular tree encounters frigid temperatures. Young plants are generally much more sensitive than plants that are well established, so the regularity and timing of extremely cold temperatures in relation to the plant's development is important. If a large tree loses a few leaves in a late frost, it will probably have enough stored reserves of energy to grow new ones. For a seedling with just a few leaves and limited food reserves, however, it is quite a different story. At Kew, experience with some frost-sensitive trees, such as certain eucalypts, showed that if their time as seedlings happened to coincide with consecutive mild winters, which allowed them to become well established, then their chances of surviving when the next hard winter came along were greatly increased. However, the reverse is also true. In southern England the winter of early 2010 was one of the harder ones of recent decades. A colleague at Kew who had ginkgo seedlings growing outside lost almost half of them.[5]

In ginkgo a further complication is that the length of the growing season affects the development of the embryo in the maturing seed and therefore the opportunities for the young plant to get established in its first season. A study by Peter Del Tredici shows how important such effects can be. Using careful observations and some simple experiments, Peter studied how reproduction in ginkgo is affected by temperature based on comparisons of plants growing at the Arnold Arboretum of Harvard University, at about 42 degrees north latitude, and in Guizhou Province, China, at about 25 degrees north latitude.[6]

At the Arnold Arboretum ginkgo pollen cones generally emerge from the winter buds around mid-May, and pollination takes place soon after. Fertilization happens about four and a half months after that, and the resulting seeds, which are generally shed from the tree in late October, germinate the following spring. On the face of it there is no problem with the reproductive cycle, and under normal conditions it seems as though ginkgo ought to be able to reproduce itself by seed in the climate of present-day Massachusetts. However, by comparing the timing of these various stages in Massachusetts with the timing in Guizhou, Peter showed that the situation in the two places is actually quite different, and that those differences could be significant for plants trying to establish themselves in the wild.

By carefully following seed development in trees growing outside, and comparing them with seeds kept in a warm greenhouse over the winter, Peter was able to show that while the tempo of the annual reproductive cycle is remarkably similar in both the Chinese and North American trees, the way in which the seeds develop is very different. Seemingly irrespective of temperature and latitude, fertilization typically occurs about 130–140 days after pollination. The time from pollination to germination was also remarkably constant at around 233–234 days for the Guizhou plants and the seeds kept in the greenhouse in Massachusetts. Maturation of the embryo from fertilization to germination was also similar, taking roughly 100 days in both cases. The embryo develops continuously without a pause, and there is no natural period of dormancy.[7]

However, seeds developing outside, without the protection of a greenhouse, experience very different conditions in Massachusetts compared with Guizhou. In both cases embryo development slows or stops completely during the winter. This helps prevent the seeds from germinating too early, which increases the seedlings' chances of surviving the cold part of the year. It does, however, increase the time to germination. In the relatively short growing seasons and long winters of the northeastern United States this means that germination is significantly delayed compared with southern China, where the winters are shorter and less severe, and the growing season is correspondingly longer.

In the northeastern United States, pollination takes place around mid-May and fertilization takes places around the end of September or the beginning of October. A month or two later, the seeds are shed. The embryo, whose maturity at the time of seed drop depends largely on local conditions, continues to develop on the ground. This means that the embryonic plant has only a month or so to develop before growth slows or shuts down completely with the onset of cold weather. As a result, development is carried over into the following season, and because of the relatively late spring and the additional time needed for the embryo to undergo its development, germination does not take place until mid- to late June. The whole process takes thirteen months from pollination to germination, and the young seedling has only about five months to become established before the next winter, with its potentially fatal cold temperatures.[8]

The contrast with the life cycle of ginkgos growing in Guizhou is stark. Here, with a much earlier start to the growing season, pollination takes place in mid-March to early April, as much as two months earlier than in Massachusetts. Seeds are shed in mid-September, and germination occurs in mid-March of the following year. The whole process goes a little more quickly, just twelve versus thirteen months, but more important is the earlier completion of embryo development, which allows germination a full three months earlier than in Massachusetts. In Guizhou, the seedlings have about eight months, rather than five, to put on good growth and become securely established before facing their first winter. And winter, in any case, is relatively mild compared to that in the northeastern United States. All these considerations show that while temperature is important in determining where ginkgo will grow, its effects are complex, and made more so by interactions with other factors, particularly soil conditions and the availability of water.[9]

One factor that does not seem to matter for the growth of ginkgo, however, is the annual distribution of light. High-latitude areas experience a strange light regime, with several months of almost complete darkness in the winter and several months of near complete light in the summer. Unless something drastic has happened to the angle at which the Earth's axis of rotation is currently tilted, which seems unlikely, this light regime at high latitudes would have been the same 55 million years ago as it is now. Fossils from the high Arctic—such as the north slope of Alaska, at nearly 70 degrees north, or Ellesmere Island or Spitsbergen at nearly 80 degrees north latitude—show that ginkgo and other trees were indifferent to spending part of the year in nearly round-the-clock sunlight and a corresponding amount in round-the-clock darkness during the winter. Peter Del Tredici's experiments also imply that the Eocene winters in these high-latitude areas were also rather mild, and certainly not cold enough to cut off a ginkgo seedling in its prime.[10]

While the control on the northward limit of ginkgo today is broadly related to different aspects of temperature, especially shorter growing seasons, frigid absolute temperatures in the winter, and the period for which those harsh conditions need to be tolerated, exactly how ginkgo is limited at the other end of the climatic spectrum is harder to understand. A quick survey in North America indicates that while ginkgo thrives in many places, it is not a characteristic part of the landscape in New Orleans or in the year-round warmth of southern Florida. Ginkgo is obviously an appropriate tree to plant at Dinosaur World in Disney's Animal Kingdom theme park in central Florida. It was certainly part of the world of dinosaurs, but along with other trees from more distinctly temperate climates, it does not look especially happy there. According to Jeff Courtney, a horticulturalist at Disney's Animal Kingdom, the trick to keeping the park's handful of ginkgos alive is to find a good spot with the right microclimates, out of the direct sun and in places where there is plenty of water.

In Mexico, ginkgo grows well at the relatively high altitude of Mexico City, where winters can be chilly and average low temperatures in January are around 43 degrees Fahrenheit, but ginkgo is not a plant that you will find growing in the lowlands of Chiapas or Oaxaca. In Brazil, ginkgo grows in the southern provinces of São Paulo, Minas Gerais, and Rio Grande do Sul, planted in outdoor plazas, and also in the traditional Japanese gardens that reflect the cultural influence of a million and a half Japanese-Brazilians. However, it does not thrive in Amazonas, or in the hot, dry tropical climates of the interior of Bahia. Similarly, in Asia ginkgo nuts are common in the Chinese cuisines of Hong Kong and Singapore, but they are all imported. Ginkgo grows happily in the warm temperate southern provinces of Guizhou and Yunnan, but not in the extreme south of China—for example, at Xishuangbanna, near the border with Laos and Myanmar. Tough as it is, ginkgo cannot tolerate the climates of the true tropics.[11]

As with its poleward limit, the factors that exclude ginkgo from the tropics, defining its southern limit in the Northern Hemisphere and its northern limit in the Southern Hemisphere, are likely to be complex. Absolute temperatures and the availability of water undoubtedly play roles, but also important is a situation well known to the world's winemakers. We tend to associate vineyards with hot summer days when we would be more comfortable taking our glass of wine in the shade rather than out in the full sun, but we should not forget the winter. Much as grapes flourish in hot summers, they do not grow well in places with year-round warmth. They need a distinct cold period as part of their normal annual cycle of growth. The same is true of many fruit trees, which also need a cold winter, a period of what gardeners call vernalization, to flower and fruit the following year. In many cases, deciduous trees from temperate

A ginkgo flourishing in Weimar, Germany, one of the historic cultural centers of Europe and once home to Goethe. Ginkgo easily withstands the cold of continental Europe and seems to thrive in its warm summers.

climates do not do well when planted in botanical gardens in the tropics. They leaf out in a seemingly uncontrolled way and eventually die. In most cases, plants from the temperate zone don't survive long in frost-free climates.[12]

The reasons behind this need for a cold period are not well understood, but they seem to be connected to the internal biological mechanisms that result in rapid and coordinated bursting of the buds, as well as flowering, in the spring. In many deciduous trees from the temperate zone, unless there is a period of about one to two months where the mean minimum temperature is about 41 degrees Fahrenheit or below, bud burst and subsequent leafing out do not seem to proceed normally.[13]

Water is also crucial, and again its effects may be subtle. In particular, it is also not just the absolute amount of rainfall that is important but also how rainfall is distributed across the year and how water is stored in the soil. Water is quickly lost from thin, sandy soils, and in these circumstances it can quickly become scarce, which can in turn affect the growth of trees. At Kew, for example, where the soil is generally well drained, trees often responded to long dry summers by shedding their leaves early, or even shedding large branches. In other circumstances—for example, where there are deep loamy soils—soil water is retained for longer periods, which helps to smooth out short-lived deficits of rainfall. Ginkgo is sensitive to extreme waterlogging; its roots cannot tolerate permanent drowning, but water is indispensable for it to really flourish.

Taken together, these aspects of temperature, seasonality, and water availability seem to be crucial factors that limit the distribution of ginkgo today, both toward the tropics and toward the poles. Similar factors potentially influenced the changing distribution of ginkgo in the past. However, while the ginkgos planted all over the world tell us something about where the tree manages to grow, this is not the same as saying that ginkgo would be able to survive in the wild under those conditions. That is altogether a taller order, requiring not just an ability to tough out the climate and set seed but also an ability to be successful in an ecosystem, alongside a varied mix of plants and animals, as well as microbes, pests, and diseases. The ability of ginkgo to survive in cultivation in many different places is only a partial test of its ability to survive in the wild.[14]

20
Retreat

It isn't so much that hard times are coming;

the change is mostly soft times going.

—Groucho Marx

Between about thirty-five million and sixty-five million years ago ginkgo was widespread across the Northern Hemisphere, but that period of great prosperity eventually came to an end as the climate began to cool. In the Southern Hemisphere, the ginkgolike plants that had persisted from the Cretaceous were still part of southern landscapes, but they too soon disappeared. The last evidence of ginkgolike plants in the Southern Hemisphere is in Tasmania about forty million to sixty-five million years ago. After that, even though there are many younger fossil floras from Australia and South America, ginkgolike plants seem to have been lost from the Southern Hemisphere until they were reintroduced, tens of millions of years later, by people.[1]

Around thirty-five million years ago the global climate not only became markedly cooler but in many places also became drier. In North America the continuing uplift of the Sierra Nevada, the Cascades, and the Rocky Mountains sucked the moisture from winds that had passed over the Pacific and intensified the rain shadow over much of the continent. This created opportunities for expansion of a new kind of landscape, the prairie. In Asia too, as the climate changed, forest was replaced with steppe. In

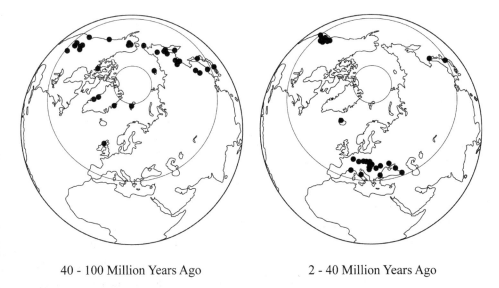

40 - 100 Million Years Ago   2 - 40 Million Years Ago

Fossil localities with ginkgo from about 40 million to 100 million years ago compared with localities
with ginkgo from about 2 million to 40 million years ago. In earlier times ginkgo was present at very
high latitudes in the Northern Hemisphere, but later it was restricted to middle-latitude regions
(maps show the present-day distributions of continents).

these new and more open habitats, new groups of plants, such as grasses, sunflowers and their relatives, and herbaceous species of many kinds flourished. There was an explosion in the evolution of flowering plants, which created much of the plant diversity that we enjoy today. At the same time, forest habitats suitable for the growth of ginkgo became steadily more restricted.

In Europe, ginkgo was certainly present around sixty million years ago, for example at Mull in Scotland, on the eastern margin of the opening Atlantic Ocean, but ginkgo leaves are absent from the many Eocene floras known from Britain, France, Germany, and surrounding countries. This was almost certainly because the climates were too warm. More puzzling is that leaves of ginkgo are also absent from some of the classic Oligocene floras from Europe, as well as from most of the rich and intensively studied Miocene and Pliocene fossil floras associated with the massive deposits of brown coal from the Rhine Basin in western Germany.[2]

There is, however, a truly remarkable occurrence of ginkgo in the Selárdalur flora from Iceland. Following up on earlier paleobotanical studies, extensive new collecting and comprehensive studies by Thomas Denk and his colleagues have shown that around fifteen million years before the present, ginkgo was growing in a forest of ancient oak, swamp cypress, coast redwood, magnolia, and grapevines on the volcanic landscapes of Iceland's west fjords, places that were perhaps not much different from those that ginkgo inhabited on Mull almost forty million years earlier.

A hotspot for the occurrence of ginkgo in Europe in the Miocene and Pliocene is in fossil floras from eastern and southeastern Europe. In this area, ginkgo appears to have been absent in the Paleocene and Eocene, but its characteristic leaves occur widely in Middle Miocene, Late Miocene, and Pliocene fossil floras, ranging in age from about five million to sixteen million years ago across a broad area from Munich in the west to Ukraine and Russia in the east. In the extreme southeast, ginkgo is known even from northwestern Greece. In this area the leaves are most common in fossil floras that reflect the vegetation growing along the banks of ancient rivers. The characteristic ecology of ginkgo recognized by Dana Royer and his colleagues in the Late Cretaceous to Early Miocene of western North America also applied in Europe and seems to have persisted.[3]

The later appearance of ginkgo in fossil floras from eastern and southeastern Europe, combined with its presence in fossil floras from the Paleocene and Eocene in places such as Scotland and Spitsbergen but absence over much of northwest Europe, suggests that ginkgo migrated into this area as it was forced south by changing climatic conditions, perhaps a combination of cooling and drying. Obviously—important in light of what happened later—this implies that its dispersal system was still working.

In North America the pattern is similar to that in Europe. Ginkgo is absent from some of the most thoroughly collected and best studied Oligocene floras, such as the Bridge Creek flora from the John Day Basin, and it is very rare in the Oligocene fossil flora from Florissant, Colorado. However, it is present in a smaller number of fossil floras, many of which are mainly, but not exclusively, western. For example, unmistakable fossil ginkgo leaves occur in the Oligocene Lyons fossil flora in western Washington, which is about the same age as that from John Day Basin. Ginkgo is also present in other Oligocene fossil floras from the Pacific Northwest, including from the Willamette flora near Eugene, Oregon. These occurrences might suggest that ginkgo was most common in places not far from the Pacific Coast. As it does today, that part of western North America may have had higher rainfall than areas inland.[4]

Similar complex patterns persist into the Miocene and Pliocene. Ginkgo does not occur in the Clarkia flora of northern Idaho, which is one of the most remarkable fossil floras anywhere in the world for the quality of its preservation. Rich collections from the Clarkia assemblage have been assembled over many years by the late Jack Smiley and Bill Rember of Idaho State University, but the unmistakable leaves of ginkgo have never been seen, even though other characteristic Chinese plants like China fir are well represented. Today Clarkia is about 370 miles from the Pacific Coast, and the area experiences only about half the annual rainfall of places like Portland and Eugene. Perhaps inland the climates were already becoming too dry or too cold for ginkgo to flourish.[5]

A classic occurrence of ginkgo from the Miocene is at Ginkgo Petrified Forest State Park, not far from Vantage, Washington. Here it is known not from fossil leaves but from its unmistakable wood. The stumps of approximately 15.5 million–year–old ginkgos occur among those of other forest trees growing on the slopes of an ancient, active, Cascades volcano, petrified by silica from the ash and preserved where they grew. Other species that were growing nearby include plants like bald cypress, sweet gum, oaks, and sycamores. This is one of the last known occurrences of ginkgo in North America until its eventual return, with the help of people, millions of years later.[6]

21
Extinction

Many rivers to cross
But I can't seem to find my way over.
—Jimmy Cliff, "Many Rivers to Cross"

Given its long fossil history, the presence of ancient ginkgo across much of the Northern Hemisphere for most of the past 65 million years is not so surprising. Ginkgo and its extinct relatives were seemingly nearly everywhere on the planet for eons, and despite their clear decline about 100 million years ago, ginkgo managed to persist in many places. However, looking back from today, the fact that ginkgo was growing wild in Bulgaria and Greece just 5 million years ago nonetheless seems strange. It reminds us that not so long ago the world was a very different place. In the grand sweep of geologic time the distribution of animals and plants on our planet has changed rather quickly; where they live and grow now bears a strong imprint of history.[1]

Fossil floras from the Late Miocene and Pliocene provide irrefutable evidence that in addition to ginkgo, there were many other plants in western North America and Europe between about five million and fifteen million years ago that no longer grow there. In terms of the trees the vegetation in these areas was much richer then than now. For example, fossils from the fill of an ancient sinkhole at Willershausen near

Göttingen, Germany, show a mix of broadleaved and coniferous forests. On richer soils broadleaved forest included species of maple, birch, hickory, beech, ash, oak, and elm among about thirty-four tree species. Conifer forest included many trees that no longer occur today in Europe but can be found growing in the warm temperate forests of eastern Asia: the umbrella pine, for example, as well as the Chinese swamp cypress, the katsura, the dawn redwood, and the hardy rubber tree. Like ginkgo, in Europe, they all disappeared relatively recently.

In North America, fossil evidence from Clarkia, Idaho, shows exactly the same pattern. Again, the Chinese swamp cypress and the katsura are both present, along with the dawn redwood and the China fir. All of these plants are today restricted to eastern Asia. At both Clarkia and Willershausen there was also the Cathay silver fir, a rare conifer discovered as a living plant only in 1955. Today it has a scattered and restricted range in southwestern China. After about five million to fifteen million years ago, these plants were never seen in Europe and North America again, but somehow they managed to persist in the East.[2]

It is hard to understand exactly when and how these species were eliminated from Europe and North America because in most cases the fossil record is not sufficiently complete to provide a detailed picture of how their distribution gradually changed from being widespread in the past to being much more restricted today. We can, however, get some idea of how they may have fared by tracing the fate of a few of their associates that have especially distinctive pollen grains. Pollen grains are produced and preserved in the fossil record in vast numbers, and when they are sufficiently diagnostic of a particular tree, and readily recognized in fossil assemblages, they can be used to get a fine-grained look at how that plant fared as global climates deteriorated.

Particularly instructive is the history of the Caucasian wingnut, a tree in the walnut family that has especially distinctive pollen grains. These pollen grains disappear and reappear through successive glacial and interglacial phases in southern Britain. After each of the first few glacial advances up to about 500 thousand years ago, pollen grains of the Caucasian wingnut reappear in the intervening warm interglacials. These plants seem to have been forced south by successive glacial advances, but they evidently migrated back again into Britain, presumably from the south and east, as the climate warmed. However, these distinctive pollen grains are last seen in Britain during the Hoxnian interglacial between about 374 thousand and 424 thousand years ago. For some reason, in the two most recent interglacials, the Eemian, which lasted from about

114 thousand to 130 thousand years ago, and the present Holocene, which began about 10 thousand years ago, the Caucasian wingnut never made it back.[3]

It would be wonderful if we could follow the history of ginkgo in a similarly detailed way, but unfortunately its pollen grains are too easily confused with those of other plants. However, the example of the Caucasian wingnut does raise a potentially important but unanswered question concerning the ecology of ancient ginkgo: having been displaced from particular places by changing climates, did it have the ability to recolonize? Colder and drier climates may have progressively restricted ginkgo's geographic range, but why did it not bounce back? Surely it should have been able to recolonize those places where it obviously grows so well today.

In most plants, the ability to colonize an area depends on the effectiveness with which seeds are dispersed. Seed dispersal provides plants with the ability to emulate an animal and move from one place to another, albeit much more slowly, generation by generation. The fruits and seeds of many plants show specializations to increase the effectiveness of dispersal, from the parachute-like fruits of dandelions that are blown along by the wind to the seeds of blackberries that are gobbled up along with the fleshy fruits in which they develop and are dispersed in the droppings of birds. A key question in the case of ginkgo is whether one of the factors responsible for its decline over the past few million years was a poor system for dispersing its seeds.

In 1982 the tropical ecologist Dan Janzen and the paleontologist Paul Martin published a provocative article with an arresting title: "Neotropical Anachronisms: The Fruits the Gomphotheres Ate." Their central idea flowed from the observation that many of the common plants in Guanacaste National Park in Costa Rica, where Janzen had worked for many years, appeared to have no natural means of dispersing their seeds. They noted that this was particularly the case for some of the plants in which the fruits and seeds were relatively large, such as guanacaste itself and another legume tree, divi-divi. Today, the fruits and seeds of these trees are eaten by horses and cattle, but these animals have been introduced by people from elsewhere only relatively recently. There are no indigenous animals that appear capable of dispersing them. Janzen and Martin argued that this mismatch arose because these plants had been dispersed in the past by animals that are now extinct. The plants had survived, but the animals capable of dispersing their seeds had not.

Janzen and Martin suggested that such plants used to be dispersed by the large mammals that once inhabited South and Central America but disappeared relatively suddenly, perhaps as a result of hunting by humans, climate change, or both factors acting together, about ten thousand years ago. These now-extinct animals would have included the gomphotheres, massive extinct relatives of modern elephants, that were obviously plant eaters and flourished in Central America for most of the past five million years. Living alongside them were other fruit eaters like ground sloths, glyptodonts, extinct horses, extinct bears, giant armadillos, flat-headed peccaries, and others. Janzen and Martin's point was that the gomphotheres, along with other extinct large mammals, probably played an important role in the ecosystems of Central America over the past few hundred thousand years and that their relatively recent extinction has left us trying to understand an ecosystem that is missing some of its most important parts.[4]

What was most important about Janzen and Martin's idea was its focus on the importance of history for interpreting the world around us. The survival of the plants, after the extinction of the gomphotheres and other animals that may have dispersed them, was an accident of history. In effect the evolutionary histories of the plants and their associated animals were now out of phase. With some slight rhetorical license, Janzen and Martin called those plants that had lost their dispersal agents the "living dead." The implication was that without the dispersers with which they had evolved, their days were numbered.[5]

Janzen and Martin's ideas proved hugely influential, and in 1984 my paleobotanical colleague Bruce Tiffney of the University of California, Santa Barbara, suggested that something similar might have happened in the history of ginkgo. Bruce argued that ginkgo, like Janzen and Martin's tropical trees, was also one of the "living dead," a plant that had lost its dispersers. He speculated that the strange and strong-smelling ginkgo seed might have been a specialization for attracting dinosaurs, or perhaps early kinds of mammals that are now extinct.

Of course, an idea like this is hard to prove, but it does begin to hint at another reason, other than local extinction due to climate, as to why living ginkgo very nearly went extinct. The apparent migration around fifteen million to twenty-five million years ago of ginkgo into eastern and southeastern Europe, areas where it was not previously present, seems to suggest that dispersal was still possible long after the demise of dinosaurs and ancient extinct mammals. However, Bruce's point was nevertheless a good one. A lack of effectiveness in the dispersal of ginkgo seeds may have played a part in its progressive restriction, and the fact that this may reflect more recent extinctions, rather than ancient extinctions at the time of the dinosaurs, is in some ways beside the point.

Unfortunately, even though its smelly seeds are one of its most well-known and distinctive features, we know very little about how seed dispersal works in living ginkgo. However, germination does improve after the fleshy seed coat has been removed—for example, by passing through the gut of an animal. In one of the potentially wild ginkgo populations in China it is also documented that the seeds are eaten by a wild cat, and in Japan they are eaten by badgers. Dogs are sometimes attracted to them too. A friend recalls his dog feasting on ginkgo seeds one autumn on the University of Minnesota campus. It would be helpful to have more information on the kinds of animals attracted to ginkgo seeds today, but even if various mammals are known to collect and eat ginkgo seeds, this is not quite the same as knowing that ginkgo has a reliable seed disperser.[6]

If Bruce is broadly correct, and sometime toward the end of the Mesozoic, or more likely during the Cenozoic, ginkgo lost the animals on which it depended for dispersal, then the effects of climatic restriction would have been greatly amplified. It would have meant that ginkgo, unlike the Caucasian wingnut, for example, was not able to easily recolonize areas from which it had been displaced. It would have continually lost ground, and its populations would have become smaller, moving it ever closer to what conservationists sometimes call the extinction vortex. Colder or perhaps drier climates would have eaten away at ginkgo's once widespread geographic range, and limited powers of dispersal would have reduced ginkgo's ability to recolonize. The effect would have worked like a ratchet; once ginkgo lost ground it was unable to take it back. In North America and Europe the impact over the past few million years may have been especially pronounced if, as seems likely from the fossil evidence, the geographic extent of ginkgo in those areas had already been reduced by climatic drying and other vegetational changes. The mountains and valleys of southern and western China may have provided a greater variety of potential refuges.

Whatever the reason, the pattern of regional extinction could not be clearer. Ginkgo has a more or less continuous record in Asia beginning with the early fossils described by Zhou Zhiyan and his colleagues more than 200 million years ago. It continues through the Jurassic and Cretaceous, to the presence of ginkgo in fossil floras from the Pliocene of Japan. However, in Europe and North America the pattern is different. Here the fossil record of ginkgo is also deep but it is abruptly truncated relatively recently.[7]

These insights provide a clear example of the importance of fossils to fully understand how our modern world came to be. The natural world is full of patterns, some of
them completely unexpected, that can be explained only by reference to history, and
as I never tire of saying to my students, if you want to understand the way anything is
today, whether it is a plant, a person, an ecosystem, an organization, or a country, then
you need to understand its history. It is a mantra that is hardly original, but one that is
easily forgotten in our modern preoccupation with the here and now. In biology, these
kinds of historical complications are the reason why we ignore evolution, and the direct historical evidence that comes from paleontology, at our peril.

In particular, the fossil record of ginkgo and similar plants helps make sense of a
somewhat enigmatic observation made by botanists since the time of Linnaeus: that
there are surprising similarities between the plants of eastern North America and eastern Asia. Highlighted at the end of the eighteenth century by the Italian botanist Luigi
Castiglioni, and then later by the American Thomas Nuttall, the full extent of these
similarities did not become clear until the work of the great nineteenth-century American botanist Asa Gray.[8]

Gray and his contemporaries were at a loss to explain how the pattern had come
about. For Darwin, writing to Gray at Harvard in 1856, this was one of the "many utterly
inexplicable problems" of botanical geography. Darwin was completely puzzled about
why there should be stronger similarities between the flora of eastern North America
and eastern Asia than between the floras of eastern and western North America. The
fossil record shows beyond doubt, just as Gray later inferred, that these seemingly
strange and widely separated occurrences are the result of regional extinction, especially in Europe and western North America, of plants that were once much more
widespread. In the case of ginkgo regional extinction went even farther; the species
was completely eliminated from Europe, from eastern and western North America,
and also from Japan. Even in China its extinction was very nearly total.[9]

22
Endurance

Here on this rugged and woody hillside has grown an apple-tree,

not planted by man, no relic of a former orchard, but a natural growth,

like the pines and the oaks.

—Henry David Thoreau, "Wild Apples"

The oldest and largest ginkgo trees on the planet occur today in eastern Asia, and it is only in China that we find the combination of really ancient trees of large size growing alongside seedlings that are regenerating naturally in what appears to be a more or less wild situation. However, whether truly wild populations of ginkgo trees still survive in China is an open question. One problem is that China is a vast country; even today it is not completely explored from a botanical point of view. Many new species of plants continue to be described from China, and in just the past decade, new and potentially wild populations of ginkgo have been discovered and have not yet been fully studied. Another problem is that it can be hard to tell whether a particular population is really wild or not.[1]

The possibility of identifying truly wild populations of ginkgo is an almost irresistible attraction to botanists interested in plant evolution. If we could identify living populations of wild ginkgo trees, we might be able to learn how their lives interconnect

*Ginkgo seedlings regenerating
naturally along a path on the campus
of EWHA University, Seoul, South Korea.
The seedlings flourish in partly shaded,
partly open areas of the woodland
understory.*

with those of other plants and animals that live in the same ecosystem. We could see whether ginkgo has natural enemies, such as insects that can feed on its leaves, and we could observe whether there are mammals that collect or eat its seeds. We could also study the bacteria and fungi that live on, or in, its roots and stems. It would be interesting to know whether some of the plants that occur at Almont and in other fossil floras still live alongside ginkgo in modern forests.

A key problem in trying to recognize natural populations of ginkgo is the difficulty of distinguishing truly wild trees from those with a history of cultivation. China has a long history of human occupation, and most of the potentially wild ginkgo populations occur in areas where people have lived for millennia. It is hard to know whether a particular tree grew from a seed that came there naturally or from one that was planted, and because ginkgo will regenerate readily from seed under the right kind of conditions, the distinction is a fine one to make. Even a wild-sown seed may have the hand of people somewhere in its background. On the campus of EWHA University in Seoul,

I have seen many ginkgo seedlings growing on a steep slope in the understory of a small piece of seminatural woodland. All of them had germinated from seeds that were naturally sown, but from seeds produced by female ginkgo trees planted along a road a bit farther up the slope. Given that ginkgo has probably been planted as a nut tree in China for centuries, the difficulties are obvious.

Western botanists first encountered ginkgo in Japan at the end of the seventeenth century. In the eighteenth century they also became aware of ginkgo and the use of ginkgo nuts from European traders in China, Japan, and Korea. However, early botanists visiting eastern Asia were mainly confined to the coasts. It was not until the mid-nineteenth century, as a result of the forced opening of trade with China, Japan, and Korea by Western mercantilist powers, that European plant explorers were able to travel more widely in Japan and Korea and to penetrate the interior of China. Large ginkgo trees were among the spectacular plants that they encountered, many of them entirely new to Western science. The accounts of the adventures of early Western plant hunters in China make fascinating reading. As a result of their explorations, and the preserved and living plant specimens that began to flow back to Europe, it quickly became clear not only that the native flora of China was exceedingly rich, but that there were also plants of commercial interest to which the European powers wished to have access.[2]

The most important of the early botanist-explorers to journey in China was Robert Fortune, an intrepid Scotsman from Berwickshire. Fortune began his career as a gardener at the Royal Botanic Garden in Edinburgh, before moving south in 1840 to work in the garden of the Royal Horticultural Society, then based in west London. Just months after the signing of the treaty of Nanjing, which brought to an end the First Opium War of the late 1830s and early 1840s, Fortune was dispatched to China by the Royal Horticultural Society and arrived in Hong Kong in July 1843. It was the first of four incident-filled visits, and Fortune was also one of the first Western botanists to travel in Japan.[3]

Fortune's adventures are vividly recounted in his several books. His botanical focus was on the collection of living plants, and his second expedition on behalf of the British East India Company had a direct economic purpose: the introduction of tea to India against the wishes of the Chinese government. By mastering the Mandarin dialect, adopting local dress, shaving his head except for a queue, the brusque Scottish botanist was able to pass himself off as a native Chinese from a distant province and gain entry into otherwise forbidden areas. Over the course of his expeditions in Asia he shipped more than twenty thousand seeds and seedlings from China and Japan, ultimately introducing more than a hundred plant species into European horticulture.

On his first journey to China, Fortune writes of ginkgo, using its then accepted botanical name: "The only tree which I met with of very large size in [Shanghai] is the *Salisburia adiantifolia,* commonly called the Maiden-hair tree, from the resemblance its leaves bear to a fern of that name. This is one of the plants which the Chinese are fond of dwarfing, and it is, consequently, often seen in that state in their gardens. Its fruit is sold in the markets in all Chinese towns by the name of '*Pa-Kwo,*' and is not unlike dried almonds, only whiter, fuller, and more round. The natives seem very fond of it, although it is rarely eaten by Europeans."[4]

After Fortune, a second wave of plant explorers gained access to China as a result of changing political circumstances, this time the Second Opium War. The resulting Convention of Beijing in 1860 greatly benefited both the French and the British. From Britain, Augustine Henry was a key link between Robert Fortune and collectors who came later, such as Joseph Rock, Ernest Henry Wilson, George Forrest, and Frank Kingdon-Ward. Also influential as great plant explorers were three French missionaries who were dispatched to China in the 1860s, of whom the first and most important was the Basque Lazarist monk Père Jean Pierre Armand David.[5]

Augustine Henry went to China as an employee of the British Imperial Maritime Customs Service and collected extensively in central China between 1882 and 1889. He sent more than fifteen thousand dried specimens and seed samples to Britain. Most were described by botanists at Kew, such as William Botting Hemsley and Daniel Oliver. Henry brought to light a wealth of exciting new species from China, many of which remain important in European horticulture and bear his name, such as the Henry Honeysuckle and the Henry Lily. He came across ginkgo on several of his expeditions. In his "Notes on Economic Botany of China," published in 1893, his comments are brief: "*Pai- kuo, Ginkgo biloba* L. Seeds eaten." However, in his seven-volume work *The Trees of Great Britain and Ireland,* published in 1906, seven pages of text are devoted to ginkgo, along with three photographs that include an ancient ginkgo beside a temple in central China.[6]

Henry is also important because of his influence on a young British plant collector who followed him, Ernest Henry "Chinese" Wilson. Through his roughly two thousand plant introductions, Wilson had a major impact on the cultivation of Asian plants in

A large old ginkgo photographed by
Ernest Henry Wilson at a temple in
Nara Prefecture, Japan, in April 1914.
In this ancient ginkgo the trunk has
split in two and another large tree
is growing in the cleft.

the gardens of the West and became one of the best-known plant collectors of all time. In the process he also made substantial contributions to knowledge about the plants of China.[7]

Wilson came across ginkgo many times during his Chinese explorations. In his book *A Naturalist in Western China with Vasculum, Camera, and Gun,* Wilson gives his impression of what he had seen of ginkgo during his Chinese travels: "This strikingly beautiful tree is associated with temples, shrines, courtyards of palaces and mansions of the wealthy throughout the length and breadth of China, and also in parts of Japan. But nowhere is it truly wild."[8]

Nevertheless, with further botanical exploration it has become clear that there are many large ginkgo trees in China that are not associated with temples, and Chinese botanists have raised the possibility that wild ginkgo continues to grow in several places in the rich forests along the Yangtze River and especially in Zhejiang Province in the area around Tianmu Mountain.

For someone wishing to get a taste of the true richness of the forests of eastern China within easy striking distance of Shanghai and Huangzhou, there is no better place to visit than Tianmu Mountain, which reaches almost five thousand feet and is one of the highest mountains in Zhejiang Province. Its varied terrain produces equally varied vegetation ranging from subtropical evergreen forests to temperate deciduous forests. The area with ginkgo and the especially rich flora is located on the south-facing slope of the west peak. The vegetation is luxuriant and includes subtropical evergreens more typical of forest farther south, mixed in with more typical temperate deciduous broad-leaves and a variety of more temperate conifers. The flora is exceptionally rich, with about fifteen hundred species of vascular plants, a number of species comparable to the entire native flora of the United Kingdom in an area of little more than sixteen square miles.[9]

The ginkgo population at Tianmu Mountain has provided important information about ginkgo and how it behaves under natural or near natural conditions. For example, fieldwork by Peter Del Tredici located 167 ginkgo trees, more than a third of which had two or more substantial trunks. It seems likely that the extra trunks were produced by the activation of the basal chi-chi. As in other plants that produce basal lignotubers that are capable of sending out shoots and roots following a disturbance to the tree, they may help to stabilize the plant growing on steep slopes.

There is, however, a nagging worry that ginkgo, like several other trees growing on Tianmu Mountain, may have been introduced from elsewhere, perhaps a rather long time ago. It may never be possible to fully understand the origin of the ginkgo population on Tianmu Mountain, but over the past few decades the discovery of other potentially wild populations in China, together with the introduction of new techniques from molecular biology, has provided further opportunities to probe whether ginkgo survives in the wild in some parts of that vast country.

23
Relic

By their fruits ye shall know them.

—Matthew 7:15

With recent rapid developments in plant molecular biology, new tools have become available to help understand the history of ginkgo in eastern Asia over the past few hundreds of thousands of years. All ginkgo trees look more or less the same from the outside, but we can now look into their DNA to see just how similar or different they really are. We can sample different ginkgo trees growing in different places to see how they might be related to one another and also how much variability there is in their genetic makeup. We can use evidence from DNA to assess whether the individual trees in a population are genetically very different from each other or whether they are all more or less the same. Armed with that information we can also develop ideas about which populations could potentially have given rise to which, based on the assumption that those populations with the most genetic diversity might be living plants not too different from those from which genetically less diverse populations in other places might have been derived.

These kinds of arguments have been widely used to infer the origins of different crop plants. The different parts of the world where particular crops were originally domesticated from wild plants, as demonstrated by archaeological, botanical, and other

The famous Five Generations Ginkgo growing in the Tianmu Mountain Reserve,
Zhejiang Province, China.

evidence, often show relatively high levels of genetic diversity, whereas other parts of
the world into which these plants have been introduced often show relatively lower
levels. This is exactly what would be expected because it is unlikely that the full genetic
diversity would be represented in a subgroup of plants that had been moved to other
places. For example, at a global scale, potatoes and tomatoes have their greatest genetic variety in the northern Andes, where wild species that still grow in those regions
were most likely domesticated in ancient times. Using arguments of this kind, although
without the benefit of modern genetics, the great Russian plant scientist Nikolai Vavilov identified the likely wild relatives and probable source areas of many of our most
important crop plants.[1]

In trying to track down wild populations of ginkgo, there has been particular interest in studying the genetic diversity among the Tianmu Mountain ginkgo trees and

comparing that with the genetic diversity seen in other potentially wild populations elsewhere in China. Early studies tended to suggest that the Tianmu Mountain population had relatively low genetic variability, but how this compared with other populations of ginkgo elsewhere in China had not been tested until a series of studies over the past decade or so by Chinese scientists at East China Normal University in Shanghai and Zhejiang University in Huangzhou.[2]

In one initial study the Chinese team sampled nine populations of possible ancient ginkgo trees from different parts of China: Guizhou in southwestern China, Henan and Hubei in central China, and Jiangxi and Zhejiang in eastern China, which included the Tianmu Mountain population. For each population they extracted DNA from between ten and thirty trees, and for each tree they used two approaches to assess the genetic variation.[3]

Their first approach used the so-called RAPD technique: an acronym for the much less user-friendly technical designation, random amplified polymorphic DNA. This technique is relatively crude compared with the level of sophistication of modern molecular biology, which could potentially "decode" the entire DNA sequence of a ginkgo tree in a few weeks, but RAPDs are nevertheless relatively quick and effective, and were once widely used. For good reason, in the modern slang of molecular biology this approach is usually known as "RAPIDS."

After extracting the DNA from individual ginkgo trees the Chinese group created large quantities of short sections of DNA, each based on a "starter" section, a so-called primer, with a precise sequence of the genetic code. The basic idea was to see how many fragments each primer would produce from the DNA of each tree and how big those fragments would be. Then the researchers characterized the resulting fragments in terms of their size by placing each sample from each tree on a gelatin-like slab with an electric current passing through it. Drawn along by the current, each sample moves along a "lane" in gel, much like a sprinter in a hundred-meter race; smaller fragments of DNA move more quickly and travel farther, while larger fragments of DNA move more slowly and do not travel as far. In total, from all 164 trees sampled, the researchers recovered forty-seven DNA fragments of different sizes, and by comparing how many of the fragments were represented in each ginkgo tree and each group of trees, they were able to come to a rough estimate of genetic diversity for the different ginkgo populations growing in different places in China.[4]

The results showed that from a regional standpoint, and in terms of the number of different DNA fragments recognized, the ginkgo populations from southwest China were the most diverse. Three of the four populations from Guizhou showed especially high levels of diversity. The ginkgo trees of Tianmu Mountain from Zhejiang were more or less intermediate in diversity along with those from Henan and Hubei. Comparison of the RAPD results from each population with every other population highlighted the ginkgo population from Jinfo Mountain in Chongqing Municipality, as the most distinct. The population from Tianmu Mountain was also relatively distinct but, on the whole, more similar to the other populations studied.

The Chinese team then followed up with a second approach to assessing the genetic diversity of the different populations of ginkgo trees. In this case, they made many duplicates of specific pieces of the DNA from the leaves of 158 trees collected from four places in China. Having obtained a sufficient quantity of each piece of DNA from all of the trees, they used three enzymes, each of which had the potential to cut the DNA in a specific place. When they analyzed the resulting pieces, they found that among the 158 trees there were eight patterns. The population from Jinfo Mountain was again the most diverse: of the eight patterns the Jinfo population showed six.[5]

So what can we conclude from these initial insights into the DNA of Chinese ginkgo populations? First, these kinds of studies provide only a crude estimate of the true genetic situation. In both cases the technique looks at only a tiny portion of the overall genetic composition of these trees, and whether that portion is really representative is an open question. There is also a slight worry, inherent in the use of both techniques, that the fragments produced, while treated the same for purposes of comparison, may actually be slightly different.[6]

Nevertheless, more recent studies by Chinese scientists at Zhejiang University and their collaborators, using more sophisticated techniques, support these earlier conclusions. Again they highlight that the populations from southwestern China are genetically more diverse than those in eastern China, but on the whole they tend to support the idea that ginkgo may have survived the Pleistocene glaciations in two refuges: one on Tianmu Mountain not too far from the Chinese coast and the other farther inland in the protected valleys along the southern margin of the Sichuan Basin.[7]

If we take the results at face value, and also recognize that they provide only a guide, rather than a definitive picture of the history of different populations, it is certainly true that a grove of ginkgo trees with low or only moderate genetic variation, like those that occur elsewhere in China, would be unlikely to be the source of a genetically more di-

*Location of two potentially native populations of ginkgo in China at Jinfo Mountain,
Chongqing Municipality, and Tianmu Mountain, Zhejiang Province.*

verse population, like that at Jinfo Mountain or even at Tianmu Mountain. However, it
does not necessarily follow that the reverse is true. Less diverse populations may have
been derived from the more diverse ones, but especially in a case like ginkgo it is pru-
dent to be cautious. There has obviously been widespread extinction, including perhaps even in historical times, through forest clearance. Both the genetically diverse and
genetically depauperate populations could be different kinds of relics from formerly
widespread and still more diverse populations that are now extinct.

Against this background, the results from the DNA analyses do not completely settle
one way or the other whether the ginkgo trees on Tianmu Mountain are relics of a natu-

ral population, or whether they were introduced by Buddhist monks, as some have suggested. They do, however, point the spotlight at Jinfo Mountain and perhaps other places in southwest China as of exceptional interest and worthy of much more detailed study.[8]

Jinfo Mountain in Chongqing Municipality lies near the border of Guizhou and Sichuan Provinces. It is only a little farther south than Tianmu Mountain, but rather than being relatively close to the coast it is deep within the interior of China. And in addition to massive ginkgo trees there are other trees in the same region, such as the Cathay silver fir and Chinese fir, that have a long fossil history and that were once much more widely distributed. In this part of China there are many small, scattered populations of ginkgo that seem to be regenerating naturally from seed. Of all the ginkgo trees in China these are the ones that seem most likely to be the tiny remnants from a time when the geographic range of ginkgo was once much more continuous.[9]

History

24
Antiquity

How cunningly nature hides every wrinkle of her inconceivable
antiquity under roses and violets and morning dew!
—Ralph Waldo Emerson, "Works"

Ginkgo now grows all around the world, but almost everywhere it has been brought
there by people; for most of us ginkgo is a plant of parks, gardens, or city streets, all
human-created habitats. These trees are also of modest size; nowhere outside eastern
Asia are there ginkgo trees of truly massive proportions. Even the Old Lion at Kew, one
of the oldest trees in Europe, has a trunk only a little more than five feet in diameter. In
China, Japan, and Korea, the situation is different; here there are some true giants, and
in a few places in China, there are huge ginkgo trees that seem to be growing wild. At
Jinfo Mountain in Chongqing Municipality on the border of Guizhou and Sichuan, a
tree recorded in the 1950s had a trunk more than twelve feet across. Another recorded
in 1999 was clearly much older than the nearby town, with a trunk more than eleven
and a half feet in diameter. In all, seventy trees at Jinfo Mountain were found to have
trunk diameters of more than forty inches, and eight had trunks more than six and
a half feet across. There are many trees in that region with trunks the size of the Kew
Ginkgo, and they are growing alongside saplings and seedlings, which show that natural regeneration is occurring.[1]

The big trees found on the forested slopes of Jinfo Mountain and Tianmu Mountain are unusual; most of the really large ginkgo trees in China, and also in Japan and Korea, are not forest dwellers. Some are associated with temples or shrines, but they are often found alone in the countryside, growing by themselves. A review of large ginkgo trees in China turned up 138 with trunks more than about six and a half feet across; many are singletons, and it is obvious that they are not growing in natural forests.[2]

Among the most impressive of these trees is the Grand Ginkgo King, which grows near the small hamlet of Li Jiawan in Guizhou, southern China, towering over the agricultural bottomland of a small valley. Perhaps it started life as part of an ancient ginkgo orchard, but equally probable, given the seemingly wild populations of ginkgo not too far away at Jinfo Mountain, it could be the last remnant of an ancient ginkgo grove that somehow survived when the forest was cleared for farming. Like the ginkgos that survived at Hiroshima, perhaps it was resistant to fire, the key tool that early farmers used to clear the land.[3]

The Li Jiawan Grand Ginkgo King is a big tree, nearly a hundred feet tall with a trunk about nineteen feet across at ground level; more massive than that of any other ginkgo so far recorded. However, even from a distance it is obvious that this tree has had a complicated history. There are four main upward thrusting leaders, each of which is more than sixty feet tall. Studies by Chinese colleagues suggest that it may have sprouted from the base at least four times in the course of its long life. Its ring of partly separate trunks surrounds a hollow center so large that it once provided shelter for a local farmer and his cattle. The hollow stem rules out a full count of the annual rings and makes the age of the Li Jiawan ginkgo hard to assess, but by one estimate it might be as old as forty-five hundred years. Such great antiquity seems improbable, but from its size alone the Grand Ginkgo King must surely be among the most ancient of all living ginkgo trees.[4]

Great ages have also been suggested for other ginkgo trees in China. One in Dongkou Xian County, Hunan Province, is thought to be more than thirty-five hundred years old. Another at Dinlinsi Temple in Juxian County, Shandong Province, is reported to be more than three thousand years old. In Zhouzhi County, Shaanxi Province, and also in Tancheng County, Shandong Province, are trees that are thought to be more than two thousand years old, while another very large tree in Fuquan City, Guizhou Province, is likely to be of similar age. Altogether there are about a hundred ginkgo trees in China that are thought to be a thousand years old or more.[5]

The estimate of forty-five hundred years for the Li Jiawan ginkgo, if anywhere near correct, would make it among the oldest of all living trees, but this seems unlikely. Individual bristlecone pines in the White Mountains of California are dated reliably as more than forty-seven hundred years old, and direct evidence from annual rings shows that several other species, especially conifers, live for more than two thousand years. Some coast redwoods are into their third millennium, while a few giant sequoias are into their fourth. But these and other methuselahs are exceptions; most trees have life spans measured in a few centuries rather than in multiple millennia.[6]

It is often hard to be confident about the great ages attributed to giant trees. The massive baobabs of Africa, some of which have trunks that exceed twenty or thirty feet in diameter, provide a classic case. The great German explorer Alexander von Humboldt never saw a baobab, but nevertheless declared it to be one of the "oldest inhabitants of our globe." The French botanist Michel Adanson, from whom the baobab takes its scientific name, had more direct experience. He came across two trees on the Îles de la Madeleine off the coast of Senegal in 1749 on which earlier sailors had carved their names and dates. Estimating the amount of wood added since that time, and taking into account the diameter of the trunk, he deduced the age of the two trees to be more than 5,000 years. Adanson concluded that they must have been alive before the Great Flood. It was a clever and perhaps slightly mischievous extrapolation that allowed him to goad the religious establishment, but recent research indicates that most massive baobabs are much younger, between 500 and 800 years old. So far, the oldest radiocarbon date from a baobab is 1,275 years, from a tree in Namibia that had a trunk about one hundred feet across.[7]

Figuring out the age of baobabs is particularly difficult because they do not have well-defined annual rings, but even for trees with rings that are easily counted, accurate age estimates are often not easy to come by. One problem, as in the case of the Li Jiawan Grand Ginkgo King, is that the oldest parts of many truly ancient trees have been lost through decay. Another difficulty is that even though foresters have developed a special tool for taking a small core from the trunk of living trees, this becomes harder to do as the trunk gets bigger. It also is not always easy to find the center of a living tree with a complicated trunk.

A further problem is that even if a reliable age can be obtained for one tree of a particular species, the relationship between girth and age is not sufficiently consistent to extrapolate reliably to other specimens. Rates of growth can vary drastically in the life of a single tree—for example, when the sapling is growing in deep shade, compared with when it has an opportunity to grow quickly into the canopy. Growth rates also vary depending on where the tree is growing. Cross sections of two stumps of Japanese cedars on display in the forest museum on the island of Yakushima, Southern Japan, nicely illustrate the difficulty.

The massive Japanese cedars on Yakushima are among the great botanical wonders of the world. Jōmon-sugi, the largest of these great trees, was discovered in the mid-1960s in the dripping wet forests about three thousand feet above sea level. It is only about eighty feet tall, but its fifty-three-foot girth is truly impressive. Age estimates for the massive tree range widely, from 2,170 to 7,200 years old. The younger suggestion may be plausible, but seven millennia seems much too old. However, part of the difficulty in narrowing the age range further is in knowing how to extrapolate from the ages of cedars felled in other parts of the island.[8]

The two stumps of young Japanese cedars in the Yakushima Forest Museum are both about two and a half feet across, and the age of each is known precisely by counting the annual rings. One, which grew at a higher elevation, is 225 years old; the other, which grew at a lower elevation, is a mere 64 years old. In the older of the two the annual rings are tightly spaced, and each is perhaps no more than about one twentieth of an inch thick. In the other, the annual rings are much thicker. Trees with good growing conditions and a longer, warmer, growing season, with access to plenty of water and light, can grow fast, and the annual rings will be broad. Trees growing in less ideal conditions will grow more slowly; their annual rings will be narrow and tightly packed.[9]

A further way to try to try understand the age of massive Asian ginkgos is to look for historical or cultural clues. But this also has its problems, because the clues most often available are legends, which are rarely reliable. A good example is the massive and much-loved ginkgo at the Tsurugaoka Hachiman-gū Shrine in Kamakura, not far from Tokyo.

When I visited the Tsurugaoka Big Ginkgo on a warm spring Saturday several years ago, the shrine was bustling with families. The grand old tree had still not leafed out, and its nakedness revealed how much its crown had been sculpted by generations of tree surgeons. Nevertheless, it was wonderfully imposing. Standing next to the steep steps leading to the main building of the shrine, and with a traditional rope made of rice straw wrapped around its massive trunk, it was resplendent in the early spring sunshine.

The Tsurugaoka Big Ginkgo at the Tsurugaoka Hachiman-gū Shrine in Kamakura, Japan, photographed in spring 2006. The great tree fell after a storm in March 2010.

The Tsurugaoka Hachiman-gū Shrine was built in Kamakura in 1180 by the first Shogun, Minamoto no Yoritomo, and in 1219 was the site of the notorious assassination of the third Shogun, Minamoto no Sanetomo, on the same stone stairs that visitors still ascend on their way to the main shrine. According to legend, the assassin hid behind the ginkgo tree as he waited to make his attack; by that measure, the Tsurugaoka Big Ginkgo would today be nearly a thousand years old.[10]

Shihomi and Terumitsu Hori, the two foremost scholars of the cultural history of ginkgo in Japan, carefully reviewed the evidence to determine whether the legend and such a great age for the tree might be true. The *Azuma Kagami,* an official record of events that occurred under the Kamakura bakufu between 1180 and 1266, reports the assassination in detail, with the date, time, weather, and even the words of the assassin as he attacked. The *Eukan-sho,* written in 1220, just a year after the assassination, by the priest and poet Jien, gives still more detail, even describing the assassin's clothes. However, in neither of these near-contemporary accounts is there any mention of the tree. Suspiciously, the ginkgo appears only in accounts written much later; the first is in the *Kamakura Monogatari,* written around 1659, more than four hundred years after the fact. It seems very likely that the ginkgo is a later addition to the legend. Rather than a thousand years, a more modest age of perhaps five or six hundred years seems more likely for the Tsurugaoka Big Ginkgo.[11]

These kinds of problems apply to the great ages ascribed to ancient trees of all kinds, including ancient ginkgo trees in China. For example, one legend associated with the Li Jiawan Grand Ginkgo King links its origin to a scholar named Bai who lived in the Tang Dynasty (618–907), while another makes a connection to a somewhat similar legend from the Ming Dynasty (1368–1644). Similarly, it is said that the massive ginkgo trees at the ruins of the ancient Huiji Temple, not far from Nanjing, were planted by Zhaoming, a famous prince of the Liang Kingdom from the early sixth century. This would make them about fifteen hundred years old, but again there is no written evidence, and an age of about half that seems much more likely.[12]

It is hard for anyone, still less a Westerner, to get to the bottom of these kinds of legends, but Joseph Needham's classic *Science and Civilization in China* is nevertheless a useful guide. Needham devoted his life to understanding the early history of science in China, and in 1986 he reviewed the history of Chinese botany. A little later, Nicholas Menzies, one of Needham's colleagues, provided a similar account of Chinese forestry. Together, these two works review some of the most important early literature on the history of ginkgo in China. Especially illuminating is the example cited by Menzies of an inscription beside an old tree at the Fu-Yen Ssu monastery "on the slopes of the sacred peak of Hēng Shan in Hu-Nan." According to the inscription it was planted by "the venerable abbot Hui Ssu, in the second year of Kuang-Ta of the Chhēn dynasty." This would mean that the tree was planted in 568. However, Menzies cites work by foresters in the Nan-Yüeh Forest who had counted its annual rings and found it to have an age of six hundred years. This was a long-lived and no doubt majestic tree, but it was less than half the age that generally had been ascribed to it.[13]

Examples like these suggest that the great ages assigned to many ancient ginkgo trees are not reliable. Legends tend to be embellished over time, and it is not easy to obtain reliable estimates of the age of a living tree. For the largest and most spectacular specimens this has rarely been done. Given these complications, it seems unlikely that even the most ancient ginkgo trees approach the three thousand years of a giant sequoia or come close to the four and a half millennia of the bristlecone pine. A more reasonable view is that the age of some might approach fifteen hundred years, but that most have ages that are best measured in centuries rather than millennia. This is also in line with data from written sources, which points toward younger, rather than older, ages for the world's largest and most impressive ginkgo trees.

25
Reprieve

A burning stick, though turned to the ground, has its flame drawn upward.
—Saskya Pandit, "Elegant Sayings"

It is not known exactly when, or why, ginkgo first became associated with people, but in *Science and Civilization in China,* Nicholas Menzies finds the two earliest references to ginkgo unconvincing. The dating to the sixth century of the old tree at Fu-Yen Ssu Monastery is known to be unreliable. The other, the mention of a fruit referred to as *phing chung* in the poem "Rhapsody on the Capital of Wu" by Tso Ssu, which dates from the Jin Dynasty in the third century, has nothing to link it to ginkgo except mention of its silvery color. Similarly, while ginkgo is sometimes identified in carvings and paintings from the fourth to the eighth centuries, these show few botanical details and are also probably cases of mistaken identity. As far as I know, there is also no firm evidence to support the sometimes-cited use of ginkgo nuts during the Han Dynasty in the third century.[1]

The earliest reliable written reference to ginkgo is from 980 in the *Ko Wu Tshu Than,* or *Simple Discourses on the Investigation of Things,* written by Tsan-Ning, a "learned monk." A little later, during the Song Dynasty in the eleventh century, the most widely cited and earliest undisputed historical reference to ginkgo is a famous exchange of poems between the early Chinese historian Ouyang Hsiu and the poet Mei Yao-Chēn.

Both refer to ginkgo by the name duck's foot, *ya chio,* but they also use the name *yin hsing,* silver apricot.

Ouyang Hsiu begins the dialogue by presenting his friend with ginkgo nuts from a tree planted in Kaifeng, one of the Seven Ancient Capitals of China. Mei Yao-Chēn responds with his thanks and reminisces about ginkgo in his native Xuancheng. The exchange concludes with a poem from Ouyang Hsiu that sets out a delightful brief history of ginkgo cultivation.[2]

> Ya chio (duck's foot) grows in Chiangnan with a name that is not appropriate. At first it came in silk bags as a tribute, and as yin hsing (silver apricot) it became cherished in the middle provinces. The curiosity and effort of the Noble Prince [Li] brought roots from afar to bear fruit in the capital. When the trees first fruited they bore only three or four nuts. These were presented to the throne in a golden bowl. The nobility and high ministry did not recognize them and the emperor bestowed a hundred ounces of gold. Now, after a few years, the trees bear more fruits.

These poems and the other early record from the late tenth century both point to a first association of ginkgo with people about a thousand years ago. There are many earlier Chinese works that deal with plants, and with cultivated plants in particular, but none mentions ginkgo. Ginkgo is also unlike many other trees, which are often first noted as growing wild and are only later mentioned for their uses. This might suggest that before it was cultivated ginkgo was a rather rare tree. From the beginning, ginkgo is treated as a cultivated plant, a tree grown for its nuts. Compared with other plants, like rice and soy, which have a history of cultivation in China that goes back several millennia, the cultural roots of ginkgo seem relatively shallow.[3]

After the exchange of poems between Ouyang Hsiu and Mei Yao-Chēn, ginkgo appears commonly in Chinese literature, and by the Yuan Dynasty, established in the late thirteenth century by the ethnic Mongols under Kublai Khan, there are many references to ginkgo being grown for its nuts. According to Menzies, in some areas the nuts became an important commodity. He cites the *Nung Sang Chi Yao* from 1273 as the earliest work to give details of how ginkgo should be cultivated, and notes that these horticultural instructions were repeated verbatim in later manuals such as the *Chung Shu Shu,* the *Pien Min Thu Tsuan,* and the *Nung Chēng Chhüan Shu.*[4]

From the beginning, it was also recognized that both male and female trees are needed to produce nuts, but then, as now, there is no easy way to tell whether a nut would produce a male or female tree. The *Nung Sang Chi Yao* offers the following advice: "The ginkgo has male and female trees. The male [seed] has three ridges, the female has two. They must be planted together. When they are planted next to a pond, they face their own reflection and so can bear fruit." There is no evidence that three-ribbed seeds, which in any case are rare compared with the two-ribbed forms, grow into male trees. However, the idea of planting female trees near water may be helpful. Ginkgo thrives when it has good access to water, and trees stressed by drought are unlikely to produce a good crop of seeds. Later horticultural treatments also recommend grafting a male branch onto a female tree, the reverse of what was done by Jacquin in his experiments. It is an easy and sensible way around a potential twenty-year wait for seeds if there are no male trees nearby.

Based on the comments of Mei Yao-Chēn reminiscing about ginkgo near his home, and mentioning that he had collected nuts from the wild, Chinese authors have suggested that the ginkgo is native to the area around Xuancheng in Anhui Province. More recently, attention has also focused on Jinfo Mountain in Chongqing Municipality, as well as Tianmu Mountain in Zhejiang. It is still not completely certain whether ginkgo is native to these areas, but it does seem likely that the cultivation of ginkgo began in the south and spread to the north. This northward movement of ginkgo is also consistent with the comment in *Shihuazonggui* by Ruan Yue: "In the capital—now Kaifong City, Henan Province—there was no ginkgo. Since the coming of Mr. Li Wienhe, the emperor's son-in-law, from the south, he introduced it and planted it in his private house. Then it yielded and was propagated and developed there."[5]

The fact that none of the largest ginkgo trees in China are from the three northern Chinese provinces of Liaoning, Jilin, and Heilongjiang, on the landward side of the Korean Peninsula, also fits this pattern. The northerly expansion of ginkgo cultivation to Korea was probably along the coastal trade routes of China and then across the Yellow Sea to the southern tip of the peninsula.

Ginkgo would certainly have been cultivated on the Korean Peninsula well before Japanese troops under Toyotomi Hideyoshi invaded at the end of the sixteenth century, and in the seventeenth century Hendrick Hamel and his shipwrecked companions are reputed to have dreamed of their escape under the massive ginkgo at Gangjin-gun in the far south of the peninsula. A list of large ginkgo trees in Korea includes twenty-one with ages estimated to range from four hundred to a thousand or more years. The age of the great Yongmunsa Ginkgo is said to be about eleven hundred years, and two others on the Korean Peninsula are reputed to be even older.[6]

In North Korea, at the Anbulsa Temple there is a huge female ginkgo, 140 feet tall and 19 feet in diameter. It was protected on orders of President Kim Il Sung during the Korean War and in 2003 was visited by his son Kim Jong Il. It is said to be 2,120 years old and to produce more than 650 pounds of seeds each year. In South Korea the ginkgo in Yeongwol, with a trunk diameter of around 15 feet, is said to be 1,000–1,200 years old. It once stood in front of the demolished Daejeongsa Temple, but the 60-foot-tall tree still gives shade to the people in the village that has now grown up around it.[7]

In Japan there are fossils of ginkgo from before the ice ages and massive living trees that are evidently of great age, but there is no evidence that ginkgo is native to the Japanese Islands. Like many other useful plants, ginkgo was introduced from mainland Asia during historic times. As in China, many of the oldest and most venerable ginkgos in Japan are on the grounds of temples, but large trees are also found in agricultural landscapes, where, like the Li Jiawan Grand Ginkgo King, they may mark the sites of former ginkgo orchards rather than former temples that have disappeared completely.

Shihomi and Terumitsu Hori have reviewed the early cultural history of ginkgo in Japan and identified eight great trees that range from seven hundred to fifteen hundred years in age according to their associated legends. The Jyonichiji Ginkgo, a large female tree on the grounds of the Jyonichiji Temple in Toyama Prefecture, is considered the oldest. According to local legend, this was already a big tree when the temple was founded in 682. The Nigatake Ginkgo, sometimes called the Chichi Ginkgo or Uba Ginkgo, at the Ubagami Shrine in Miyagi Prefecture, is thought to be almost as old. The Hōryō Ginkgo in Aomori Prefecture, which is thought to be eight hundred to twelve hundred years old, is said to have been planted to commemorate the founding of the now lost Zenshoji Temple in the Heian period.[8]

It is hard to probe the veracity of these legends from Korea and Japan, but as in the case of the Tsurugaoka Big Ginkgo, such extreme ages are probably not correct. In Japan unambiguous written evidence for ginkgo appears even later than in China, and ginkgo is notably absent from classic early literature, where it might be expected to have appeared. Traditional Japanese thirty-one-syllable poems called *waka* have a history that goes back about thirteen hundred years. The *Man'yōshū*, the oldest anthology of such poems, was compiled at the end of the Nara period, during the late eighth century, but while there are many references to yellow leaves in these poems, these could belong to many other plants. Similarly, the Horis conclude that the word *chi-chi* mentioned in the poems of Otomo-no-Yakamochi, which date from the middle of the eighth century, more likely refers to the widely grown Japanese fig.[9]

There are no unequivocal references to ginkgo in the anthologies of waka collected by Imperial Command in the tenth and thirteenth centuries, either. Nor does ginkgo feature in *The Tale of Genji* by Murasaki Shikibu, one of the oldest novels in Japan, or in the *Makura no Soshi,* a collection of ancient essays written by Sei Shonagon in the late tenth to early eleventh centuries. Considering how prominently nature is featured in this literature, and how often ginkgo features in later works, its absence may be significant.

The first indisputable reference to ginkgo comes in two Japanese dictionaries and a textbook from the mid-fifteenth century. The dictionary *Ainosho* from 1446 by Gyoyo is published in a characteristic style, also used in early Chinese literature, which poses direct, specific questions and then uses previous references to answer them. To the question "What is ginkgo?" the answer comes back "There is no name in Wamyo," which is believed to be an earlier dictionary, the *Wamyo-sho,* published about 934. The *Ainosho* seems to set a lower and upper limit for the introduction of ginkgo into Japan between 934 and 1446.[10]

In the *Ainosho* and also in the textbook *Sekiso Orai,* written by Ichijo in the fifteenth century, the Japanese name for ginkgo is written in kanji characters as "silver apricot," exactly as in earlier Chinese sources. But it is also accompanied by the phonetic pronunciation in katakana script that gives the pronunciation as *icho.* In the *Kagaku-shu,* a dictionary from 1444, ginkgo is listed in the same way, but "duck's foot," another older Chinese name, is also mentioned, along with an explanation of why that term became the name of a tree. It is a reference to the similarity between the shape and diverging veins of the ginkgo leaf and the webbed feet and diverging toes of the mandarin duck, a lone bird revered for its symbolism. In Japanese literature the earliest reference to ginkgo appears in the travel diary of the poet Socho in 1530, who writes of giving a gift of beautiful yellow autumn ginkgo leaves.

Taken together, evidence from historical documents indicates that ginkgo was in cultivation in China by the eleventh century, or perhaps the late tenth century. It perhaps spread through the gifting of seeds and cuttings, which grew increasingly common in China from the tenth century onward. It clearly came to Korea and Japan somewhat later. In Japan there is no reliable evidence for ginkgo before the early part of the fifteenth century, three or four hundred years after ginkgo is well-established in China, probably with an association with Buddhism. The implication is that any ginkgo tree in Japan is unlikely to be older than about seven hundred or eight hundred years; none is likely to approach an age of a thousand years.[11]

The link between ginkgo and Buddhism may have been part of the motivation for its transfer to Japan, but ginkgo was perhaps even more important for food and medicine. And these motivations may not have been mutually exclusive; as in Europe, the early development of medicine in eastern Asia is closely linked to the development of religious beliefs. Buddhist monks would have been among the most important early practitioners of Chinese traditional medicine in medieval Japan.[12]

As to the likely place of ginkgo's introduction to Japan, it seems most plausible to look to the west and the south. In particular, the spotlight settles on Kyushu, the southernmost and westernmost of the four main Japanese islands, and a place with especially strong historical connections with mainland Asia. For centuries through medieval times, there was a thriving trade through the Kyushu area, in which ceramics played an especially important role. Beginning toward the end of the Heian period in the twelfth century, and continuing through the Kamakura and Muromachi periods in the thirteenth through sixteenth centuries, ceramics were greatly valued by the elite and were imported into Japan in large quantities. Colonies of immigrant Chinese became established in Japan and helped fuel trade in everything from precious metals to inkstones and water droppers for calligraphy. They were also importers of medicinal and other useful plants from the mainland; ginkgo may well have been among them.[13]

26
Voyages

He goes a great Voyage, that goes to the Bottom of the Sea.

—Thomas Fuller, *Gnomologia*

In May 1975 a fishing boat working in an area of high tides and strong currents off the northwest coast of Jeungdo Island, off the southwestern Korean Peninsula, dredged up in its nets six pieces of light green Chinese celadon ceramics and white porcelain: the first hint of the Shinan Ship, a discovery that ranks alongside the *Vasa* of Sweden and the *Mary Rose* of Britain among the most remarkable in underwater archaeology. At first the Korean authorities were uncertain how to proceed, but when the site attracted looters in the autumn of 1976, a high-profile salvage project was launched by the Korean Cultural Heritage Administration with the support of the Korean Navy. This massive nine-year effort opened an extraordinary time capsule from the fourteenth century and a treasure trove of exquisite objects that provided new perspectives on trade between China, Korea, and Japan in medieval times.[1]

The circumstances under which the Shinan Ship sank are not known. It was probably blown over in a storm, and as it went down, with the crew aboard, it crashed on underwater rocks before settling on the muddy seabed. Over time, in about sixty-five feet of water, the ship was partly buried by mud and sand. The parts left exposed were destroyed or washed away, but those parts of the cargo and hull protected from decomposition and shipworms remained in excellent condition. In 1990 the Marine Antiques Preservation Center was established at Mokpo on the mainland to help deal with the huge amount of material recovered and to oversee the ship's preservation.[2]

Excavating the wreck of the Shinan Ship was a formidable task. The Korean authorities had never undertaken an underwater excavation, and the navy divers found themselves working in darkness in muddy, fast-flowing waters. Initially, operations were aimed mainly at protecting the site, preventing looting, and assessing whether further work was warranted. However, the nearly two thousand pieces of ceramics and more than six thousand coins recovered by those early efforts made it clear that a careful, large-scale excavation would be well rewarded.

Salvage started in earnest in July 1977 with the support of two naval vessels and about sixty deep-sea divers. The initial work recovered tens of thousands of objects, many still preserved in the wooden crates in which they were originally packed. There was no doubt that the Shinan Ship was a trading vessel, and at ninety feet long, twenty-five feet wide, and about 260 tons, it was a relatively large one for its time, packed to the brim with merchandise of all kinds.[3]

Over the following summers until September 1984, the excavations recovered a further vast collection of objects, as well as the remains of some of the crew. An area of more than half a mile around the wreck was also checked with dragnets to scour up any last evidence about the ship, what it was carrying, and where it was going. The ship's timbers were also removed from the seabed for study, restoration, and exhibit, and many small objects were scooped or siphoned from the mud of the seabed in those final stages of the excavation. Among the plants recovered there was a single but unmistakable seed of ginkgo.[4]

Altogether, more than twenty thousand objects were brought to the surface after six and a half centuries on the seabed. The bulk was pottery of various kinds. There were about five thousand pieces of white porcelain and about three thousand pieces with varied glazes, but the majority, more than twelve thousand pieces, was distinctive green celadon. There were figurines, water droppers, incense burners, teapots, cups, cup stands, mortars and pestles, and even a ceramic pillow, as well as a variety of beautiful dishes, bowls, and vases. The preponderance of Chinese celadon makes it obvious that the Shinan Ship was on its way from China when it sank. Only seven pieces of Goryeo celadon were from Korean kilns.[5]

Some of the plant materials recovered from the Shinan vessel were part of the cargo,

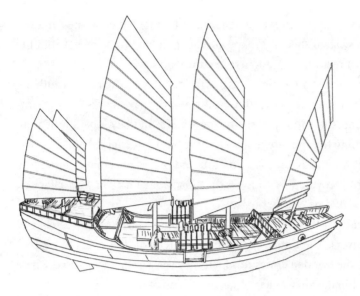

*Reconstruction of the
Shinan Ship, a trading
vessel that sank in June
1323 on its way from China
to Japan. A seed of ginkgo
and other exotic plants
were among its cargo,
which consisted mainly
of fine ceramics from
Chinese kilns.*

and had probably come from southeastern Asia, perhaps the Malay Peninsula, or even farther afield. There were more than a thousand pieces of sandalwood, most cut into pieces about six feet long. This exotic aromatic hardwood, which was used to make high-quality furniture or burned to produce fragrant smoke, could have come from southern China or southeast Asia, but it may have been brought from as far away as India. Also from the tropics was a large amount of black pepper in a rectangular storage box, as well as massive quantities of croton fruits, which are used medicinally as a purgative.

Other plant materials were found in much smaller quantities, but of sixteen species identified, fourteen were used in traditional Chinese medicine. They included two different kinds of ginger root as well as cinnamon sticks. There was also charcoal powder, made from dense logwood, which is used to help stop bleeding and treat dysentery. Along with the ginkgo seed, there were pieces of lychee, betel nut, apricot, peach, walnut, and hazelnut. All may have been part of the medicine chest on board the ship, a conclusion also supported by other objects discovered in the wreck, such as millstones for grinding medicines, as well as spoons and scales suitable for measuring and weighing herbal ingredients.

The celadon pottery showed that the Shinan Ship was carrying a cargo from China, and that the bulk of these ceramics were from the Longquan kilns. Other bluish white and white porcelain came from the Jingdezhen kilns. Both are in the hinterland of the important port area of Qing Yuan Lu, which is now the area around Ningbo. Further evidence that Ningbo was the port of departure comes from one of the scale weights, which is inscribed *Qing Yuan Lu,* as well as some of the curious 364 wooden tags recovered from the wreck. Like some of the other fragile objects from the Shinan Ship, these were recovered mainly during the final phases of the excavation from the mud in and around the wreck site.

Each tag is a short flat piece of wood up to about six to eight inches long with a hole or two notches at one end that were evidently used to tie the tag to the cargo. Remarkably, some of these tags identified the owner of that piece of the cargo, and included the owner's signature, the number of units or weight of the cargo, and its type. The tags not only confirm the port of departure as Qing Yuan Lu but also tell us when the ship left on its ill-fated voyage. More than a hundred tags record the shipping date. One is labeled April 20, six are labeled April 23, thirty-seven are labeled May 11, one is labeled June 1st, nine are labeled June 2, and fifty-eight are labeled June 3. And on the eight tags where the year is mentioned it is 1323. The ship probably left port in early June 1323 and may have been at sea for several weeks before it went down in a summer storm or perhaps an early typhoon.

The tags also provide direct evidence of who owned the cargo and where it was to be delivered. A hundred and one tags were labeled *Gangsa,* the title of a Buddhist monk responsible for aspects of temple administration. Forty-one tags were labeled for the Tōfukuji Temple in Kyoto, while others included the Hakozaki Shrine and Chojuan, one of the smaller branch temples belonging to the Jotenji Temple, both in Fukuoka. Historical records for this time, during the thirteenth and fourteenth centuries, show that some of the larger temples in Japan dispatched licensed trade ships to China, especially from Fukuoka on Kyushu, where there was a naturalized Chinese population. Among the tags designating the names of private cargo owners, about twelve seem to belong to Buddhist monks; a further twelve are clearly Japanese. The Chinese or Korean nationality of the others is uncertain.[6]

There would have been many possibilities for the movement of ginkgo from China to Korea and Japan through well-established trade routes plied by craft like the Shinan Ship. The ginkgo nut from the Shinan Ship, which dates from the early fourteenth century, is also consistent with the documentary evidence for the presence of ginkgo in Japan by the 1440s. The trees probably would have needed to have been in Japan and yielding nuts for a few decades before they attracted sufficient interest to be included in a dictionary. This suggests that ginkgo may have been in Japan by the 1300s.[7]

By the time of the first Western contact with Japan in 1543, ginkgo had probably been assimilated into Japanese culture for a century or two. By then there is also abundant evidence for an early association between ginkgo and Buddhist temples. The *Okazari no sho* from 1523 records the furniture, tools, and utensils that the Shogun Ashikaga Yoshimasa kept in the great Ginkaku-ji Temple, the Temple of the Silver Pavilion, that he founded as a private villa in Kyoto in 1460. Among them is mention of *icho-guchi,* a "ginkgo-mouthed flower vase." Also, then as now, the distinctive leaves of ginkgo proved an irresistible motif. From the late fifteenth century or early sixteenth century there is a long box (*nagafubako*) used for the storage and transportation of scrolls that is decorated with a crest (*mon*) of five ginkgo leaves. The *Kenmon Shokamon,* a book of heraldry written in the fifteenth century, includes some 260 family crests, one of them composed of three ginkgo leaves. From the Azuchi-Momoyama period at the end of the sixteenth and beginning of the seventeenth century, there are kimonos decorated with ginkgo leaves. One of the national treasures in the National Museum of Japan in Tokyo is a man's silk hip-length jacket (*kodofuku*) decorated with embroidered ginkgo leaves. By late medieval time the Japanese elite had already recognized ginkgo as something special.[8]

27

Renewal

The only real voyage of discovery consists not in
seeking new landscapes but in having new eyes.
—Marcel Proust, *À la recherche du temps perdu*

After millennia of decline, the reprieve that ginkgo found through its association
with the cultures of eastern Asia was followed in the eighteenth century by an even
more marked turnaround in its fortunes. This renewal began on September 25, 1690,
with the arrival of the German-born physician-botanist Engelbert Kaempfer in what
is now the heart of the Japanese city of Nagasaki. Kaempfer stayed just two years, but
what he learned, and what he later wrote, distinguish him in the West as "the first interpreter of Japan." He was also the first to introduce ginkgo to Western science.[1]

Located on the western shore of Kyushu, the farthest west and south of the four
main Japanese islands, Nagasaki stands at the head of a fine natural harbor where
the ancient trade routes that connect mainland Asia and Japan come together. To the
north, there is a short crossing to the Korean Peninsula via the ancient port of Hirado
and the island of Tsushima. To the south, the western shore of Kyushu leads to the
gentle arc of the Ryuku Islands and on to Okinawa and Taiwan. From there, Canton,
Hong Kong, and Macau are not far to the west. To the south, across the Luzon Strait,
the Philippines are the gateway to the islands of southeast Asia.

From Nagasaki the most direct route to China is west to the Goto-retto Archipelago, on to Jeju Island, and then southwest, out across the East China Sea to the Zhoushan Archipelago just off the Chinese coast, and from there to the ancient Chinese cities of Ningbo and Huangzhou. This was the lucrative trade route that the Shinan Ship was taking and that connected the rich ancient cultures of China to Korea and Japan. By the time Kaempfer arrived in Japan at the close of the seventeenth century, ginkgo had been assimilated into Japanese culture for perhaps two hundred or three hundred years, and by the 1690s Kaempfer records that it was grown "almost everywhere in Japan." This makes it likely that he was not the first Westerner to see it.[2]

In discussing trees from China, which were then very poorly known, John Evelyn, who published his great *Silva: a Discourse of Forest Trees* in 1664, mentions two great trees; one "a certain Tree called Ciennich (or the Tree of a thousand years) in the Province of Suchu near the City Kien, which is so prodigiously large, as to shrowd two hundred sheep under one only branch of it without being so much as perceived by those who approach it," the second "a greater wonder . . . in the Province of Chekiang, whose amplitude is so stupendiously vast, as four score persons can hardly embrace." There are several possibilities for the identity of these trees, including figs, which are often striking and achieve a great size, but there is just a chance that they could also be ginkgo.[3]

The first contact between Japan and the West had come even earlier with the arrival of the Portuguese in 1543, nearly 150 years before Kaempfer landed at Nagasaki. The Portuguese came from the south, bringing goods from Europe, as well as ceramics and other valuables from China, especially after establishing a trading base in Macau in 1557. European guns, first introduced by the Portuguese in the sixteenth century, later proved decisive in internal power struggles within Japan. The Portuguese also brought bread and tobacco, most likely chilies too, as well as new words that found their way into the Japanese language. The Japanese word *tempura* comes from the Portuguese word *tempero*, "seasoning." However, the most important Portuguese introduction into Japan was Christianity, which, intertwined with trade, was instrumental in the establishment of Nagasaki as a major port. Just a few years after first contact with the Portuguese, Jesuit missionaries began to arrive, and widespread conversion to Christianity followed. By the late 1580s the Catholic influence in southern Japan was widespread and growing. Steadily and predictably, it became an increasing cause of friction.[4]

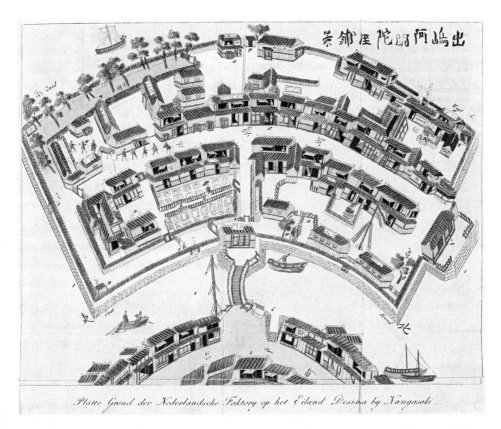

出嶋阿蘭陀屋舗景

Platte Grond der Nederlandsche Faktory op het Eiland Desima by Nangasaki.

The fan-shaped man-made island of Deshima in Nagasaki Harbor where Engelbert Kaempfer, Carl Peter Thunberg, and Philipp Franz von Siebold were based while working for the Dutch East India Company.

In 1587, concerned with the growing influence of the Jesuits in local politics, the feudal ruler Toyotomi Hideyoshi, the great unifier of Japan in the late sixteenth century, ordered the expulsion of the missionaries. Nine years later, on suspicion of a potential invasion, he ordered the crucifixion of twenty-six Catholics in Nagasaki, but the tension continued. In 1614 the Shogun Tokugawa Ieyasu finally banned Catholicism and followed through on the expulsion of the missionaries.

In 1636, in a further attempt to curb the influence of the Portuguese, Tokugawa Ieyasu confined them to the man-made island of Deshima constructed in Nagasaki Harbor. From Deshima, their activities could be monitored more readily and the lucrative trading activities could be more easily controlled. When asked in what shape the island should be made, Tokugawa Ieyasu is reputed to have simply spread out his fan. Deshima was thus created by cutting a canal through a small promontory and fashioning the resulting island as an arc linked to the mainland by a single bridge. This tiny island served as the main point of contact between Japan and Europe for more than two hundred years. It was the cultural and commercial conduit through which ideas and materials flowed in both directions.

While Deshima was founded as a base for the Portuguese, it soon became home to the Dutch. In April 1600, the Dutch ship *Liefde* arrived off the coast of present-day Usuki after a catastrophic voyage through the Straits of Magellan and across the Pacific. On board, one of the few lucky survivors was the Englishman William Adams. He became a key figure in early Western contact with Japan and nearly four hundred years later was the model for John Blackthorne, the fictional Englishman at the center of James Clavell's novel *Shōgun*.[5]

Adams flourished in Japan through the patronage of Tokugawa Ieyasu. He became the Shogun's personal adviser, was established as a samurai, and is best known for assisting the Japanese in the development of Western-style sailing ships and new trading ventures. Adams helped strengthen faltering early trade between Britain and Japan, but more important, Adams assisted the Dutch East India Company in breaking the near monopoly enjoyed by the Portuguese. With his help the Dutch began trading from Hirado, also on Kyushu, in 1609. With the expulsion of the Portuguese after the Shimabara Rebellion of 1637 and 1638, the Dutch became the new occupants of Deshima; this relationship became the most important European influence on Japan until the mid-nineteenth century.[6]

In the seventeenth century Deshima was a far-flung outpost of the Dutch trade network, but the writings of several of those who were stationed there became the main source in the West for early information on Japan. In Japan it also became known as a center of learning about certain branches of science and technology. It was a magnet for scholars of *Rangaku*, or "Dutch studies," and through the inquisitiveness of a succession of three physician-botanists, Kaempfer, Thunberg, and Siebold, all in the employ of the Dutch, Deshima also became the route through which Western science first learned about the plants of East Asia. Almost certainly it was also one of the ports from which ginkgo first embarked on its journey from eastern Asia to Europe.

Engelbert Kaempfer, the first of the Dutch physician-botanists to be based at Deshima, grew up in Lemgo, in northern Germany, about midway between Amsterdam and Berlin. As the second son of the pastor of Saint Nicholas church, he came from an educated family, and from 1674 to 1676 he studied languages, history, and medicine at the University of Cracow. This was followed by four years at Königsberg in Prussia studying natural sciences and medicine, and in 1681, Kaempfer moved to Sweden and the University of Uppsala.

The journey that took Kaempfer to Japan had an unlikely beginning in Stockholm at the court of Charles XI. It then followed an even more unlikely route, overland via Russia to Iran when Kaempfer was appointed secretary to the Swedish Legation to the Persian Court. The legation traveled to Isfahan in central Iran by way of Moscow—where Kaempfer had an audience with Peter the Great and his half-brother Ivan V—and it eventually arrived at the court of the sultan in March 1684. Ultimately, the Swedish mission was not a success, but Kaempfer then took employment with the Dutch East India Company, working for a time as a physician at the Dutch trading base at Bandar Abbas on the Persian Gulf. While there, he studied the local plants and also made careful observations of the date harvest, about which he published a detailed account on his return to Europe.

In June 1688, in the employ of the Dutch, Kaempfer left Persia on a voyage that took him to Muscat, the coast of India, Sri Lanka, and then on to the Dutch East India Company headquarters in Batavia, close to present-day Jakarta in Indonesia. He arrived there in September 1689, and in early May of the following year he left for Japan. In his posthumous work *The History of Japan,* Kaempfer recounts his voyage from Indonesia to southern Japan, including a month spent in Siam, modern-day Thailand.

When Kaempfer was at Deshima, the Dutch were virtual prisoners; their activities and especially interactions between them and the Japanese were strictly controlled. Over the winter Deshima was home to only a few dozen Dutch residents supported by Japanese who worked as everything from interpreters to cooks, but when the ships arrived, the port would have swarmed with both Dutch and Japanese as the cargo was unloaded, trades were made, orders were taken, and the new cargo was stowed for the long journey back to Europe.

As the Dutch contingent posted on Deshima came and went, the year followed a regular cycle. In the late summer and early autumn the boats would arrive from Batavia loaded with goods from Europe and also from the tropics. During the early autumn they would unload and be reprovisioned in time for departure before the weather deteriorated and winter set in. In the spring, the head of the Dutch mission, treated like a Japanese feudal lord (*daimyo*), would lead a delegation to Edo, modern Tokyo, to pay tribute and present his respects and gifts at the court of the Shogun. This was the same journey that Siebold took more than a century later. It took about a month in each direction, and was the best opportunity for the Dutch to see and learn more about Japan.[7]

In his two years in Japan, Kaempfer made the journey to Edo twice. In *The History of Japan* he gives a vivid account of "the Author's two journies to the Emperor's Court at Jedo, the City of his Residence," describing the preparations for the journey and the buildings and other structures that he saw. He describes how he traveled, the accommodations in which he stayed, the people that he encountered, and what took place along the way. He also describes in detail his stay in "the city of Jedo, its Castle and Palace, with an account of what happened during our stay there; our Audience and Departure." He writes: "I went to the Emperor's court twice myself to my very great satisfaction: the first time, in the year 1691 with Henry von Butenheim, a gentleman of great candour, affability and generosity . . . the second time, in 1692, with Cornelius van Outhoorn, brother to the Governor General at Batavia, a gentleman of great learning, excellent sense, and well skilled in several languages . . ."

Kaempfer continues:

> In the first place we set out from Nagasaki, to go by land across the Island Kiusju, to the town of Kokura, where we arrive in five days time. From Kokura we pass the streights in small boats going over to Simonoseki, which is about two leagues off where we find our abovemention'd barge riding at anchor and waiting our arrival, this harbour being very convenient and secure. . . . At Simonoseki we go on board our barge to proceed from thence to Osacca, where we arrive in eight days, more or less, according as the wind proves favourable or contrary.

He goes on:

> Osacca, is a city very famous for the extent of its commerce, and the wealth of its Inhabitants. It lies about thirteen Japanese water leagues from Fiogo, which we make in small boats, leaving our large barge at that place to await our return. From Osacca we go again by land, over the continent of the great Island Nipon, so far as Jedo, the Emperor's residence, where we arrive in about fourteen days or more. The way from Osacca to Jedo is by the Japanese call'd

Tookaido, that is, the Sea, or coast-way. We stay at Jedo about twenty days, or upwards, and having had our audience of his Imperial Majesty, and paid our respects to some of his chief ministers and great favourites, we return to Naga-saki the same way, compleating our whole journey in about three months time.[8]

Kaempfer made good use of his journeys to Edo and his time in Japan. When he boarded his ship on the last day of October 1692 to return to Batavia and then to Europe, he took with him a rich collection of specimens, books, and cultural objects: the first effort by any European to take samples reflecting the environment and culture of Japan. By May of the following year Kaempfer was in the Cape, and on October 6, 1693, he arrived in Amsterdam.

Kaempfer had intended to devote his time to writing an account of his travels, but his book *Amoenitatum Exoticarum* was not published until 1712, twenty years after he had left Japan. It is an extraordinary work, full of original observations and the story of his journey overland to Persia and then to southeast Asia. He also provides accounts of cultural and natural items that caught his interest, such as Japanese tea, and describes the plants that he had encountered, many of which were completely unknown to Western science. Included are the first Western descriptions of many Japanese plants, including many, like ginkgo, that had been introduced from China.[9]

On page 811 of the *Amoenitatum Exoticarum,* in the section dealing with fruit and nut trees, Kaempfer gives an illustration of a twig of ginkgo with its distinctive leaves inscribed "杏銀 *Ginkgo.* vel *Gín an,* vulgò *Itsjò.* Arbor nucifera folia Adiantino." This is the first Western illustration of ginkgo and probably was drawn by Kaempfer himself. The original sketch is among the Kaempfer papers in the British Library.

Decades later, in 1775, the position held by Kaempfer was taken by Carl Peter Thunberg, a talented and inquisitive student of Linnaeus. Building on what Kaempfer had accomplished, Thunberg's *Flora Japonica* was the first detailed account of the plants of Japan. A few years later, in 1779, Isaac Titsiugh arrived as *Opperhoofd* of the Dutch trading station. As a surgeon, scholar, and trader, he became a key figure in late-eighteenth- and early-nineteenth-century European relations with Japan and China.[10]

Throughout its history the export of ceramics was a lucrative part of trade passing through Deshima, and some of these early-eighteenth-century ceramics give direct evidence that the aesthetic possibilities of ginkgo had already caught the eye of artisans who were creating high-quality porcelain in southwestern Japan. Ginkgo leaves ap-

The drawing of ginkgo published by Engelbert Kaempfer in his Amoenitates, *1712: the first illustration of ginkgo by a Western botanist.*

pear occasionally among the many plant motifs that decorate the extraordinary bowls, dishes, and many other objects that were created as special gifts for the shogunate from the kilns of the Arita region. In the Kyushu Ceramic Museum one such white bowl shows three ginkgo leaves arranged in a circle. Still more striking is a beautiful large shallow dish with two stout upright ginkgo branches picked out in a cobalt-blue underglaze and covered with ginkgo leaves. Strong horizontal elements, a little above the midpoint, represent mist passing through the branches. It was produced in the Okawachi Kiln in Hizen between about 1700 and 1730, not long after Kaempfer was in nearby Nagasaki. Several large ginkgo trees still growing in that part of Japan would have been alive during the time of Kaempfer, Thunberg, and Siebold.[11]

28
Naming

If you do not know the names of things, the knowledge of them is lost too.

—Carl Linnaeus, *Philosophia Botanica*

Perhaps the most curious thing about Kaempfer's introduction of ginkgo to the West is the word itself, and much has been written about how Kaempfer came up with this seemingly strange name. To understand how and why Kaempfer fastened upon the name *ginkgo* it is important to understand how the plant was referred to in Japan around the time he was there in the late seventeenth century.[1]

When ginkgo was introduced into Japan, perhaps in the thirteenth or fourteenth century, the names most commonly attributed to it in China probably came with it. Among them was the name still used there today, silver apricot. In Chinese characters this is written 銀杏, and this is also how ginkgo is written in the Chinese character form (kanji) of Japanese. The characters must have seemed strange to Kaempfer, but he carefully copied them into his book. He wrote "silver apricot" as 杏銀 rather than 銀杏 because at that time in Japan the characters were written from right to left. Kaempfer then followed those characters with the transliteration of the two words that his Japanese translators were using to refer to the tree: *Gín an* and *Itsjò*.

When ginkgo first appears in Chinese historical texts it has two names: 銀杏, silver apricot, and 鴨脚, duck's foot. Later, beginning in the Yuan Dynasty, other names also began to be used: 白果, white fruit; 公孫樹, grandfather-grandchild tree; and 白眼, white eye. Among these, *pei kuo* or *bai guo,* meaning "white fruit," is the most widely used colloquial name in China, although 銀杏, silver apricot, is used most often in writing. In modern Japan, however, as in Kaempfer's time, when ginkgo is written in Chinese characters, 銀杏, it can be pronounced in two ways, *icho* and *gin'nan.* Shihomi and Terumitsu Hori reviewed how these pronunciations arose, and they suggest that they are corruptions of ways that the two original Chinese names of ginkgo, duck's foot and silver apricot, were pronounced. In standard Chinese these characters are pronounced *yajiao* or *yinxin,* but in the dialect of southern Jiangsu and northern Zhejiang, the part of China from which the Shinan Ship came and with which Japan had strong links through trade, 鴨脚 is pronounced *ai cho* and 銀杏 is pronounced *nin an.*[2]

The links between what Kaempfer wrote as "Itsjò" and the modern Japanese "icho," and between Kaempfer's "Gín an" and "gin'nan," are clear enough, but where did "ginkgo" come from? Shihomi and Terumitsu Hori investigated this carefully with the help of a distinctive feature of the Japanese language. Frequently, when a Chinese character is used, it may be accompanied by a phonetic spelling, a so-called *kana.*

By examining the kana associated with the Chinese characters 銀杏, silver apricot, in various Japanese dictionaries and books published between the fifteenth and eighteenth centuries, the Horis made an extraordinary discovery. In almost all the works they looked at, the pronunciation of 銀杏 is generally given either as "icho" or "gin'nan," or a clear variant of one of these two possibilities. However, in two Japanese pictorial dictionaries from the seventeenth century that were in use around the time Kaempfer was in Japan, the pronunciation is given as "ginkyo." In the *Kagaku-shu,* a dictionary published between 1617 and 1619, the pronunciation of 銀杏 is given as "icho" and "ginkyo," while in the *Kinmo Zui,* published in 1666, the pronunciation is given as "gin'nan" and "ginkyo."

In the 1940s A. C. Moule carefully examined Kaempfer's manuscripts, now preserved in the British Library, and found about ten mentions of the plant to which Kaempfer gave the name *ginkgo,* and noted in particular that "Gin'nan 32" is referred to as *Catalogus Plantarum Iaponicarum in Kin mo chjju i.* The Horis point out that this almost certainly refers to the *Kinmo Zui,* and they make a further important connection by realizing that Kaempfer obtained copies of this book while in Japan. There can be little doubt that Kaempfer referred to the *Kinmo Zui* both when he was in Japan and when he was writing the *Amoenitatum Exoticarum,* and that this is where the name *ginkgo* came from. With its later endorsement from Linnaeus it has become perhaps the most internationally recognized of all seventeenth-century Japanese words.[3]

There is one final puzzle, and perhaps the most difficult one: why did Kaempfer write "ginkgo" rather than "ginkyo"? This is particularly intriguing because in the preface of his book Kaempfer explains that he went to great lengths to transcribe the Japanese language as accurately as possible. He listened carefully to how words and sounds were pronounced, and developed a system of rules to commit them to paper. Because Kaempfer was so meticulous, Shihomi and Terumitsu Hori point out, the plant names that he transcribed into the Roman alphabet in *Amoenitatum Exoticarum* vividly capture how those names were pronounced by the Japanese people at the end of the seventeenth century. So if Kaempfer heard "ginkyo," why did he transcribe it as "ginkgo"?

Some have thought that the spelling of ginkgo was a simple mistake, a misspelling by Kaempfer or an error by the typesetter, but the Horis think that *ginkgo* is the spelling Kaempfer intended. They suggest that the mysterious second *g* in his spelling of *ginkgo* is a legacy of his own roots in northern Germany. In northern German dialect, the sound of *j* is often written as *g*. For example, Kaempfer might have said "jut" but would have written it "gut." The Horis suggest that in the context of his own background, Kaempfer transcribed "ginkyo" just as he heard it.[4]

Whatever the origin of the second *g* in *ginkgo*, the broader significance of Kaempfer's journey to Japan was widely recognized after the publication of the *Amoenitatum Exoticarum,* and after his death on November 2, 1716, his collections were put up for sale. They were quickly snapped up by Sir Hans Sloane, a wealthy and acquisitive patron of the sciences during the Enlightenment. It was Sloane, more than anyone else, who ensured that Engelbert Kaempfer and his work were not forgotten, and it was through Sloane that significant parts of Kaempfer's unique collections, which document one of the earliest Western encounters with ginkgo and with Japan, were saved for posterity and came to be in London.[5]

Sloane was an inveterate collector with a particular interest in plants. He collected plants in southern England and also in the south of France, but he will be forever associated with the plants that he collected during his fifteen months as physician to the governor of Jamaica, the Duke of Albemarle. Sloane came back from the West Indies in 1689 with a collection of eight hundred dried plant specimens and enough original observations for a large two-volume book; he also found that while he was away, the Catholic King James II had been replaced by the Protestant William of Orange.[6]

Returning to London, Sloane established a lucrative medical practice, based in Bloomsbury, which catered to some of the most influential and wealthy figures of the day. As a result Sloane too became well known, wealthy, and increasingly influential. He was elected to the Royal Society in 1685, became its secretary in 1693, and in 1696 published his *Catalogus Plantarum,* a list of all the plants that he had encountered while in Jamaica. His better-known two-volume work *Natural History of Jamaica* appeared in 1707 and 1725. Volume 1, which deals mainly with the plants, contains the first record in English of the use of chocolate. In 1719 Sloane became president of the Royal College of Physicians, a post he held until 1735. He succeeded Sir Isaac Newton as president of the Royal Society in 1727.[7]

One of Sloane's great scientific passions was acquiring important collections of plants being gathered during the European exploration of different parts of the world at that time, and it was this that preserved the legacy of Engelbert Kaempfer. Sloane's extraordinary collection of collections is the most extensive and historically important accumulation of early botanical specimens anywhere and vividly documents the rapid expansion of scientific knowledge about the diversity of plants during the seventeenth and early eighteenth centuries. All told, the Sloane collection consists of 337 separate collections preserved in 265 huge luxuriously bound books with green leather covers that are today housed in the former Department of Botany, now part of the Department of Life Sciences, in the Natural History Museum in London.[8]

Sloane collection 211 is labeled "H.S. 211, ff Volumen Plantarum in Japonia collectarum ab Engelberto Kempfero M.D. annis 1691 et 1692. Additae sub finem Plantae aliquot ab eodem in Persia et Insula Ceylan repertae." It is one of the most consulted herbaria in the Sloane collection. No other herbarium contains specimens of such an early date collected from Japan, and it is therefore a key resource for the scientific study of Japanese plants. For this reason, Sloane's Kaempfer herbarium has been studied by some of the greatest figures in the history of botany, including Daniel Solander, who accompanied Sir Joseph Banks on Cook's first voyage to Australia, and Robert Brown, the great botanist who first observed "Brownian motion." It was also consulted by Carl Peter Thunberg and Franz von Siebold, Kaempfer's successors on Deshima.

The Kaempfer collection was also well known to J. E. Smith, who, in 1797, in one of the more curious episodes in the history of scientific study of ginkgo, took it on himself to rename ginkgo *Salisburia adiantifolia,* "in honour of Richard Anthony Salisbury, Esq., F.R.S. and F.L.S., of whose acuteness and indefatigable zeal in the services of botany no testimony is necessary in this [the Linnean] society, or in any places which his writings have reached." *Salisburia* is used in favor of *Ginkgo* in some of the nineteenth-century botanical literature, but it never really caught on, and under the modern international rules of botanical nomenclature the name *Ginkgo* takes precedence. *Salisburia* is considered an invalid synonym.[9]

In addition to obtaining Kaempfer's collection of plants, Sloane also obtained Kaempfer's papers, and Sloane manuscript 2914, which is now in the British Library, is titled in Kaempfer's own hand "Delineato Plantarum Japonicarum manu Engelberti Kempferi." It is a volume of 217 folio-size drawings of Japanese plants, accompanied by an alphabetical list of names in Japanese with references to his *Amoenitatum Exoticum* and other works.

Realizing the significance of what Kaempfer had written, and also how much Kaempfer had not covered in his *Amoenitatum Exoticarum,* Sloane gave Kaempfer's papers to John Gaspar Scheuchzer, his personal librarian, to be translated from Kaempfer's High German dialect. The work that Kaempfer was preparing for publication, *The History of Japan,* was completed and published by Scheuchzer in 1727.[10]

Ginkgo is mentioned specifically in Kaempfer's chapter 9, which is titled "Of the Fertility of the Country as to Plants," and along with short accounts of "Mulberry trees," "Paper-tree," "Varnish-tree," "Bay-tree," "Tea-shrub," and many other kinds of trees and useful plants, Kaempfer comes to a nut called "ginau": "Another sort of Nuts, call'd Ginau, as big as large Pistaches grown very plentifully almost everywhere in Japan, on a fine tall tree, the leaves of which are not unlike the large leaves of an Adianthum. The Japanese call it Itsionoki. The nuts afford plenty of Oyl, which is also much commended for several uses. As to a more accurate description of this Tree, I refer the Reader to the Amoenitates Exotic."[11]

The History of Japan proved to be an enormously influential work. According to some, Jonathan Swift's Gulliver is something of a composite of Kaempfer and the sailor William Adams. Some of the incidents in Gulliver's journey seem to connect directly to similar descriptions and incidents in the Scheuchzer translation of the *The History of Japan.* Perhaps Sloane circulated Scheuchzer's translation among his friends at the Royal Society, in which Jonathan Swift was also a fellow.[12]

29

Resurgence

At your return visit our house; let our old acquaintance be renewed.

—William Shakespeare, *Henry IV, Part 2*

Through his *Amoenitatum Exoticarum,* Kaempfer was the first to bring ginkgo to the attention of Western science, but it is unlikely that he was the first to introduce living seeds or plants to Europe. This probably did not happen until later, perhaps not until three or four decades into the eighteenth century. The evidence for this time lag between Kaempfer's recognition of ginkgo and its cultivation in the West is mainly negative, but in this case it has at its center the most reputable source of information on all the plants known to science at that time: the Swedish physician and botanist Carl Linnaeus.[1]

Linnaeus, one of the greatest of all cataloguers and describers of new plants, spent the years 1735 to 1738 in Holland. Then a young man, he was employed by the wealthy Anglo-Dutch banker and great patron of the sciences George Clifford. Clifford was one of the directors of the Dutch East India Company, which at that time had been in existence for more than a hundred years. A few decades earlier Clifford's predecessors had been Kaempfer's employers. Clifford, Linnaeus, and their scientific circle were well connected and would have been on the lookout for interesting exotics from the East, but there is no mention of ginkgo in Linnaeus's catalogue of the plants growing in Clifford's garden, which was published in 1738. There is also no specimen of ginkgo in the

Clifford collections, which were acquired by Sir Joseph Banks and are now preserved in the Natural History Museum in London.[2]

Kaempfer's work was published in 1712, and Linnaeus was certainly aware of its importance. He owned a copy of Kaempfer's book and commemorated Kaempfer by naming a genus of tropical ginger after him. However, even by 1753 Linnaeus was still apparently unaware of the existence of ginkgo or, more likely, unwilling to commit it to print until he had seen a specimen with his own eyes. In his monumental work *Species Plantarum,* a comprehensive compilation of all the world's plants known to him at the time, there is no mention of ginkgo.[3]

A key figure in the circle of people in correspondence with Linnaeus at that time, and also someone with an interest in procuring plants from all over the world, was John Ellis, a leading scientist of his day and a prominent member of the Royal Society. In 1767 Ellis received the Copley Medal, the society's most prestigious award, for his work showing for the first time that corals are animals rather than plants. Ellis was also an excellent networker who used his connections to advance both science and Britain's commercial interests. In 1770 he published an account of how seeds and living plants should be transported from overseas. It was Ellis who played a key role in bringing ginkgo to the attention of Linnaeus.[4]

In a letter dated April 25, 1758, Ellis writes to Linnaeus:

> If you want a correspondent here that is a curious gardener, I shall recommend you to Mr. James Gordon, gardener at Mile End London. This man was bred under Lord Petre and Dr. Sherard, and knows systematically all the plants that he cultivates. He has more knowledge on gardening than all the gardeners and writers on gardening in England put together; but he is too modest to publish anything. If you send him anything rare, he will make you a proper return. We have got a rare double jessamine (*Gardènia flórida*) from the Cape, that is not described: this man has raised it from cuttings, when all the other gardeners have failed in the attempt. I have lately got him a curious collection of seeds from the East Indies, many of which are growing, but are quite new to us. He has got the ginkgo (*Salisbùria*), which thrives well, and when he has increased it, he will dispose of it.[5]

The anticipated arrival of ginkgo took another nine years. In a letter dated July 3, 1767, Ellis wrote to Linnaeus, "James Gordon intends you a plant of the Ginkgo of Kaempfer. I shall send your specimen of the *Siren lacertina* at the same time."[6]

James Gordon owned the then famous Mile End nursery in east London, and this was the conduit for many new plants into horticulture. According to Smith, he was already growing ginkgo in his nursery at Mile End in 1754. Further evidence of the early cultivation of ginkgo by James Gordon comes from the notes of Dr. John Hope, king's botanist to Scotland and professor of botany at Edinburgh. His trip to London in the summer of 1766 is an important source of information about the plants growing there at that time.[7]

Gordon was not only among the first to introduce ginkgo into British horticulture but he was also the first to provide Linnaeus with a living plant, probably as a cutting rather than as seed. It was probably the first ginkgo specimen that Linnaeus ever saw. Gordon wrote to Linnaeus in 1769, just a few years before Linnaeus's death:

> London, Mile End
> October 26, 1769
>
> My Lord
> Being many years much beholden to your Lordship's most ingenious and learned Labours, please to accept the underwritten plants. If there is any thing particular, that is acceptable, which falls in my way it will be the greatest pleasure and honour done me, to furnish your Lordship with it
> I am my Lord,
> Your Lordship's obliged servant
> James Gordon
>
> 1 Ginkgo kaempferi
> 1 Magnolia acuminata
> 1 Andromeda mariana
> These stand in the natural ground, and bears the severest frost of our winters, without damage.

Linnaeus wrote back thanking Gordon for the ginkgo plant that he had sent earlier that year. Since the plant was not yet mature, Linnaeus could not assign to it a definite systematic position. A specimen, probably preserved from the plant that Gordon sent Linnaeus, taken after the plant had been established at Uppsala, is now in the Linnaean herbarium of the Linnean Society of London. This is the specimen on which Linnaeus based the scientific name *Ginkgo biloba*. It is a typical long shoot with deeply bilobed leaves of the kind seen especially clearly on leaves near the tip of the shoot. If the specimen had come from a more mature plant, perhaps Linnaeus would not have chosen the

12923·2

The ginkgo specimen in the collection of Carl Linnaeus on which the name Ginkgo biloba *is formally based. Linnaeus almost certainly made this specimen near the end of his life from a living plant sent to him in 1769 by the British nurseryman James Gordon.*

specific name *biloba*. Linnaeus first published the name *Ginkgo biloba* in 1771, almost at the end of his life, in his *Mantissa Plantarum Altera*. Linnaeus, who was by then in poor health, published little after that; he died in Uppsala in 1778.[8]

Gordon was clearly a key figure in propagating ginkgo from limited early sources, but as far as I know, Gordon's records no longer exist, so it is not certain how he obtained the plants, or where they came from. One possibility is that through the connections of his wealthy friends and patrons he was able to obtain cuttings from the very few other trees that were in cultivation in Europe by then. It seems less likely that he was able to obtain seeds from Japan. However, there is also the strong possibility that he obtained material from China, and again it may have been John Ellis who provided the key connection.[9]

As well as Linnaeus's specimens, the collections preserved in the Linnean Society of London also include his extensive personal library. Pasted into the cover of Linnaeus's copy of Kaempfer's *Amoenitates* is a single handwritten sheet of paper headed: "Remarks on Kaempfer's Amonitat. Exotic by John Bradby Blake Esq. Canton 1770."

John Bradby Blake, a Londoner from a seafaring family, was born in November 1745 and was the son of Captain John Blake, who worked for the British East India Company and later ran a successful business supplying fresh fish to markets in London. Blake followed his father in working for the East India Company and first sailed to China when he was twenty-one, in 1766 as a supercargo. On one of his visits home he brought with him the Chinese boy later painted by Sir Joshua Reynolds. Eventually the company based him more permanently in China.[10]

While in Canton, Blake pursued a wide range of scientific interests. He sent back specimens of the kaolin and petuntse used in the making of Chinese ceramics. There are references to the "Chinese porcelain materials sent from China by Mr. Blake" in the experiment books of Josiah Wedgwood. He also sent back seeds of all the useful plants that he came across. According to Henry Laurens, writing in 1773: "Mr. Blake has an excellent opportunity by the aid of his Son who resides SuperCargo at Canton in China of introducing many things from thence which may become peculiarly beneficial to Carolina. He Seem'd most benevolently disposed to do all the good he can in that way, & only wants the assistance of minds a kin to his own, on our Side of the Water. . . . Severaral [sic] of the articles from Mr. Ellis were derived from Mr. Blake."

From Ellis or his associates some of these plants found their way to the Americas. For example, "the fine Cochin China rice" was distributed to growers in Jamaica, South Carolina, and the present-day Dominican Republic. It seems likely that Blake transmitted living ginkgo seeds too.[11]

The note written by Blake in 1770, and now in Linnaeus's copy of the *Amoenitates,* lists information on eight plants, referring to them by their numbers in Kaempfer's book.

811
Ginkgo vel Gin-nan pronounced Maulon ging:hang
Pao=Zuo
from the Northward—grows in gardens
but does not flourish at Canton the nuts are eat in

various ways: and are brought down for sale dried
by fire in great quantities to Canton. Procuring seeds
this hint must be particularly alluded to as also in
many other seed.

It seems unlikely that the original ginkgo plants grown by James Gordon in his Mile End nursery came from John Bradby Blake. Blake was not sent to Canton until 1766, eight years after Ellis's letter to Linnaeus proves that Gordon was already growing it, but it is not impossible that Gordon obtained living ginkgo in the 1750s from one of Ellis's other correspondents, perhaps even Blake's father. A few years later, Gordon, and perhaps others in England, may well have received ginkgo collected in China by Blake via John Ellis. After all, we know that Blake had been sent out for the express purpose of collecting living plants, and from the note in Linnaeus's copy of Kaempfer we have direct evidence that he was focused on making sure his seeds were alive. Confirmation that Ellis was being sent seed also comes from an aside in the account of ginkgo written by Loudon in his *Arboretum et Fruticetum Britannicum; or, The Trees and Shrubs of Britain:* "The nut, when examined by Sir J. E. Smith, from specimens in his possession, which were sent from China to Mr. Ellis, was found to be larger than that of the pistachia . . ."[12]

From this, it seems likely that the enigmatic piece of paper from John Bradby Blake, pasted into Linnaeus's copy of Kaempfer, also came from the Ellis correspondence which he gave to the Linnean Society of London. The note from Blake also makes it likely that the early ginkgo plants grown in Europe were the result of several separate introductions from diverse sources, from both China and Japan. The introduction of ginkgo into Europe almost certainly happened more than once and over a long period of time. Whatever the ultimate source of the early ginkgo trees growing in Europe, the evidence suggests that living ginkgo was rare there at least until the late 1750s, but only a decade or two later it was widely distributed.

It is also clear that James Gordon played a key role in its propagation and spread. He was probably the source of the ginkgo planted in the Royal Estate at Kew, now the precious Old Lion. He was growing ginkgo at the Mile End nursery already in 1754. There is also a tree at Blaise Castle, Henbury near Bristol, which is reputed to have been planted in 1762, and there is another at Whitfield, Warmbridge in Herefordshire, which is reputed to have been planted in 1776. Gordon's nursery may have been the source of all of them.

Use

30
Gardens

The garden suggests that there might be a place where we can meet nature halfway.

— Michael Pollan, *Second Nature*

Ginkgo owes its resurgence in historical times not just to its utilitarian value but also to some kind of irresistible biological charisma that has taken hold in both Eastern and Western cultures. In the East, ginkgo may have made the transition from wild forests to gardens almost by accident. Buddhist and Taoist priests have long nurtured tracts of forest with ancient trees around their temples. Chinese and Japanese Buddhism in particular believe that achieving nirvana, a state of spiritual liberation, is available to all life, including trees. With their unusual leaves and bizarre chi-chis, ancient ginkgos perhaps embodied an element of Buddha nature. Adopted also by Confucianism and Shintoism, ginkgo became widely revered as a symbol of vitality, longevity, and resilience.[1]

In the eighteenth century, ginkgo expanded from the forests, temples, and orchards of the East to the botanical gardens and grand estates of the West. Only a few decades after Kaempfer's first description, living ginkgo, probably as seeds but perhaps as seedlings or cuttings, began to arrive in Europe. Exactly when this took place is uncertain, but that it was no earlier than about 1700 and no later than about 1750 fits the evidence from many different sources. This means that there is no living ginkgo outside China, Japan, and Korea that is more than about three hundred years old.

Given the strong Dutch connections with Japan from the early seventeenth century onward, the earliest introductions of ginkgo into Europe may have been into the Low Countries, and two trees, little more than a hundred miles apart, vie for the title of the oldest living ginkgo outside of Asia. The largest is an old female that stands next to the church in Geetbets, near Hasselt in Belgium. With a trunk now about five feet across, it was reportedly planted around 1730. It is said to have been brought to Belgium by missionaries returning from China. Unfortunately, there is no corroborating evidence, but from its size alone this tree must be an early introduction, certainly from the eighteenth century. Whether it dates from as early as 1730 is less certain, and that age would be unusual for a female tree. Most early ginkgos in Europe are males, probably propagated as cuttings from one or a few original sources; the first record of a ginkgo producing seeds in Europe is not until 1814. A female tree planted in 1730 ought to have been mature in the 1760s or 1770s.[2]

In competition with the Geetbets tree for the distinction of being the oldest in the West is the large old tree in De Oude Hortus, the old Botanic Garden, close to the center of Utrecht. This tree is a male with a trunk about four feet across, roughly a foot less than the Geetbets tree, but still large. It is unquestionably among the oldest ginkgo trees in Europe, but again its exact age is uncertain. It is not mentioned by Linnaeus, who was in Holland between 1735 and 1738, would have visited the Utrecht garden, and had an eye for unusual plants. It is also not mentioned in the catalogue of the Utrecht garden from 1747 or in the list of plants growing at Utrecht sent to André Thouin at the Jardin de Roi in Paris in 1780. However, by the end of the eighteenth century the Utrecht ginkgo was flourishing. In 1787 the Swiss naturalist Frederick Ehrhart visited the garden and mentions that the tree was thirteen feet tall.[3]

In 1838 John Loudon, who covered ginkgo in his comprehensive account of the trees and shrubs grown in the British Isles, quotes the director of the Utrecht Botanic Garden in the 1830s: "Professor Kops informs us in a letter dated December 7, 1835, that it is a branchy tree and still continues to grow vigorously. He adds that, when he succeeded to the directorship of the garden in 1816, it was then calculated to be between 70 and 80 years of age; and hence, it must now (1837) be between 90 and 100 years old; and, if so, it must have been planted at Utrecht before the tree was introduced into England."[4]

A literal reading places this ginkgo as having been planted somewhere between about the mid-1730s and mid-1740s, but its absence from the 1747 catalogue may suggest that it dates from the 1750s or 1760s, around the time when we know that James Gordon already had ginkgo growing in his London nursery. Perhaps this ginkgo came to Utrecht together with fine ceramics on one of the annual Dutch shipments from Deshima, but it is also possible that it was sent by Gordon from London. Whatever its true pedigree, the Utrecht tree is a living monument to early Dutch contact with Japan and reminder of the key role of both countries in the introduction of plants into the world's gardens.[5]

In Britain it is part of garden folklore that the oldest ginkgo at Kew was among a select group of five specimens, the so-called Old Lions, introduced from the estate of the third Duke of Argyll, who had extensive nurseries, not far along the Thames at Whitton, near Twickenham. According to the story, when the duke died in 1761, his nephew John Stuart, the third Earl of Bute, moved some of the best young trees to the Royal Estate at Kew, bringing them down the Thames by barge.[6]

The reality is less romantic. Among the Old Lions, only a black locust and a plane have documented links with the Argyll Estate. The big Japanese elm, planted in 1760, is unlikely to have been acquired by Princess Augusta because it sits beyond the boundaries of the original arboretum. One of the others, the Pagoda Tree, is thought to have come to Kew in 1760 from the nursery of James Gordon, who introduced the tree into Britain in 1753. Given that Gordon was also an early source of ginkgo in Europe, it seems likely that he was also the source of the Old Lion ginkgo at Kew. It might be that both the Pagoda Tree and ginkgo were part of the same batch of seeds that Gordon somehow received, probably via John Ellis, from China or Japan.[7]

Early in the second half of the eighteenth century in Europe, ginkgo was already being propagated from cuttings and passed from garden to garden. In 1785 a ginkgo was planted in the Hortus Botanicus at Leiden, the oldest botanical garden in the Netherlands. In 1787 Giorgio Santi, director of the Botanical Garden of the University of Pisa, planted a ginkgo in the new arboretum, which still stands today just a five-minute walk from the Leaning Tower. Conrad Loddiges, who ran a prominent English nursery specializing in exotics, supplied a plant to the royal estate of Schönbrunn just outside Vienna in 1781, and this may also have been the source of the old male tree in the botanical garden in Vienna.[8]

Ginkgo first found its way to France in the late 1770s. Loudon cites the planting of a specimen in Rouen in 1776. Translating the account given by the French botanist André

Thouin, he also gives an account of the eccentric introduction of ginkgo into the gardens of Paris, which also explains the origin of the unusual name of the tree in French, *l'arbre aux quarante écus:*

> In 1780, a Parisian amateur, named Pétigny, made a voyage to London, in order to see the principal gardens; and among the number of those that he visited was that of a commercial gardener, who possessed five young plants of *Ginkgo biloba,* which still rare in England, and which the gardener pretended that he then alone possessed. These five plants were raised from nuts that he had received from Japan; and he set a high price on them. However, after an abundant déjeuné, and plenty of wine, he sold to M. Pétigny these young trees of Ginkgo, all growing in the same pot, for 25 guineas, which the Parisian amateur paid immediately, and lost no time in taking away his valuable acquisition. Next morning, the effects of the wine being dissipated, the English gardener sought out his customer, and offered him 25 guineas for one plant of the five that he had sold the day before. This however was refused by M. Pétigny, who carried the plants to France; and as each of the five had cost him about 120 francs, or 40 crowns (quarante écus) this was the origin of the name applied to this tree in France, of "arbre aux quarante écus"; and not because it was originally sold for 120 francs a plant.[9]

According to André Thouin, almost all of the early ginkgo trees in France were propagated from these five plants, one of which was given to the Jardin de Plantes in Paris. There are, however, other well-documented ginkgo introductions, for example, the ginkgo sent by Sir Joseph Banks to Pierre Marie Auguste Broussonet in 1788, which probably came from Gordon's stock. Broussonet gave the plant to Antoine Goüan, who planted it at the Jardin des Plantes in Montpellier.[10]

Around the turn of the eighteenth century the first generation of ginkgos planted in Europe began to mature; in horticultural vernacular they began to "flower," even though ginkgo has nothing resembling a flower in the normal sense of that word. The first is thought to have been the Old Lion at Kew, which produced pollen cones in 1795. Male trees also "flowered" in Pisa in 1807, and at Montpellier and Rouen in 1812. When a female ginkgo was eventually recognized in Europe in 1814, perhaps sixty years after the first introductions, it created a stir.[11]

The botanist Augustin Pyramus de Candolle from the Geneva Botanical Garden was the first to notice the female reproductive shoots on the ginkgo at the Bourdigny Estate just outside Geneva, but in the absence of males the seeds did not mature. However, cuttings from the female tree were circulated to botanical gardens across Europe and grafted onto many male trees, including in Strasbourg and Kew. The first viable seeds produced in Europe were borne on a male tree in the Montpellier Botanic Garden in 1835 that had been "covered . . . with grafts."[12]

Ginkgo made the leap across the Atlantic to the New World with the help of William Hamilton, a wealthy late-eighteenth-century botanist, plant collector, and landscape designer. Inspired by a year spent in England, Hamilton wrote to Thomas Parke on September 24, 1785: "The verdure of England is its greatest beauty and my endeavours shall not be wanting to give the Woodlands some semblance of it. . . . Having observed with attention the nature, variety and extent of the plantations of shrubs, trees and fruits and consequently admired them, I shall . . . endeavour to make it smile in the same useful and beautiful manner."[13]

His efforts resulted in the Hamilton Woodlands, a country estate on the Schuylkill River south of Philadelphia, one of the first American landscape gardens designed in the "natural" style of English estates of the time. Groves of native and exotic trees were set among pathways, meandering streams, and carefully maintained lawns. We can thank Hamilton for several familiar trees that he introduced into North America; the elegant Lombardy poplar, the tough Norway maple, and, less happily, the aggressively invasive tree of heaven. In 1784 Hamilton also cultivated the first ginkgo trees to grow on North American soil in several million years.

As the winter of 1785 approached, Hamilton wrote to his private secretary somewhat anxiously on November 2: "Secure the tender plants from the severe weather, otherwise all my pains will have been to no purpose. The Cistus's, the Heaths, eleagnus, Ginkgo, Laurus's, Tamarisks, Yucca glorioso, the Carolina mahogany, Zantoxylon sempervirens &c, should be secured by skreens of Dry straw or some other means, but by [no] means let dung be put to their roots for it will inevitably kill them."[14]

Hamilton's ginkgo was one of several acquired from England, perhaps through a connection to Peter Collinson, a British scientist, gardener, and fellow of the Royal Society. Through Hamilton's social connections, young ginkgo plants found their way to the gardens of some of the most prominent early American botanists. He gave one to his cousin and neighbor the prominent early American naturalist John Bartram, also a correspondent and friend of Collinson's. This male tree, which still survives in Bartram's Garden just a few miles from the former Woodlands estate, is now the oldest living ginkgo in North America. Another of Hamilton's cuttings went to the botanist and physician David Hosack. This is probably the ultimate source of the large ginkgo on the Vanderbilt Mansion Estate at Hyde Park that Hosack acquired in 1828.[15]

Within a few years of the arrival of ginkgo in the northeastern United States, André Michaux, a diplomat and botanist to the King of France, also introduced the tree into the southern United States. Having first attempted to establish a botanical garden in New Jersey, where the climate proved too harsh, Michaux moved to Charleston, South Carolina, in 1787 and developed what became known as the French Botanical Garden. Charleston's agreeable climate allowed him to cultivate and tend his plants year round, and it was here that he introduced many Old World plants to America, including the mimosa or silk tree, crepe myrtle, and camellia, and planted many others, including ginkgo. With the interruption of the French Revolution the garden was abandoned, but Michaux's son François, returning in 1802, recalled: "I found in this garden a superb collection of trees and plants that had survived almost total neglect . . . some of which were in the most flourishing state. I principally remarked two *Ginkgo biloba,* that had been planted about seven years, and which were then upward of thirty feet in height."[16]

Besides their interest in gardens and exotic plants, pioneering ginkgo growers in the United States shared another attribute: political prestige. André Michaux moved in the circles of Thomas Jefferson, Benjamin Franklin, and George Washington. John Bartram had cofounded the American Philosophical Society with Franklin in 1743. Peter Collinson had frequently exchanged correspondence with Franklin about electricity and other matters, and supported the American Philosophical Society with gifts of books. David Hosack was the founder and first president of the New York Horticultural Society, the first horticultural organization in America, in which John Adams, Thomas Jefferson, and James Madison were honorary members. On the Fourth of July in 1788, the official acceptance of the United States Constitution was celebrated with a parade that ended with seventeen thousand of the young country's citizens picnicking at the Woodlands.[17]

Given the company kept by some of the early ginkgo growers in the United States, it is perhaps not surprising that another prominent politician, Henry Clay, probably played an important role in the early cultivation of ginkgo through the American South. Several stately ginkgos that grow today in Kentucky are rumored to have arrived in Washington from Japan as seedlings and were subsequently sent by Clay to

Kentucky. Attempts to document this connection have so far failed, but a group of large and undoubtedly early ginkgos survive on Clay's former estate, namesakes for the estate's Ginkgo Tree Cafe. Two such trees also grow on the grounds of the former Kentucky Military Institute. Magnificent large ginkgos at Cave Hill Cemetery in Louisville, the former Cave Hill Farm, seem also to be of similar age, and they too may have connections to Henry Clay.[18]

Whether or not the ginkgos planted at the Kentucky Military Institute came from Henry Clay, they are significant as the first to bear seeds in the United States. Nurseries had distributed ginkgo as cuttings since 1810, but the availability of abundant seeds allowed ginkgo to be cultivated more widely. The Arnold Arboretum at Harvard received seeds from Kentucky on January 7, 1878. In 1890 William R. Smith, the curator of the United States Botanic Gardens in Washington, reported that the Kentucky female was the garden's "chief source of supply" of ginkgo seed: "Up to a recent date the Ginkgo was a very rare tree, nurserymen asking £1 apiece for them. The first Japanese Embassy brought over some seeds and presented them to the Botanic Garden and the trees raised from those seeds are now bearing seed."[19]

By the late nineteenth century, ginkgo was well established in the preeminent botanical gardens across the United States. In 1859 the Missouri Botanical Garden was founded by Henry Shaw, a British merchant from Sheffield; he planted several ginkgos that are now large trees in his magnificent garden. The New York Botanical Garden, founded in 1891, in the Bronx in New York, boasts massive old ginkgos growing near the Botanical Garden station, which was once part of the garden's estate. The largest, planted in 1898, is seventy-seven feet tall, with a crown spreading to fifty-seven feet. The garden's records indicate that by the 1890s ginkgo was readily acquired as small trees from nurseries.

Further enthusiasm for planting ginkgo came with the new excitement about its evolutionary significance following Hirase's discovery in 1896. As an icon of turn-of-the-century botanical science, it was planted on many university campuses. At the University of California, Berkeley, the elegant old ginkgo that stands next to Giannini Hall dates from around this time and is one of the most beloved trees on campus. Its rare display of autumn color among the otherwise dominant California evergreens has inspired poetry from professors with offices that face it. At the City College campus of the City University of New York, one of the most diverse colleges in America, students

*A carefully pruned espaliered ginkgo growing on the wall of the Department
of Plant Sciences at Cambridge University.*

complain about the smelly seeds of several old ginkgos in more than ninety languages. At Cambridge University, a carefully tended and unusual espaliered ginkgo cloaks the south side of the Plant Science Building on the Downing Street Site.[20]

The large female ginkgo that stands beside Botany Pond at the University of Chicago was planted almost at the founding of the university and is embedded in the memory of thousands of University of Chicago students and staff. A recent lighthearted competition among University of Chicago alumni to find the best haiku resulted in many entries about Botany Pond and its towering ginkgo. This was one of the best:

> Rustling gold ginkgo,
> Languid koi circling below
> In Botany Pond.[21]

As the meaning of *ginkgo* has changed in gardens of the world, so too has the plant itself. In the wild, new variants arising by mutation or unusual genetic combinations are quickly eliminated by natural selection. However, such novelties—for example, with an unusual growth habit or strange leaves—are easily perpetuated in gardens using cuttings or grafts. In this way, astonishing diversity has emerged from the gene pool of those few native ginkgo populations that survived into historical times in China. Today more than 220 different horticultural cultivars have been documented, and at least 28 cultivars are distinguished based on the details of their nuts.[22]

A landscape gardener recommended Princeton Sentry for the limited space in our own front garden in Oak Park. Tall and slender varieties like Mayfield and Tremonia can be also squeezed into narrow spaces. The varieties Golden Globe and Globosa have crowns as round and full as an apple tree's. Fastigiata, with its pyramidal form, looks a bit like a Christmas tree, while varieties like Pendula, Umbrella, and Horizontalis sprawl and weep like willows and Japanese maples. Mariken and Troll are dense, squat, and shrublike, barely resembling trees at all. Also much sought after are cultivars with unusual leaves. Saratoga, named for the Saratoga Horticultural Research Foundation in California, where the variety was developed, has drooping, triangular leaves with frayed edges like a fishtail palm. Variegata has striking variegated leaves with white streaks. Tubifolia has leaves that are almost entirely fused into trumpetlike funnels.

Ginkgo bonsai are also in high demand. They have the same dramatic fall color and abrupt leaf fall seen in full-size ginkgo trees, and are especially prized for these reasons. Even more coveted are chi-chi ginkgo bonsai, which begin life as cuttings taken from the descending woody chi-chi of old ginkgo trees. Planted upside down, chi-chi bonsai have an unusual conical form: the stalactite becomes a stalagmite, but with roots and protruding branches. Since there are few ginkgos outside of Asia that are old enough to produce chi-chi, these especially strange bonsai are available in the West only as imports and at great expense.[23]

31
Nuts

Sweetest nut hath sourest rind.

—William Shakespeare, *As You Like It*

Long before it became popular in gardens, or as a memory-enhancing supplement, ginkgo was valued for its edible nuts. With a plump, soft, partly creamy, partly waxy white "meat" not much bigger than a peanut, the ginkgo nut has a taste that has been variously described as like "mild Swiss cheese," "pine nuts," "potatoes crossed with sweet chestnuts," "green pea crossed with Limburger cheese," or just "fishy." Today, despite their enigmatic flavor, ginkgo nuts are common in the cuisines of China, Japan, and Korea, and they can be bought almost everywhere that people from those parts of the world have settled.[1]

It is possible that people have collected ginkgo nuts from wild trees for thousands of years. However, the first written accounts of ginkgo and its value as a nut tree appear much later. During the Song Dynasty, in the early eleventh century, when the poet and historian Ouyang Hsiu presented his friend Mei Yao-Chēn with ginkgo nuts, he was prophetic about how the association between ginkgo and people was likely to grow:

> In the past Chang Chien (second century B.C.) introduced grape and pomegranate (from central Asia). We can imagine that when these first came, they

must have been similarly highly valued as these nuts. But now these plants are common all over China, growing along fences and walls. The very things are still the same, but human nature changes in time. Someone should record the beginning so that future generations can know its origin. This is thus not only continuing your verse, but also contributing to history.[2]

Chun Chu Chi Wen, another early source, describes four large trees in Kaifeng, the ancient capital of China in eastern Henan Province, which produced several bushels of ginkgo nuts each year. A big female ginkgo can produce seeds in huge quantities, and such bounty may have been what first attracted the attention of people. Peter Del Tredici mentions a tree a hundred feet tall in Yang Tang village, Zhejiang Province, China, which produced more than eight hundred pounds of cleaned nuts in a single season. Today, production of ginkgo nuts is focused in ginkgo orchards of much smaller trees. In China production extends to more than twenty provinces and cities. By one estimate, there may be as many as 800,000 trees in China that produce an average of up to seven thousand tons of dried nuts each year.[3]

Collecting nuts for food is something that we inherited from our nonhuman ancestors. Nuts provide proteins, carbohydrates, vitamins, and minerals, and in the hands of an intelligent primate the hard shell makes them easy to keep and transport. Some of the earliest human tools may have been used to free the meats of nuts from their hard shells. At the same time, both consciously and unwittingly, people have spread nut-producing plants everywhere they have gone. The Old Testament tells us that Joseph's brothers carried pistachios down to Egypt. In northern Spain, the expansion of beech-dominated forests after the last glaciation seems to have followed human settlement and the use of beech nuts in animal husbandry. Native Americans may have used fire to maintain and increase populations of important nut trees like chestnut, hickory, and oak, and after the first European encounters with the New World, peanuts quickly found their way to China, now the world's largest peanut producer. Ginkgo is just one success story among many of the nut-people symbioses.[4]

In strict botanical parlance, a nut is a hard-walled fruit with a single seed inside, a definition that does not apply to ginkgo. Ginkgo is different; its hard shell is part of the seed rather than the fruit wall. However, the function of the "meat" in ginkgo seeds and the true nuts of other plants is the same: to provide food for the embryonic plant. Packed with carbohydrates, fats, and proteins, the "meat" helps sustain the young seedling as it develops until it can become self-sufficient. Ginkgo nuts are rich in starch and protein but low in fats. Compared to pine nuts, for example, ginkgo nuts contain about one-twentieth of the fat and a third of the calories. They also contain about 6 percent sucrose, which may account for the faint sweetness in their flavor.[5]

One of the Chinese names for ginkgo is *kung sun shu,* meaning "grandfather-grandchild tree." The name alludes to the patience needed to grow ginkgo from seed: the tree you plant today yields nuts in your grandchildren's generation. However, the time from seed to seed is generally much shorter, usually about twenty to thirty years. Even so, commercial growers are not prepared to wait. Grafting gets the trees into production earlier, often after only about ten years. It also brings the advantage that grafts can be chosen from trees known to produce large nuts.

Most of the ginkgo nuts consumed in China, Japan, and Korea are grown in commercial orchards often operated by small holders using methods that have changed relatively little over hundreds of years. In Japan, the area around Sobue, west of Nagoya in central Honshu, is a traditional region for ginkgo nut production. I visited on a bright, cold, early spring day in 2007. Around Sobue the land is flat, low lying, and often wet underfoot, with scattered houses, rice fields, and small farms; everywhere there are ginkgo trees.

The life of a ginkgo tree in the nut orchards around Sobue begins with a three- or four-year-old rootstock grown from seed, onto which several branches from female trees are grafted, usually in March and early April, just before the leaves appear. A useful by-product of grafting is that when normal branches from "mother trees" are grafted to the rootstock, they somehow "remember" how they were growing and continue to grow in a more or less horizontal fashion. They seem to have lost their ambition to produce the strong upward growth of a normal ginkgo tree.

This phenomenon, a so-called topophytic effect, is well known to horticulturalists and has been exploited in some species of conifer to produce the low-growing prostrate forms that are useful for gardeners. In ginkgo, the effect is to accentuate the normally spiky form of a young tree and produce the characteristic shape seen in ginkgo orchards: small trees with low, spreading branches. Such trees would be very inconvenient if planted along narrow streets where there is little space, but in a ginkgo orchard keeping the trees small and the branches low makes for easier harvesting.

Around Sobue even the largest ginkgo orchards consist of only a few hundred trees, and most of the seeds are produced by small-scale farmers who tend a few dozen trees on small plots of land close to their homes. Trees in orchards are pruned more regularly and kept in a uniform shape. Most are no more than about twenty or thirty feet tall, with a crown of spiky branches from the top of the trunk six to ten feet above the ground. Each tree is carefully tended and is often perched on a low mound of fertilizer and soil that is covered by a mat of straw to keep the fertilizer in, the weeds down, and the roots moist.

Whether the production of ginkgo nuts is on a small or large scale, the objective is the same: to produce good yields of the large nuts that command the best prices. In both China and Japan female trees that produce larger than normal seeds are much prized and often allowed to grow to reach their massive natural size. Over several centuries selection for larger nuts has produced a modest increase in size. Nuts from the Chinese cultivar King of Dongting Mountain are typically a little more than an inch long, while those from a normal female ginkgo are two-thirds that size. Many cultivars are distinguished by the size and shape of the nuts.[6]

Nuts start to ripen on the trees in August and can then be collected by hand. Usually, though, they are left longer to fill out, to give the shell more time to harden and to help ensure that the nuts are not damaged during harvesting. Timing is critical; the nuts cannot be left too long: they need to be gathered before the young embryo inside grows too much.

Traditionally, after collection, the seeds are buried to encourage decay of the smelly flesh, then dug up, washed, dried in the sun, and made ready for sale. Alternatively the seeds are collected in a bucket of water, where the pulp is allowed to partially rot until it can be rubbed off in changes of water until all of the flesh is gone. However, growers wanting to get paid more quickly are even more direct. In the countryside around Sobue, seeds are loaded into small vats that are stirred by large screwlike blades that help break up and remove the fleshy seed coat. The putrid flesh is then washed off as a gut-wrenchingly malodorous slurry. Active depulping needs to be done carefully, to avoid damaging the shells and squirting the noxious juices.

When the flesh is gone and the shell is exposed, the nuts are sorted by size, dried quickly in the sun, and sent on the way to the packers and distributors the next day. The nuts cannot be kept. Often the shell is removed and the "meats" are vacuum-packed before they find their way into the shops. More traditionally, ginkgo nuts are air-dried for one or two weeks, stored in a cool place, and then shipped to market, sometimes after having been roasted.[7]

The annual ginkgo harvest is also carried on every autumn by Asian families in the West, in streets and parks from Montreal to San Francisco. In New York City, a Harlem homeowner speaks of a new and unwelcome tradition: waking in early November to find people climbing the gangly ginkgo in front of her home, knocking down seeds with long poles. Other New Yorkers encounter the seed collectors under the old ginkgos that accent the northern boundary of Central Park. These foraged urban seeds mainly find their way onto local dinner tables rather than to the markets of Chinatown.[8]

Recently, a new group of ginkgo gatherers has appeared at the autumn harvests. Motivated by an ethos of eating local, seasonal food, intrigued urban "locavores" are on the lookout for new experiences, and recount their exploits in magazines, on food blogs, and over dinner parties. Sara Crosby, a writer for *Gourmet* magazine, went door to door among Chinatown restaurants toting a reeking bag of uncleaned ginkgo seeds, seeking advice on their preparation. After attempts at boiling, salting, and roasting all produced a snack she found too bitter to be palatable, her southern instincts came to the fore and she deep-fried them: ginkgo seeds, like most things, taste good that way. For curious cooks unwilling to delve into putrid pulp for the full experience, there is also the option of visiting the local Chinatown. A pound of plump, bleached, cream-colored *bai guo* sold in bulk, or a vacuum-sealed package of cooked and shelled "semen ginkgo," can be had for under two dollars.[9]

Interest in ginkgo nuts has mainly centered on their value as a food, but they have also been used in ritual and medicine. In sixteenth-century China, the nuts were often used at weddings and feasts, sometimes dyed red and substituted for lotus seeds. In traditional Chinese medicine raw ginkgo nuts are used to "send down adversely rising *chi*, remove toxic substances, and destroy parasites." Generally, however, it is not advisable to eat the nuts raw. They contain toxins that are partly broken down by heat, so ginkgo nuts are generally eaten only after cooking. Even then, they should not be eaten in quantity like almonds or peanuts. Cooked inside the shell, they emerge slightly yellow in color, with a waxy texture, and a slightly bitter taste.[10]

As Engelbert Kaempfer noted in the *History of Japan,* ginkgo nuts have also been used in Japan to produce oil. They may also have been used in this way in China. In ancient times plant oils provided an important alternative to animal fats for lighting and cooking, and this may have encouraged widespread ginkgo cultivation. The modern use of mineral oils and the introduction of gas and electricity for lighting and cooking

Ginkgo nuts baked in their shells at a banquet in Nanjing, China.

have pushed this use of ginkgo aside. Nevertheless, a long list of edible oils from nuts such as almond, cashew, hazel, peanut, pine nut, and walnut are still relatively easily obtained for various culinary, manufacturing, and medicinal uses. The fact that ginkgo oil is not among them may suggest that in the time of Kaempfer it was used mainly for nonculinary purposes.[11]

Culinary uses of ginkgo nuts in Asian cuisines vary by region and occasion; the nuts may appear in savory dishes and desserts, in formal banquets, and in day-to-day fare. In China, they may be enjoyed in many traditional dishes, boiled into soups, fried with celery and lily bulbs, or served in sweet dessert soups along with Chinese dates. At a dinner in Singapore not long ago, the dessert menu offered the options of "Teochow yam paste with gingko nut" and "Chilled snow fungus with lotus seed and gingko nut." The spelling was not quite right, but the taste for these nuts has evidently followed the Chinese diaspora to places well outside where ginkgo trees can be grown. Most often I have come across ginkgo nuts still in their shell, just baked on a tray or roasted in the oven in tinfoil. This is the easiest way to prepare them, but thorough cleaning of the seed is vital: as with poison ivy, sensitive people can have an allergic reaction if they are caught in the smoke.

In Japan, as in China, the culinary uses of ginkgo nuts were recorded soon after the

*Vendor selling roasted ginkgo nuts at the Tsurugaoka
Hachiman-gū Shrine, Kamakura, Japan.*

first written records of the plant itself. Ginkgo nuts are listed as a fruit or as tea cakes in *Shinsen Ruiju Orai,* a text written sometime between 1492 and 1521, and ginkgo nuts are also mentioned in tea ceremony records written between 1533 and 1596. In the *Miyoshi-tei Onariki* from 1561 ginkgo nuts are listed as sweets and a dessert on the menu of a meal served to Shogun Ashikaga Yoshiteru and his followers.[12]

Kaempfer, Thunberg, and Siebold may all have eaten ginkgo nuts during their time at Deshima, perhaps in traditional dishes little different from those of today. Edward Morse, a Harvard archaeologist and naturalist who was among the first Westerners to teach at Tokyo University in the late 1870s, considered himself an enthusiast of traditional Japanese food, but he remarked after a dinner at a Japanese tea house: "There were many things I tasted for the first time. The bulb or root of the lily was an excellent substitute for the potato; there were a number of water plants similar to the watercress; a preparation of fish, like macaroni; the nut of the gingko tree, which I did not like, and a preparation of tea, which I did."[13]

In Japan, as early as 1785, a book on traditional Japanese cuisine entitled *Kaiseki-ryori-cho* mentions ginkgo nuts as a side dish for sake drinkers, a tradition that continues. It is sometimes claimed that ginkgo seeds help prevent drunkenness and are an effective hangover cure. Research provides just a hint that such optimism is not entirely wishful thinking. An enzyme in the seeds seems to speed the breakdown of alcohol. In one study, laboratory animals were given enough alcohol to get them nicely inebriated: those fed ginkgo nuts beforehand were able to better clear the alcohol from their blood.[14]

Perhaps the best way to eat ginkgo nuts is out beneath the trees themselves; roasted ginkgo nuts are often sold by street vendors, like traditional European or North American chestnuts. To me, they taste a little like chestnuts, too. I saw them for sale one sunny spring weekend at the Tsurugaoka Hachiman-gū shrine in Kamakura not far from Tokyo. The brightly colored open-air stall advertised them with a big sign in Japanese, "delicious." The vendor was roasting them in a large wok over a gas burner, selling them in small bags and calling out in Japanese to the visitors to the shrine, "Freshly roasted ginkgo nuts, easy to eat, ready to eat." However, extracting the meat from the shells of roasted seeds can be a trial for heat-sensitive fingers. At home, my Japanese friends crack open the shells with pliers.[15]

On my visit to Sobue, we had lunch at a restaurant specializing in the use of ginkgo in Japanese cuisine. Our guides were enthusiastic for us to sample the local produce. There were ginkgo nuts in just about every dish of our traditional Japanese lunch. They appeared as a garnish with sashimi, for example, on a plate of finely sliced octopus. In *nabe-ryori* they were boiled with vegetables, fish, and meat in a hot miso-like soup. They were also deep fried in the tempura batter around shrimp. However, most memorable was the *chawan mushi*, a savory, pale yellow, custardlike dish with one or two ginkgo seeds at the bottom. This steamed, lightly set egg soup usually containing pieces of chicken, fish, or vegetables is one of the most common places to encounter ginkgo nuts. They are similar in color to the soup itself, and their flavor, like the flavor of the whole dish, is delicate and subtle: not quite sweet, but not fully savory either. This typical Japanese dish, like many others from Asia, extends the normal experience of the Western palate. Like the ginkgo tree itself, the versatile yet mysterious nuts continue to serve up more questions than answers.

32
Streets

I had to compete. In the concrete, in the jungle.

—Tower of Power, "Back on the Streets Again"

Ginkgo is among the most widely planted street trees in the world. With most of the world's people now living in cities, it is seen by millions of people every day. Along with other street trees, ginkgo has a role to play in sustaining connections between people and the natural world. In the United States alone, the trees growing in backyards, streets, parks, and the urban reserves comprise about 74.4 billion trees that account for about 8 percent of the total national tree canopy. These trees are important in the lives of those three-quarters of Americans who work and live in metropolitan areas.[1]

Interest in planting trees in cities has never been higher. The United Nations Environment Programme's Plant for the Planet: The Billion Tree Campaign in only five months secured pledges for a billion new trees globally. Two years after the program was launched in 2006, more than 1.8 billion trees had already been planted. In London the British charity Trees for Cities organizes its work around five themes: Trees for Food, Trees for Learning, Trees for Play, Trees for Streets, and Trees for Volunteering. It has planted nearly a quarter of a million trees around the world in the past few years. Initiatives to plant millions of new trees have taken off in New York, Los Angeles,

Memphis, Miami, Denver, and Philadelphia. Delhi has a target to increase its tree cover by 13 percent. These large-scale efforts have been aided by the simple fact that planting trees can be a good investment. They make people feel good about their environment, and within a few years economic and other returns more than make up for the costs of their planting and maintenance.[2]

Trees in cities contribute to floodwater reduction, temperature moderation, pollution abatement, energy savings, and improved property values. The crown of a mature tree can intercept as much as 100,000 gallons of rainfall each year, allowing much of it to evaporate before it flows into overloaded storm drains. Trees also take up substantial amounts of water through their roots and moderate temperature by breaking up and shading concrete and asphalt surfaces that absorb and intensify heat. Urban heat islands may have temperatures up to 10 degrees higher than places nearby. A well-placed tree that provides shade in the summer and shelter in the winter can reduce household energy costs by a third. In front of a home, a single tree can increase property values by about 6 percent. Studies also show that we are more likely to linger, and open our wallets, in business districts where the hard landscape and hard selling have been softened by trees.[3]

At the USDA Forest Service Center for Urban Forest Research, Greg McPherson and his team quantify the benefits of urban trees with their computerized valuation program i-Tree Streets. Using this tool, property owners, urban forestry groups, and city governments can quantify the benefits and cost savings that trees provide. The program uses data on the different kinds of trees and their sizes, together with local and regional information on climate, species growth rates, property values, energy prices, water prices, air pollutant emissions, stormwater costs, and costs of tree maintenance.[4]

Results from i-Tree show that the economic performance of ginkgo as a street tree varies substantially according to location. In Minneapolis, the average annual value of each of the city's 5,002 ginkgo trees at the time of inventory was $11.52. In San Francisco the savings were double that, at $23, reflecting differences in property values and costs of energy and water. With 16,184 ginkgo trees at the time of the i-Tree survey, not only are the ginkgos in New York City more numerous than anywhere else in the country, but they are also more valuable. Even so, $82 a tree is low compared to the New York City street tree average of $209. Ginkgo's lower value reflects in part a "penalty" for its relatively modest leaf surface area compared with other trees of similar height and spread. On the positive side, ginkgo scores high for its aesthetic value; people like it. Greg McPherson, the developer of i-Tree Streets, was quick to confess, "Ginkgo is my favorite city tree."[5]

Street trees also help us in ways that do not translate easily into monetary values. Neighborhood trees increase the time that children spend playing outdoors and the amount of supervision they receive. Parents and children spending more time outdoors helps to strengthen community bonds and reduce crime through improved vigilance. Tree-lined streets also help decrease road rage and improve the attention of drivers. By creating the illusion of a narrower street, trees prompt drivers to drive more slowly. Trees and nature generally also affect us psychologically in ways we can't always describe or explain. In one study, patients whose hospital room looked out on to a view of trees and water had shorter stays and more positive evaluations and needed fewer painkillers than a group of similar patients with windows facing a brick wall.[6]

In New Haven, Connecticut, the Urban Resources Institute, a nonprofit university-community partnership based in the Yale School of Forestry and Environmental Studies has worked for more than a decade to foster community-based land stewardship, environmental education, and urban forestry. Its work has touched every neighborhood in New Haven and has engaged thousands of residents, including unemployed ex-convicts and proud homeowners, most of whom might previously never have given trees a second thought.[7]

The scale of current campaigns for urban forestry is unprecedented, but the idea that trees are important for healthy cities has a long history. More than three thousand years ago in Egypt, Ramses III had trees planted along the streets for promenading and recreation, and in ancient Greece trees were planted to beautify cities and shade the pathways leading to market. In the sixteenth century, as the influence of the Italian Renaissance spread across Europe, the idea of a garden allée was widely embraced and soon moved into the city. In 1615 Amsterdam became the first city to formally incorporate buildings, transportation, and trees in its Plan of Three Canals, and King Louis XIV, in 1670, ordered that the walls around Paris be destroyed and replaced with new tree-lined boulevards for pedestrians and carriages.[8]

Aside from their aesthetic and recreational roles, urban trees were also thought to cleanse the city air and prevent the "miasmas" that were believed to cause illnesses and disease. The first public greening campaign in the United States was in Philadelphia, which passed an ordinance in 1700 that every owner of a house "should plant one or more trees before the door that the town may be well shaded from the violence of the sun . . . and thereby be rendered more healthy." In 1732 the Assembly decreed that "walks may be laid out and trees planted to render [the town] more beautiful and commodious." In 1792 citizens of Philadelphia petitioned to have trees planted in public areas because "It is an established fact that trees and vegetation . . . contribute to . . . the increased salubrity of the air." In 1872 in New York, the city commissioner of health declared that street trees should be planted to mitigate heat and reduce the death rate among infants.[9]

By 1773 Savannah already had a blueprint for avenues lined with live oaks, and in 1791, when George Washington appointed Major Pierre L'Enfant to design the capital city that would bear the first president's name, tree-lined avenues were planted throughout the city. In the early 1870s ninety ginkgo trees were planted as an avenue on the grounds leading to the Department of Agriculture, one of the first major uses of ginkgo as a street tree. A 1929 article gave the D.C. ginkgo trees a hearty endorsement that foreshadowed their broad acceptance in cities elsewhere:

> Visitors to Washington, D.C., are always much impressed with the beauty of the avenues of Ginkgo trees that line the approaches to the Department of Agriculture and that ornament the city in many other places. There is no good reason why Washington should be the only city in the country especially favored with this famous tree, sacred to the Chinese and Japanese and grown for centuries in their temple courts. It does very well in all parts of the United States where the winters are not too severe and can at least survive as far northwest as central Iowa. . . . A great virtue of the Ginkgo is the almost complete freedom from the fungus disease and insect pests that bedevil practically all of our other ornamental trees.[10]

One problem with the use of ginkgo as a street tree in Europe and North America is the disagreeable smell of the seeds. I have heard of several urban homeowners who have taken matters into their own hands and tried to kill unwanted female ginkgos on streets outside their homes. Every year arborists in Washington, D.C., spray large quantities of the herbicide chlorpropham on female trees to prevent them from seeding.[11]

Today, reputable nurseries circumvent the problem of smelly female ginkgos by selling only male trees propagated as cuttings. In New York City, the Department of Parks and Recreation, as a formal policy, hasn't planted a female ginkgo tree in twenty

*Three young ginkgo trees among
the many thousands growing on the
streets of Seoul, South Korea.*

years. Ordering trees from nurseries that propagate only cloned males is one way to ensure that the gender ban is enforced, but older trees that were planted before such regulations and that now are grandfathered into city tree heritage laws complicate the removal of healthy old female trees. Moreover, females often sneak back, sometimes through the planting efforts of less cautious residents and sometimes with approved permits.[12]

However the issue of female trees is handled, ginkgos remain common in cities from Beijing to Berlin. Often they are a defining feature of the streetscape. In London ginkgos are among the plantings around the Tower of London; on the other side of the city they accent the streetscape near Imperial College and outside the nearby Natural History Museum. In Manhattan, ginkgo accounts for 10 percent of the urban forest and is the third-most-common of all street trees. Whether on the streets of Chelsea, on Fifth

A young ginkgo being nursed back to health after having been transplanted during the renovation of the streetscape in Sejongo, Seoul, South Korea.

Avenue, along the northern border of Central Park, or through Harlem and Washington Heights, the spiky forms and distinctive leaves of ginkgos are ubiquitous.[13]

In Japan, ginkgo accounts for around 11 percent of all street trees and is the most widely planted tree in the country. Its use was nurtured there in the late nineteenth and early twentieth century as the country underwent rapid modernization. In Tokyo in particular, a 1907 plan focused on ten fast-growing and resilient trees, of which ginkgo was one. After the fires from the Great Kanto Earthquake in 1923 and the bombing of the Second World War, which destroyed close to half of the 270,000 street trees planted along the roads of Japan, city planners again turned to ginkgo to enliven and soften their streets. Today, more than half a million ginkgo trees are planted along Japanese roadsides.[14]

The same attributes that have allowed ginkgo to survive for thousands of millennia may have also contributed to its success as a hardy and resilient street tree. The lives of street trees are "nasty, brutish and short": on average, they survive a mere seven to thirteen years, compared with the sixty years the same species might expect to enjoy in a park and the hundreds of years it might live in its native forest. Deborah Gangloff, executive director of San Francisco's nonprofit group American Forests, explained some of the reasons why: "They're stuck in a concrete box, get bikes chained to them, with dogs relieving themselves and cars hitting them. . . . It's a hard life." From the salt that attacks their roots in winter to the ozone that assaults their leaves in summer, the constant barrage of chemicals faced by street trees is well beyond the level of abuse that evolution designed them to endure.[15]

Street trees also face challenges belowground. Urban soils are often made of landfill rubble, building material, and other contaminants. Such soils are highly variable in nutrient content, low in organic matter and fertility, and generally lacking in soil microorganisms needed for healthy plant growth. The ginkgo trees wedged into the pavement of my old neighborhood in the South Loop area of Chicago were pushing their roots through the remains of the old railway tracks that once entered Dearborn Station, but they did well nevertheless.[16]

When you next see a large street tree with an elegant crown, try to imagine what it looks like underground. An average tree has only about a fifth or sixth as much biomass belowground as above, but for a decent-sized tree that is still a huge amount of roots. With impermeable pavement running almost to the base of the trunk, those roots receive only a tiny fraction of the oxygen and rainwater that they would receive in the wild. Typically the roots of successful street trees are able to make do with little water and oxygen. Many of the hardiest and most widely planted, such as London plane, sweet gum, swamp cypress, and red maple, are floodplain plants with roots that are used to getting by with little access to oxygen. The long association of ginkgo with rivers may give them an edge as well.

The extensive networks formed by tree roots below sidewalks, driveways, and roads often torment homeowners and local officials. A former colleague in the United Kingdom made a successful living as an expert in the identification of tree roots. His services were much in demand from individuals and insurance companies wishing to find out whose tree was undermining the foundations of whose house. The problems flow the other way too. Most of the fibrous roots, which are the most important for absorbing water and nutrients, occur in the first foot of soil. When they are crushed by concrete, vehicles, and pedestrians, the aboveground parts of the tree begin to die.

Ginkgo tolerates root compaction better than most trees, but sometimes prevailing in a battle of wills with the sidewalk triggers other consequences. In the spring of 2007 in Everett, Pennsylvania, a pedestrian tripped on a pavement upended by the roots of a sixty-five-foot-tall ginkgo in front of the local library. Removal of the tree was recommended to the library board. However, it didn't take long for citizen advocates to launch a "Save the Ginkgo" campaign, with a petition circulated by the local high school, a benefit concert, and T-shirts, as well as bowls and wine stoppers made from the ginkgo's trimmed branches. Their efforts not only saved their prize ginkgo but also raised the $15,000 needed for pavement repairs.[17]

Along the way, renewed interest in the tree revealed that it was truly a survivor. It was the last of three ginkgos planted in 1861 to honor the three sons of the founder of the city, then called Blood Run, as they left to fight for the Union in the Civil War. The tree also turned up in old photographs of the Fourth of July parade on Main Street during in the 1920s, and could also be seen rising above the waters of the infamous flood that swamped the town on Saint Patrick's Day in 1936. It also turned out that the Everett ginkgo had also been saved once before, in 1985, from the city's attempts to widen its street. A group of elderly ladies had threatened to chain themselves to it. Resilience may be only part of the story of ginkgo's success, with charisma making up the rest.

33
Pharmacy

In all natural things there is something wonderful.

—Aristotle, "On the Parts of Animals"

One Friday afternoon at Kew, at the end of a harrowing week, my long-suffering secretary confided that she was among the estimated ten million Europeans regularly taking ginkgo leaf extract. "Oh yes," she said, "I've been taking ginkgo for my memory for quite a while—when I remember." Ginkgo is now a common herbal medicine in the West. In the East, interest in the health-giving properties of ginkgo goes back much farther. For almost as long as ginkgo has been prized for its nuts, it has also been valued in medicine.[1]

According to some sources, the medicinal use of ginkgo dates back to 2800 B.C., to the first pharmacopeia of traditional Chinese medicine, attributed to the perhaps mythical figure Shen Nung. However, the first undisputed written records of ginkgo come much later, and no original copies of Shen Nung's work survive. Ginkgo first appears in copies of the Shen Nung pharmacopeia around the eleventh and twelfth centuries, about the same time as it begins to appear in other written records. It is certain that ginkgo has been used in medicine for nearly a thousand years, but probably not for three or four millennia.

In the legends and folklore of China, Korea, and Japan, ginkgo is often associated

Stalactite-like zhōng-rǔ, *or* chi-chi, *bound with prayer ribbons on an ancient ginkgo at the Huiji Temple in Tang-Quan County, not far from Nanjing, China.*

with health and longevity. For example, in South Korea, the legend of the two ginkgo trees in Myeoncheon makes a link to Bok Ji-gyeom, one of the founders of the Goryeo Dynasty. It portrays him suffering from an incurable disease that has led his daughter to climb Ami Mountain and pray for one hundred days. There she meets a Taoist hermit, who tells her, "Brew Dugyeonju [wine from azalea flowers] . . . and drink the wine." Then after another hundred days, he tells her, "plant two ginkgos, dedicate your entire mind to praying, and it will cure your father." His daughter having dutifully followed the instructions, Bok Ji-gyeom recovers.[2]

Ginkgo is also often linked with fertility, and here the chi-chi, an eastern manifestation of herbalists' "doctrine of signatures," are especially important. Sometimes, chi-chi in the shape of a phallus become a particular focus for the red prayer ribbons that are often tied to the parts of old ginkgo trees. More often, the link is made to the nursing of infants. In Japan, the word *chi-chi* refers directly to breasts. Legend has it that the massive Nigatake Ginkgo in the city of Sendai was planted as the dying wish of Byakkonni, a wet nurse of the Emperor Shoumu. Women with problems in producing their own milk still worship at the tree.[3]

In traditional Chinese medicine, the seeds of the ginkgo are used most often; the best are said to be "large, dry, white, full and heavy." Seeds from Guangxi Province in southwest China are considered to be of especially high quality. They are used in the treatment of lung and respiratory ailments, sometimes combined with other plants such as the dried stems of ephedra, licorice root, and mulberry bark. They have also been used to treat a broad range of conditions from nocturnal seminal emissions and vaginal discharge to cavities, ringworm, scabies, and sores. Today ginkgo seeds continue to be used as an antitussive, expectorant, and antiasthmatic, as well as in the treatment of bladder infections.

Different medicinal uses call for different modes of preparation: dried unprepared seeds, generally known as *yin hsing* (silver apricot), are used to clear phlegm and kill parasites, whereas dry, fried, or baked seeds known as *cha bái guo* (charred white fruit) are pulverized and used to treat wheezing and vaginal discharge. To treat sores, fresh ginkgo seeds are cut in half or a poultice is made from the powdered seed. A colleague from Singapore also tells me that ginkgo seeds are good for the complexion.[4]

The chemistry of ginkgo has been studied for nearly two hundred years, and more than 170 different chemicals have been extracted and described from the seed and leaves. Some of these underpin the use of ginkgo in medicine, while others are responsible for some of its less desirable attributes. In 1927 the Japanese scientist Kawamura separated three novel, allergenic chemicals, ginkgolic acid, ginkgol, and bilobol, from the fleshy seed coat, and found them chemically similar to the compounds responsible for the allergic reactions produced by poison oak and poison ivy.[5]

Also problematic is butyric acid in the fleshy seed coat. This is what has given ginkgo the nickname "ginkgo stinko" among some American urbanites and led to the outlawing of female trees in many Western cities. Descriptions likening the smell of fallen ginkgo seeds to vomit and rancid butter are completely accurate; butyric acid is the main volatile compound in all three sources. Nevertheless, in spite of the smell, the fleshy coat from ginkgo seeds has sometimes found a use. In ancient China, it was mixed with lye to make soap, and it has also been used by fishermen who apply it to their bait to catch carp. In February 2007 the marine conservation group Sea Shepherd protested against a Japanese whaling expedition by spraying butyric acid at the whalers. Extracts from the smelly flesh show activity against disease-causing fungi, drug-resistant strains of bacteria, and even the snail hosts of the parasitic schistosoma fluke. The true value of the unwelcome waste product from ginkgo seeds remains to be seen.[6]

Notwithstanding their culinary and medicinal uses, an unfortunate truth is that ginkgo nuts are potentially toxic. In adults, ginkgo poisoning is rare, because generally only eating a large quantity would trigger a reaction. However, there is no authorized safe dose, and deaths have been reported from consumption of as few as fifteen or as many as 574 nuts. It is especially recommended that children under the age of six limit their consumption. However, the risk of ginkgo poisoning is very small, and I have enjoyed ginkgo nuts prepared in many ways; they are an essential part of the experience of being in eastern and southeastern Asia.

The effects of ginkgo seed poisoning have been known since at least 1709, when a case was first described in *Yamato-Honzo*, an old Japanese scripture. Symptoms range from irritability and vomiting to convulsions and loss of consciousness, which may begin one to twelve hours after ingesting. The toxin interferes with the body's ability to absorb Vitamin B6, which is crucial to maintaining functional nervous and immune systems, as well as many other vital processes. During food shortages in Japan between 1930 and 1960, reports of ginkgo poisoning increased significantly.

Despite a long history of selection and cultivation by people, the toxic compound ginkgotoxin, also known as MPN, has not been reduced or eliminated. Nevertheless, the toxin is water soluble and can be reduced by soaking. Levels of ginkgotoxin are also reduced by cooking; concentrations can be over forty times greater in raw seeds than in their cooked equivalents. In Yunnan Province, the Naxi ethnic group first soak ginkgo nuts, then sauté them with onion, garlic, apple cider vinegar, soy sauce, sesame oil, chili pepper, black pepper, and salt.[7]

As the medicinal use of ginkgo has spread from the East to the West, it has taken a surprisingly different trajectory. In the East the seeds have been used most widely, but in the West attention has focused almost exclusively on an extract from the leaves, which has been promoted mainly as a memory enhancer. There are early references from China to medicinal uses of the leaves. *Dian Nan Ben Cao*, written by Lan Mao in 1436, recommends the use of the leaves to treat freckles as well as skin and head sores. Slightly later, the medical text *Ben Cao Pin Hue Jing Yaor* recommends ginkgo leaves for internal use. In this and other Chinese materia medica, the leaves are described as used for treating dysentery, asthma, and cardiovascular problems. However, these uses have not been widely adopted in the classic texts on traditional Chinese medicine.[8]

The active compounds in ginkgo leaves come from an impressive battery of chemicals produced as a normal part of the plant's growth and include two main classes of chemicals, terpenoids and flavonoids. Flavonoids are the diverse chemicals responsible for the colorful pigments of certain flowers, absorption of potentially harmful ultraviolet radiation, protection against pathogens, and many other functions. Flavonoids are often suggested to have health benefits because they are an important source of antioxidants in food.[9]

Ginkgo leaves contain more than forty different flavonoids. Generally the leaves are harvested just as they begin to change color in the autumn, when the flavonoid content is highest. Green and yellow ginkgo leaves harvested in the autumn hold three times the amount of one key flavonoid than spring and summer leaves. On the other hand, the green spring and summer ginkgo leaves have a more potent content of other flavonoids and terpenes than the yellow ones, and also command a greater price in China as a tea.[10]

In Europe the main conditions for which ginkgo leaf extract is prescribed are peripheral vascular diseases, or the narrowing of arteries surrounding the heart and brain, resulting in reduced blood supply. In particular, ginkgo leaf extract has been widely applied to the symptoms of "cerebral insufficiency" often seen in elderly people, including difficulties of concentration and memory, absentmindedness, confusion, lack of energy, and other symptoms.

Standardized ginkgo leaf extracts developed by Schwabe Pharmaceuticals in Germany first appeared in 1964. Standardization is especially important for products from a plant like ginkgo in which the content and concentration of flavonoids and other chemicals may vary significantly through the season or depending on where the plant is growing. The extraction process consists of twenty or more steps to enrich the active and desirable substances in the leaves, while eliminating or greatly reducing the inactive or potentially harmful substances. The first commercialized ginkgo product was made available by Beaufour Laboratories in France in 1973, and was introduced to the market in 1975 by the subsidiary group IPSEN under the registered name of Tanakan. Soon after, INTERSAN and Schwabe laboratories released the ginkgo products Rökan and Tebonin forte on the German market.[11]

By 1988 doctors in Germany were writing more prescriptions for drugs containing

The Ginkgo Museum, a haven for ginkgo lovers in Weimar, Germany.

ginkgo extract than for any other plant-derived drug. Such use for symptomatic treatment of deficits in memory, concentration, and certain kinds of depression is approved by the public health insurance system. Currently ginkgo leaf extracts are among the leading prescription medicines in both Germany and France, accounting for 1 and 1.5 percent of total prescription sales, respectively.[12]

Annual global sales of crude ginkgo leaf extract were around $1 billion in the late 1990s, mainly from sales in Germany, but also from elsewhere in Europe, as well as in the United States and Asia. It is estimated that two billion daily doses have been used over the past twenty years. Recently, ginkgo has become a top-selling herbal medicine in the United States, despite the lack of the U.S. Food and Drug Administration's approval for the standardized purified extract. There are currently dozens of products based on ginkgo extracts, which can be administered intravenously, ingested in liquid form, or taken as tablets.[13]

According to some studies, the clinical effects of ginkgo, taken in the form of the standardized purified extract, include improved memory and learning capacity, increased brain tolerance to low oxygen, and improved circulation and microcirculation. Negative side effects appear to be few, and some of those occasionally reported, such as skin reactions and stomach upset, may be related to residual ginkgolic acids. Headaches may be a product of increased blood flow. Studies that reported significant improvements on one or more of the outcomes measured generally used dosages between 120 and 300 mg administered daily for between three and twelve weeks. When the extract was being taken to affect physiologic functions, such as memory or mood, four to six weeks were needed before positive results were noted. A review of 188 different studies conducted with humans and animals and in the laboratory found ginkgo extract to be promising in demonstrating a range of neurological and physiological improvements, including some that can take effect in a matter of hours.[14]

In spite of such positive indications and its widespread use, the effectiveness of ginkgo leaf extract remains a controversial subject. As with all herbal medicines, the use of ginkgo leaf extract raises eyebrows among researchers from a more analytical scientific tradition. They want evidence, based on extensive trials, and at the same stringent levels required of synthetic pharmaceuticals. In the United States, because ginkgo leaf extract is classed as an herbal treatment, manufacturers are not required to test the drug's safety and effectiveness, as they would be if ginkgo leaf extract were regulated by the FDA. The headline from a 2003 article in *Scientific American,* "The Lowdown on Ginkgo biloba," set a fairly typical tone from the orthodox end of the medical practitioner spectrum: "This popular herbal supplement may slightly improve your memory, but you can get the same effect by eating a candy bar."

However, the assessment of the efficacy of ginkgo leaf extract in the article that followed was balanced, fair, and based on a published review in a reputable scientific journal by three respected neuroscientists. They found "evidence that Ginkgo enhances cognitive functions, albeit rather weakly, under some conditions." Overall, though, they felt that more information was needed "to state conclusively whether Ginkgo does or does not improve cognition." They pointed out that too few experiments of the right kind had been conducted, and generally at too small a scale to provide definitive results. They summed up their feelings on the current state of research into the effectiveness of ginkgo leaf extract as follows: "there are enough positive findings to sustain our interest in conducting further research on Ginkgo," but they added an important caveat. "Many years of experience with investigations of new drugs have demonstrated that the initial positive results from studies involving a small number of subjects tend to disappear when the drugs are tested in larger numbers of subjects from diverse populations, so the true test of Ginkgo's efficacy lies ahead."[15]

Future

34
Risk

We see in many cases . . . that rarity precedes extinction; and we know that
this has been the progress of events with those animals which have been
exterminated, either locally or wholly, through man's agency.

—Charles Darwin, *On the Origin of Species by Means of Natural Selection*

Perhaps more than any others, Dave Raup and Jack Sepkoski, working at the University of Chicago in the 1980s and early 1990s, initiated the modern quantitative study of extinction in the fossil record. Dave led science at the Field Museum when I arrived there in 1982, but his move to the University of Chicago later that year gave new impetus to the development of that university's distinctive tradition of paleontology, which still continues. Along with Steve Stanley, at Johns Hopkins University, Dave was the driving force behind the development of a fresh and analytical way to study the fossil record of animals. As it developed in the late 1980s and early 1990s the approach that Dave, Jack, and others fostered came to be known, sometimes with affection, and occasionally dismissively, as the Chicago School of Paleontology.[1]

Jack was also a creative and brilliant thinker, and much of his best work emerged from his analysis of the massive database that he developed over more than two decades. In effect Jack built a giant table summarizing the history of occurrence of nearly every kind of animal through its entire fossil record based on specimens described from anywhere in the world. From this mind-numbing compilation he was able to chart and analyze the appearance and disappearance of different kinds of creatures over about the past 550 million years. The main focus was on the history of life in ancient oceans, and his database eventually summarized the fate of about 3,500 different families of marine animals. It was a herculean undertaking, but Jack's ambition went even farther. He developed a second database summarizing the comings and goings of 11,800 different genera of animals across the same vast expanse of geologic time.[2]

The work of assembling such a huge mass of data was tedious in the extreme, but also required careful attention to detail. Jack spent untold hours trawling through libraries to pull together thousands of fragments of information, often from obscure publications written in many different languages, but in the process he had to assemble the data in a consistent way, as far as possible standardizing different treatments of geology written by different authors at different times in different parts of the world. He also did what he could to untangle the complexities that come from different authors using different names for the same kinds of fossils.

Out of Jack's synthesis came strong evidence, assembled with new rigor and analyzed with new insight, of large-scale patterns in the history of life, and with Dave Raup also contributing his expertise, what emerged was the first well-founded attempt to quantify changing levels of extinction and origination through geologic time. One result, beguilingly simple but underpinned by a vast amount of data and huge effort, was a bar chart showing the first and last appearances of families of animal fossils at different times in the past. It revealed many ups and downs in the history of life, but five intervals stood out as seemingly cataclysmic moments in which an especially large number of families were lost never to appear again.

One such spasm of extinction took place around the boundary between the Ordovician and Silurian periods about 444 million years ago; a second occurred about 75 million years later in the Late Devonian, and this was followed by a third, more dramatic than any other, around 251 million years ago at the transition between the Permian and Triassic periods. The fourth great extinction occurred at the end of the Triassic, about 200 million years ago, while the fifth, the so-called K-T extinction that extinguished the dinosaurs and about which so much has been written, occurred at the boundary between the Cretaceous and Cenozoic, about 65 million years ago. Jack and Dave distinguished these five extinction episodes from the background extinction that is an inevitable part of the evolutionary process, and they also showed that while there have been twelve significant extinction events over the past 250 million years, the "big five" were of an entirely different order.[3]

The ancestors of living ginkgo, perhaps plants like glossopterids, evidently survived the mass extinction at the end of the Permian. This extinction, the so-called Great Dying, which wiped out perhaps 96 percent of all the different kinds of animals living in the oceans, may be the closest that life has come to going completely extinct. Plants very similar to living ginkgo also survived the extinction at the Triassic-Jurassic boundary, as well as the mass extinction at the end of the Cretaceous. Nevertheless, despite its resilience, ginkgo very nearly succumbed to the massive ecological reorganization that has occurred over the past few million years and especially during the Quaternary ice ages. Ginkgo came within a whisker of becoming one of those plants, like so many others, that we know only from the fossils they left behind.

There is direct evidence from fossils that the climatic trauma of the past few million years resulted in the loss of ginkgo from both western North America and Europe; a similar loss probably also occurred in eastern North America, although we don't have the fossils to prove it. The cause was probably complex. Drier climates may have been one problem. Repeated assaults by colder conditions, harder winters and shorter summers, and eventually the advance of glaciers from the north certainly constituted another.

Young ginkgo seedlings were probably the first to succumb, cut off before they could become established and start producing seeds of their own. Mature ginkgo trees may also have been killed as they struggled to find water or were eventually ground down by relentless frigid winters, shorter growing seasons, successive late frosts, and perhaps also by temperatures that occasionally went so low that the young leaves were killed inside the buds. If this increased mortality was coupled with limited capabilities for dispersal, as seems likely, the overall effect would have been to reduce the geographic range over which ginkgo could grow, and steadily deplete the number of living trees. As our world emerged from the last ice age, the remaining living populations of ginkgo in China may have been very small indeed.[4]

When speaking about extinction, Dave Raup would often use a memorable rhetorical flourish. He would ask, "Did they go extinct because of bad genes or bad luck?" It was a gentle rebuttal to the common assumption that animals or plants become extinct because they are somehow poor performers: that species lose out to other species that are superior ecological competitors. Dave's point was that such groups need not necessarily have been competitively inferior, they could just have been unlucky: perhaps a small population in the wrong place at the wrong time, or a large population that had an unfortunate encounter with an asteroid. In the case of ginkgo, losing the animals that dispersed its seeds might also be thought of as "bad luck."[5]

At the same time, Dave's favorite poser also had another subtext; it emphasized the importance of what are sometimes thought of as "random" events in evolution. At one end of this spectrum are certain mass extinctions, a form of extreme "bad luck"; it could hardly be the fault of dinosaurs or ammonites that they fell afoul of an asteroid from outer space. But in the same way, the cumulative impact of many small chance events, building contingently and relentlessly one upon the other, might also be significant. As Jimmy Stewart's character finds out in *It's a Wonderful Life,* Frank Capra's classic film, the present is the result of many small, seemingly insignificant, and sometimes chance episodes that over a lifetime add up and make a difference.

Both ideas speak to the significant role of chance and contingency in evolution. Stephen Jay Gould explored these ideas in his book that played off the title of Capra's film. Gould's particular focus was the early evolution of animals and how that early diversification influenced what came later; but his broader point was that chance and contingency have influenced much of the history of life. With regard to our own place in biological evolution, he posed the provocative question: "Would we appear at all if we could rewind the tape . . . and let it run again?"[6]

As Darwin made clear, as Dave Raup emphasized, and as modern-day conservationists also recognize, the risk of extinction by simple "bad luck" is greatly increased when species are reduced to just a few individuals. The risk is increased still further when those last remaining holdouts live in just one place or a few places where they could be wiped out by a single catastrophic event. The International Union for the Conservation of Nature (IUCN) formalizes such thinking in the different categories of threat that it recognizes in its so-called Red List. To qualify under the IUCN scheme as critically endangered, which is defined as facing "a very high risk of extinction in the wild," a species must meet several formal criteria, but basically they boil down to whether or not the species is known from just a small number of individuals or from only one or a few restricted locations.[7]

In the current IUCN Red List the most complete data are for the 5,488 known species of mammals, all of which have been carefully assessed based on strict criteria; 188 are critically endangered. More than one in ten of all mammal species are either critically endangered or endangered; seventy-six are thought to have gone extinct since 1500. The scimitar oryx and Père David's deer are extinct in the wild and survive only in captivity. The situation is similar for all other vertebrate animals and also, as far as we can tell, for plants. A recent assessment of all 800 different kinds of conifers and cycads in the world suggests that more than a third are threatened with extinction.[8]

Another important consideration in Red List assessments is whether there is already evidence that the population size of a species is declining. Measuring such declines is central to the Living Planet Index, a different approach to assessing the health of biological diversity. Developed by the World Wide Fund for Nature, the index is based on a compilation of data on the changing size of populations of 1,313 animal species, including fish, amphibians, reptiles, birds, and mammals. Since 1970 the Living Planet Index has fallen by about 30 percent, mirroring the gloomy picture from the IUCN Red List. Whichever way one looks at the current state of biodiversity, the news is not good; much diversity has already been lost, and the trends are consistently downward. The situation is urgent. It raises important questions about conservation priorities and what the most effective approaches should be.[9]

35
Insurance

They took all the trees and put them in a tree museum,

And they charged all the people a dollar and a half just to see 'em,

Don't it always seem to go that you don't know what you got till it's gone?

—Joni Mitchell, "Big Yellow Taxi"

In a famous phrase from *A Sand County Almanac*, Aldo Leopold, the best-known graduate of the Yale School of Forestry and Environmental Studies, argued: "If the biota, in the course of aeons, has built something we like but do not understand, then who but a fool would discard seemingly useless parts? To keep every cog and wheel is the first precaution of intelligent tinkering." The message for modern conservation is clear, but too often we are forced to make choices, and exactly how conservation priorities should be determined is a complicated issue on which there are many different views. Nevertheless, focusing conservation attention on those species at greatest risk is just common sense, and this means devoting particular effort to species that are known only from small and threatened populations. Doing what we can to protect those species where they occur, reducing the rate at which they are dying, and encouraging their reproduction are all critical steps toward the same basic goal: stabilizing or increasing the number of living individuals of that species in the wild.[1]

Zoos recognized long ago that the threats to some animal species in the wild were so severe, and the chances of their succumbing to simple "bad luck" or to the poacher's gun were so great, that the best hope was to rear them in captivity. In some cases it has proved possible to introduce these animals back into the wild. In the same way, for plants, conservation through cultivation is an important part of the toolkit needed to preserve the variety of plant life for the future. Everything possible should be done to conserve plants where they grow, as part of the broader ecosystem in which they evolved and the broader ecological processes in which they play a part; this should always be our main objective. However, as in other spheres of life where the risks of irreversible loss are great, it also makes sense to take out insurance. In the world of plants the actions taken to assure the long-term survival of the Wollemi pine are instructive. Like ginkgo, it is one of those species for which an encounter with people has helped make it more, not less, secure.[2]

The initial discovery of the Wollemi pine in 1994 in the Blue Mountains, just west of Sydney, was dramatic. David Noble, a park ranger working for the New South Wales National Parks and Wildlife Service, rappelled down into an otherwise inaccessible gorge and came across a peculiar species of tree with strange-looking leaves and "bubbly" bark. Specialists at the Royal Botanic Gardens, Sydney, quickly saw that this was a new species, but almost equally quickly they also saw that it was vulnerable to what Dave Raup would have called "bad luck." The total population comprised only about 110 individuals, all restricted to three localities that were very close together. The "bad luck" of a bush fire in the wrong place or the introduction of a stray pathogen on the boot of a visiting hiker could be enough to extinguish it. To the authorities in New South Wales the potentially catastrophic consequences of such chance events added up to a compelling case for swift conservation action.

The first step was to protect the area itself. The exact location of the Wollemi pine populations is kept secret, and access, even for the most legitimate reasons, is highly restricted. It was also decided not to intervene in a pristine habitat that was seemingly untouched by direct human influence. However, right from the start, a major effort was mounted to bring the Wollemi pine into cultivation and to distribute it to other gardens. Some of the first plants outside Australia were grown at Kew and Wakehurst Place. Tens of thousands of Wollemi pines, all ultimately clones from the hundred or so wild trees, were distributed by the National Geographic Society for $99.95 each. With seedlings now available from many sources, this unique and distinctive tree, un-

Seed cones, seeds, and shoots of the dawn redwood, with its distinctive opposite leaves and branching. The dawn redwood is a rare plant that was recognized as a fossil before being discovered in 1946 living in the forests of Sichuan Province, China. Like ginkgo and the Wollemi pine, it is now widely grown all over the world.

known less than twenty years ago, is increasingly common in the world's gardens. In the coming decades and centuries, even if the few remaining trees in the Blue Mountains should succumb to fire, disease, or climate change, the species is now safe. Its long-term survival has been secured by a combination of effective vegetative propagation, clever marketing, and high-profile publicity.[3]

At the end of the last ice age the number of ginkgo trees living in China may not have been quite as small as the number of Wollemi pines in the Blue Mountains, but its situation may have been almost as precarious. However, ginkgo has also been fortunate; through its association with people, the population of ginkgo trees growing around the world has been vastly increased. Even though many of the ginkgo trees in cultivation may all be genetically rather similar and could be more or less equally susceptible to a new pest or disease, ginkgo is now so widespread that there is little chance it will go extinct simply through "bad luck." By bringing the tree into cultivation in large numbers and in many different places around the world, we have greatly increased its chances of long-term survival, and we are all among the beneficiaries.

This kind of *ex situ* approach to plant conservation is needed because everywhere the world's trees are under threat. According to the Global Trees Campaign, led by Sara Oldfield at Botanic Gardens Conservation International, more than eight thousand species, approximately 10 percent of all known trees, are threatened with extinction. More than seventy are thought to be extinct, about eighteen are now known to exist only in cultivation, and these depressing figures are almost certainly an underestimate. Like ginkgo, each of these trees has its own story to tell; each helps fill in part of the grand evolutionary puzzle.[4]

The drautabua is a small tree restricted to steep narrow ridge tops high in the mountains of Viti Levu, Fiji. Like the Wollemi pine, it is one of the world's most endangered conifers; fewer than ten small and widely dispersed populations are known, and one has already been lost. Fossils similar to the drautabua occur in Australia and Antarctica, but the only other living species grows on New Caledonia. The drautabua is on the brink of extinction. It is under threat from a nearby copper mine as well as from the impacts of climate change in the special mountain habitats where it lives. It is not easy to grow, but the few plants in botanic gardens provide invaluable insurance against its loss in the wild.[5]

The Mulanji cypress, the national tree of Malawi, faces a similar crisis. It grows only on Mount Mulanji and has been decimated by overcollecting for its highly prized, decay-resistant, timber. It is no longer reproducing in the wild. Its close relatives, the Clanwilliam cedar, one of the most majestic trees of the Western Cape of South Africa, and the Willowmore cedar, restricted to the Eastern Cape, have similar problems. Three of the four species in the genus are in trouble. The Millennium Seed Bank at Kew and its partners in Malawi and South Africa are collecting seeds, working to improve the survival of seedlings, and bringing the plant into cultivation as a prelude to its reintroduction into the wild.[6]

There are many similar examples. Franklinia, named for Benjamin Franklin and discovered in what was then Britain's American colony of Georgia by John and William Bartram in 1765, is another tree that is widely cultivated but no longer known in the wild. The Hawaiian cotton tree is known from only four trees growing in their original habitat; many more are conserved in botanic gardens. It is a small tree with spectacular red hibiscus-like flowers; pure good luck brought its ancestor to Hawai'i, perhaps three million years ago; a run of bad luck could easily wipe it out. The Robinson Crusoe cabbage tree, known only from the Juan Fernández Islands, was down to just three surviving plants in the 1980s; the Café Marron, known only from Rodrigues, survives as a single wild plant; the Saint Helena ebony is known from just two individuals; the Toromiro tree, from Easter Island, is now extinct in the wild. All have a safe haven in the living plant collections at Kew and other gardens around the world.[7]

These examples provide instances of welcome insurance against extinction provided by cultivation, sometimes combined with the protection of the species in a seed bank. *Ex situ* conservation, by itself, is assuredly not enough. It cannot preserve the processes that maintain species in their natural habitat, nor can it sustain the ecological services provided by the community of which the species is part. However, *ex situ* conservation is a key tool to preserve plant diversity for the long term. Both *in situ* and *ex situ* conservation are needed, and must be integrated, to ensure the long-term survival of species that might otherwise be lost.[8]

Some contend that efforts to encourage *ex situ* conservation undermine efforts to conserve plants in their native habitats or draw funds away from *in situ* conservation. Such concerns are understandable, but fail to account for the realities of our current predicament or different opportunities for conservation funding; in conservation, as in many other areas, we should be careful that the perfect does not become the enemy of the good.[9]

Ex situ conservation, as zoos have long recognized, is worth doing and brings many other benefits. It draws attention to the plight of species that are threatened, it creates opportunities to assess the state of populations in the wild, and it helps build collective capacity around the world to preserve biological diversity. Ginkgo and the Wollemi pine are excellent examples from the world of plants. Both are known from the paleontological record, but surely everyone must agree that we are fortunate to know them as living trees rather than only as fossils. As living plants we can continue to enjoy them, study them, and better understand their secrets.[10]

Ginkgo and plants like it illuminate World History on the grandest scale, what some have called Big History. Just as history taught solely from a Western perspective is incomplete, history limited to a narrow time horizon only partly captures the contingencies through which our modern world came to be. This is not just about our delight or enjoyment; studying species and where they came from is one of the most important windows through which we can understand the world we live in and our place in it.

It is sometimes said that the reasons to preserve species differ little from the reasons to save a great work of art. It is a helpful analogy; it reminds us of what we stand to lose when species go extinct. It underlines the cultural, emotional, and ethical dimensions of animal and plant conservation. But it is also incomplete. Conserving species is not just an indulgence for a few "tree huggers" who like "wild nature."

For me, there is a fundamental difference between losing something created by human genius over days, weeks, or years and losing something created by nature over millennia. Either would be a tragic loss, but comparing the loss of species with the loss of human achievements, equating human creativity with the creativity of nature, somehow fails to capture the enormity of the situation we face. I prefer another analogy that one also hears from time to time: letting species go extinct when we have the power to intervene is like letting the library burn just as we are learning how to read the books. It is a waste of information and a loss of knowledge about our world. Losing species is a wasted opportunity to better understand the past, and understanding the past will be necessary for managing the future. Preserving ginkgo and other species preserves information on our own origins, our own history, and the history of the biological and geological systems of which we are part. Extinction destroys the evidence of how our world, and everything in it, came to be.[11]

36
Gift

In communities drawn together by gift exchange, "status," "prestige,"

or "esteem" take the place of cash remuneration.

—Lewis Hyde, *The Gift*

In June 1992 the nations of the world gathered in Rio de Janeiro at the first Earth Summit. Attended by 172 countries, and more than a hundred heads of state, as well as 2,400 representatives of nongovernmental organizations, it was one of the largest United Nations conferences ever convened. The aim was to address increasing concern about deterioration of the global environment. Greater awareness of pollution and declining environmental quality, as well as their impacts on human health, had been building since the 1950s and 1960s. Rio was a culmination of that process and a key moment in the growth of the global environmental movement.[1]

One focus at Rio was desertification, the loss of vegetation in the drylands of the world, and the threat that poses to poor people in developing countries. Another was climate change, and it was in Rio that the United Nations Framework Convention on Climate Change was introduced to the world; it led to the Kyoto Protocol and much later to the climate change conferences in Copenhagen, Cancun, Durban, and elsewhere. Also opened for signature in Rio was the Convention on Biological Diversity, and in December 1993, ninety days after ratification by the thirtieth country, the convention entered into force. All the countries of the world, except Andorra, the Holy See, South Sudan, and the United States of America, are now parties to the convention.[2]

The Convention on Biological Diversity (CBD) has as its aim the conservation and sustainable use of biological diversity. It was a response to growing misgivings about the fate of the animals and plants with which we share our planet. From Rachel Carson's warning about the devastating impacts of widespread synthetic pesticide use, to the attention brought by Chico Mendes to the loss of the rain forest and abuse of its peoples, to the coining of the term *biodiversity* itself in the mid-1980s, the CBD was the result of a multidecade process through which the nations of the world sought a common approach to reduce the destruction and degradation of biodiversity. The Convention on International Trade in Endangered Species had already come into force in 1975, in an effort to reduce threats to species from overexploitation, but there was a strong feeling that more international action was needed.[3]

Early in the negotiations leading to the CBD it became obvious that biological diversity was not evenly distributed across the world's surface. The richest places on the planet for plant species are the tropical regions of Central and South America, especially in the foothills of the Andes in both Colombia and Ecuador, as well as the rain forests of central Africa and southeast Asia, especially Malesia. Important outliers of high species diversity also occur in a few other so-called hot spots, including the southern tip of Africa and the extreme southwest of Australia. The comparison with other parts of the world is stark. In the Atlantic rain forest of eastern Brazil, only 5 percent of the original forest remains, yet it contains approximately twenty thousand plant species, of which nearly half occur nowhere else in the world. The native flora of the island of Madagascar totals about twelve thousand plant species, about 90 percent of which occur only there. The native flora of the United Kingdom is meager by comparison. It comprises only about fourteen hundred native plant species, and almost all occur elsewhere in Europe.[4]

The striking pattern is that many countries that are relatively rich in biodiversity are relatively poor economically and struggle with massive day-to-day problems to improve the lives of their people. As a result, concerns about the fate of animals and plants, emanating especially from the wealthier countries of the world, often seem to collide with more immediate concerns about the well-being of people in poorer countries. One outcome of this tension was that the process of negotiating the CBD became highly charged and deeply politicized.

A specific manifestation of this political reality was a perceived conflict between the conservation of biological diversity and its use by people. While there are concerns over the long-term fate of species of animals and plants, poverty, malnutrition, infant mortality, and the many other pressing problems highlighted, for example, in the Millennium Development Goals are critical issues of basic human justice that must be addressed. Nevertheless, while it is easy to portray this tension as a simple dichotomy, nature versus people, the reality is much more complicated. The fates of people and the fates of their environment are inextricably interlinked. The real question is: how can the Millennium Development Goals be met while also preserving biological richness and the ecological goods and services that ensure healthy and sustainable supplies of energy, food, medicines, and water, as well as a good quality of life?

It was not the role of the CBD to provide a definitive answer to such a complex question, but it tried to find middle ground by establishing the principle that the conservation and use of biological diversity are two sides of the same coin; conserving biological diversity enables us to continue to use it in sustainable ways. Implicit was the idea that positioning conservation and use as complementary activities strengthens the case for conserving biological diversity by emphasizing its value. It is a position that pivots around a simple point; biological diversity should be conserved because it is useful.

In broad terms this makes sense: we keep what we use, and we are more likely to conserve what we value. This key principle of the CBD takes an important step with the emphasis that utilization should be sustainable. By any sensible biological definition this means that the total number of individual living plants or animals should be maintained, not depleted. Populations of a species need to be allowed to reproduce and replace themselves at a rate equal to or greater than the rate at which they are being lost.

However, on top of these principles the political context of the CBD negotiations was muddied by the arguments by overenthusiastic conservationists that a key reason to protect natural habitats, most iconically the rain forest, was that the species living in those places were an Eldorado of riches waiting to be exploited. It was a well-meaning argument that went further down the utilitarian route, but it had unforeseen consequences. Because it held out the promise of substantial future profits, it fueled nationalistic and protectionist tendencies in what one might have hoped would have been more principled negotiations about the future of life on Earth.

One outcome of these complexities was that while the text of the convention acknowledged that the conservation of biological diversity "is a common concern of human kind," it nevertheless enshrined into international law for the first time the principle that biological resources within the borders of individual countries are a national patrimony. At one level, this simply reaffirmed practices already implemented in many countries, where the use of animals and plants living within their borders was subject to national laws, but when the issue was raised in international negotiations leading to the CBD, the question of under what terms biological resources should be shared internationally was given new prominence. Inevitably, it inspired new nationalistic sensitivities and further complicated the political landscape.[5]

The struggle with these complexities resulted in another key provision of the CBD: "the fair and equitable sharing of the benefits arising out of the use of the genetic resources inherent in living organisms." Again, a seemingly reasonable argument was made that countries that are engaged in the conservation and sustainable use of biological diversity within their borders ought to share in any benefits that might arise from its utilization. Heightened expectations of financial reward, concerns about possible "biopiracy," and a general lack of trust took the convention still farther down this road, to the point of linking the sharing of benefits not just to the commercial utilization of biodiversity but also to access to it. Legally binding "access and benefit sharing agreements" were one result.[6]

In these ways the convention sought to place the onus on countries to conserve the biological diversity within their borders, while also seeking to provide economic incentives by leaving open the option for sustainable utilization and introducing the principle of benefit sharing. It was a seemingly sensible approach, but it had at its heart one fundamental consequence: no longer were the plants and animals bequeathed to all of us by 4.6 billion years of planetary evolution part of our common human patrimony; instead, they became the proprietary interests of nations. Trees, birds, flowers, and all other kinds of organisms from insects to bacteria, and the genetic material they contained, were taken as property by the people living inside the borders of individual countries.

The ramifications of this fundamental shift are still being played out in never-ending negotiations around the implementation of the CBD, but one important practical effect is that many countries are now very sensitive about sharing their perceived biological wealth with others. Their overriding concern is not to give away what they regard to be valuable natural assets. One unintended consequence is that by limiting access through complex permitting regulations the CBD actually helped stifle commercially oriented work that could potentially generate revenue. Another is that many countries greatly restrict scientific access to their native plants and animals, even for noncommercial work in collaboration with in-country scientists that helps support animal and plant conservation. A third consequence is that the CBD unwittingly created a serious issue regarding how the genetic diversity of the world's most important crops, which was once freely exchanged among countries, should be treated.

In essence, the International Treaty on Plant Genetic Resources for Food and Agriculture, which was adopted by the FAO conference in November 2001 and came into force as of June 29, 2004, is an international agreement that excludes sixty-four of the world's most important crop and forage plants from some of the more unfortunate provisions of the CBD. Ironically, for our most important plants this international treaty seeks to restore the preexisting situation by attempting to create a multilateral system through which crop genetic resources may be shared for purposes of breeding new varieties. Despite broad agreement about the potential shared benefits it would bring, in the atmosphere created by the CBD, the international treaty took seven years to negotiate, failed to reach agreement on several key crops, remains highly contentious, and is being implemented halfheartedly if at all by certain countries, even those who are signatories. Terms can be revisited only by consensus among all 127 of its contracting parties. It is symptomatic of where we have come to in seeking to manage biological diversity in the twenty-first century: something is wrong in the way that we think about the natural world and the common interests of humanity.[7]

The reality, two decades after Rio, is that the CBD has struggled to find its place among national and international priorities. It has also not been well funded by those governments that have ratified it. It would not be hard to argue that despite huge investments of time and effort, and vast amounts of money spent mainly on the associated meetings, the CBD has so far produced little revenue for countries that are rich in biodiversity; nor has it resulted in a real advance toward its overriding goal of the conservation and sustainable use of biological diversity.[8]

For the future of ginkgo, the Convention on Biological Diversity has been neither a help nor a hindrance. The convention does not apply retroactively, and ginkgo was already ubiquitous before we embarked on our clumsy global approach to conserving the variety of plant and animal life. Nevertheless, an interesting question is, if a "new ginkgo" were discovered today in some far-off land, would the CBD make it more secure or more vulnerable?

Saplings of Wollemi pine for sale at
Wakehurst Place, the country garden
of the Royal Botanic Gardens, Kew
near Ardingly, West Sussex.

The answer would depend entirely on the attitude of the country in which this new plant was discovered. Under the current regime of the CBD and the atmosphere that it has created in many countries around the world, a "new ginkgo" would be jealously guarded. Unfortunately, that is not the same as saying that it would be well protected or that its long-term future would be secure. Other countries, however, would perhaps take a more enlightened approach, and the Wollemi pine provides an encouraging example through which to explore how such a thought experiment might turn out.

Australia, or more strictly the government of New South Wales, to which implementation of the CBD is delegated under Australia's federal system, took a sensibly pragmatic attitude to ensure the long-term future of the Wollemi pine. In effect the government sought to gain some short-term financial reward to assist conservation of the species while also trying replicate what occurred with ginkgo through its interaction with people. The authorities in New South Wales decided that in addition to protecting the tree in the wild, they needed to get it into cultivation—and, crucially, not only in Australia. They did not hold out for some illusory long-term economic advantage.

First, they entered into a contract with a company to handle the propagation and distribution of the Wollemi pine. The company raised thousands of young plants that were ultimately distributed for sale around the world. The government then mounted a successful and well-coordinated publicity campaign, which included selling off some of the first plants to be propagated at a high-profile international auction at Sotheby's in New York. The funds raised from the auction and sales flowed back through the company into conservation of the remaining small population in the natural habitats.

The Wollemi pine provides an excellent example of an integrated approach to conservation of a rare plant. Protection of the wild population has been enhanced and this unique species has been insured against "bad luck" by being brought into cultivation. The Wollemi pine is now growing successfully in gardens all over the world, and in the coming decades and centuries it is inevitable that the plant will be propagated by others. Step by step the Wollemi pine, as it is grown in gardens far from its home in the wild, will join many other wonderful gifts of Australia to the world.

To the extent that the CBD is an impediment to such effective *ex situ* conservation, is perceived as such, or is used as a reason to prevent such exchanges of plant material, it can do more harm than good. In the garden world there is a well-worn adage: "If you have a rare plant; give it away." The idea is simple: increase the number of plants being grown and eliminate the risk that something valuable might be lost just through "bad luck." Unfortunately, the CBD steers us in the opposite direction, away from that simple principle of common sense.

It is also an interesting question, for a plant like ginkgo, to think about what the CBD really means. It would be hard to argue with the idea that ginkgo in some sense belongs to China, but in an equally valid sense ginkgo belongs to us all. As the last of a line of plants that was once much more diverse and grew on every continent, it is part of the shared natural heritage that binds all people together. Ginkgo is both a gift of the world to China, and a gift of China to the world. When we see a ginkgo on the street in London, New York, or Tokyo, we should realize that it is a gift for which we all should be grateful. It is a gift that enhances the esteem in which China is held by others; even more so because it is a gift that has been freely given.

This line of thinking asks us to look beyond science and economics for guidance about how we should view the natural systems of which we are part, and how we should

seek to manage them. It takes us into a realm where ethics, moral values, spiritual relations, and a broader sense of our place in the world, and indeed the universe, come
into play. In many parts of the world, religions are central to the way that people think
about these issues and play important roles in constructing the moral frameworks
through which we interact with other people and the environment.[9]

In some religions, however, ideas about the natural world are peripheral rather than
central. For example, the cultural historian Thomas Berry offers the view that Christianity, with its emphasis on the individual, and that individual's relationship with his
or her own God, in many ways directs us inward. It predisposes us to a worldview that
is fundamentally centered on the relation of humans to God and that too often turns its
back on the natural world. At the same time, Berry argues that a purely secular, mecha-
nistic, and scientific worldview has narrowed our perspective, and blunted our sense
of wonder. Fortunately, for many people around the world, and not just botanists, gardeners, or conservationists, trees still have the power to inspire, to evoke awe, and to
lift our spirits. They embody the intrinsic value of nature. Thomas Berry would have it
that in some sense we are part of them and they are part of us. Trees meet a basic bio-
philic need embedded deep within us: a need that we have inherited from our ancestors and that is increasingly starved by our highly urbanized and dominantly indoor
existence.[10]

The international approach that resulted in the CBD, with its parochial emphasis on
benefits and commoditization, brings into sharp focus fundamental questions about
our currently unsatisfactory relationship with the natural world. Does it really make
sense to try to manage the global environment on a country-by-country basis? Are we
comfortable with a view that so clearly asserts that nature is simply there for human
benefit? Is it morally or ethically right for the demands of people always to trump long-
term survival of species of plants and animals? And is it really in our long-term interest
to further extend our hegemony over nature? The ways in which such questions are answered will be important for the future of all of humanity. If we take a broader view of
the history of our planet, and recognize that we have evolved over millennia as part of
complex global system of which we still have only limited knowledge, placing humans
so explicitly at the center seems arrogant, shortsighted. It might also be risky. To borrow a phrase from my friend Paul Falkowski, "Our destiny lies in understanding that
humility leads to enlightenment and that hubris leads to extinction."[11]

37
Legacy

In the empires of usury the sentimentality of the man with the soft heart

calls to us because it speaks of what has been lost.

—Lewis Hyde, *The Gift*

Tony Kirkham, who runs the arboretum at Kew, has become something of a celebrity. He featured prominently in the BBC programs *A Year at Kew*, which appeared on British television between 2004 and 2006. He then graduated to a television series of his own: *The Trees That Made Britain*. With his colleague John Hammerton he explored the contribution of trees to British history, landscape, and culture, from the preservation of ancient yews in rural churchyards to the use of their wood in the medieval longbow. He sampled cider made from the British apples in Somerset, went sailing in ash-frame coracles, and investigated the sunken timbers of Britain's sixteenth-century warship the *Mary Rose*.[1]

However, Tony's "day job" is to care for the nearly fifteen thousand trees on the Kew estate. He makes sure that they are healthy, that they grow well, and that information relating to their planting, growth, and management is up to date in the massive database used to keep track of Kew's collections of living plants. It is also Tony's responsibility to make sure that none of his trees is a danger to Kew visitors. Every tree is monitored carefully every year. Tony is also an expert in the potentially dangerous business of climbing trees, and he recently revised the classic book on pruning trees and shrubs written by one of his predecessors in the Kew Arboretum. These days, his team does most of the ropework needed to remove dead limbs from high in the canopy.[2]

Tony also has responsibility for adding to Kew's collections of living trees. With his colleagues, especially Mark Flanagan, who is keeper of the Savill Gardens and the Valley Garden Arboretum in Windsor Great Park, he has collected seed of different species of trees in Japan, Taiwan, South Korea, and China. Back in Britain this seed is the source for the new saplings planted out every year at Kew and in other gardens and arboreta around the country. Tony has seen many of Asia's ancient ginkgos on his travels. In 2006 he visited Lengji, near Luding in western Sichuan, to check on the great ginkgo photographed by Ernest Henry Wilson on August 1, 1908. It is still there, with a small shrine at its base, crowded between the houses in the middle of the village, in the valley of the Dadu River.[3]

Tony is a tree person through and through. Outside his family he has no greater passion. Like a doctor he does his best to avoid emotional involvement with his patients, for from time to time he needs to perform major surgery, or even remove a tree completely, but he has not lost the sense of awe and wonder that drew him to trees in the first place. He also knows that he must plant many more trees than he takes down to allow for losses to disease and storms as his saplings grow to maturity. In 2009 on the 250th anniversary of the founding of the Royal Botanic Gardens, Kew, he led the planting of 250 new trees on the Kew estate. In May 2009 the last two trees in Tony's yearlong effort were a Wollemi pine planted by the Duke of Edinburgh and a ginkgo planted by Her Majesty the Queen. Tony takes the long view; he understands that he is planting for those who will come to Kew a hundred, or perhaps two hundred, years or more in the future.

The Old Lion is one of more than fifty ginkgo trees under Tony Kirkham's care. He keeps a close eye on all of them, but the Old Lion is special. In the past few years he has cleared the shrubs that once grew around its base; he saw no need to make it compete with plants of lesser importance. He has stripped away the nearby grass and mulched a broad circle around the base of the trunk. He has removed part of the tarmac path that ran beneath the tree to allow more air and water to the roots; and he has used compressed nitrogen to break up the compacted soil around them.

The care that Tony lavishes on his trees, and the concern that he feels for the Old Lion and the other long-lived charges under his care, is paralleled not just in Britain

The Old Lion ginkgo at the Royal Botanic Gardens, Kew: one of the oldest and most celebrated ginkgo trees in the United Kingdom, planted around 1761. It is a living link to the eighteenth century, when King George III often spent time with his family on his Kew estate, but it also connects us to the glorious and powerful deep history of plant and animal life on our planet.

but everywhere around the world. In Oak Park, just outside Chicago, when the ginkgo outside the Frank Lloyd Wright Home and Studio showed signs of stress, the tree surgeons were called in, dead branches were removed, the tree was carefully pruned, and it emerged rejuvenated. Concerned that the Yongmunsa Ginkgo in Korea might be struck by lightning, the temple authorities built a steel lightning tower close by to ensure that this would never happen. When the great ginkgo at the Tsurugaoka Hachiman-gū Shrine in Kamakura, Japan, fell after heavy rains in March 2010, work began immediately to propagate it so that the tree could live on at the same spot. When the Mizufuki Ginkgo at the Nishi Honganji Temple in Kyoto, Japan, showed signs of decline, careful pruning, soil renewal, and protection from the feet of thousands of temple visitors brought it back to health.[4]

In the mid-nineteenth century, when the ancient Huiji Temple in Tang-Quan County, not far from Nanjing, China, fell into disrepair and was lost, it was the two great ginkgos that grew close to where it once stood that kept its memory alive. They also came to preside over the small ginkgo orchard planted beneath them. When I visited in August 2008, the temple was being rebuilt and both ginkgos were wrapped in red ribbons; some had wishes of all kinds written on them. Other visitors had expressed their hopes more directly in neat vertical columns of Chinese characters written on small bare patches of wood where the bark had been worn away. These trees are important in their community; local people still go there to pray. Late in the afternoon, as we left to return to Nanjing, a young woman arrived. She looked nervously behind her to be sure that we were leaving, then made her way to one of the two great ginkgos and stood alone in a few moments of quiet contemplation. Many others, from previous generations, had stood there before her.[5]

In rural Aomori Prefecture in northern Honshu, the Hōryō Ginkgo sits on just a sliver of uncultivated land sandwiched between commercial forests and agricultural fields. It is approached with reverence down an aisle of closely spaced, moss-covered stepping-stones. Local people visit it regularly. They lovingly clear away the twigs and branches that fall in the winter snows, they care for the rice straw rope wound around the trunk, they explain the tree's legends to local schoolchildren, and they work to spread word of its importance. This tree was a friend to their grandparents; it will probably also be a friend to their grandchildren.

These scenes are repeated many times over, in many ways, in many parts of the world. People feel connected to trees; old trees in particular are objects of affection that have earned and deserve our love and respect. Despite the comings and goings of nations, times of conflict and times of peace, years of austerity and years of plenty, the great ginkgos of Asia, and other ancient trees around the world, provide continuity in communities with the passing of generations. It is not only "plant people" who are moved by the grandeur of a coast redwood, or humbled by the antiquity of bristlecone pines. These feelings resonate among anyone who takes the time to reflect on the true meaning of ancient and long-lived plants. In cultures around the world trees place our own ephemeral existence in proper perspective.

When I was growing up in the Midlands of England, the countryside around my home was dominated by majestic avenues, each composed of hundreds of elms. They were all planted more than two centuries before by a wealthy local landowner, the second Duke of Montagu, sometimes known as John the Planter. It is said that his plan was to plant an avenue of elms and lindens all the way to London, but thwarted by the difficulties of persuading other landowners to join in the project, he satisfied himself with planting his great avenues around and across his own estate. Altogether the avenues he created were more than seventy miles long. Then, sometime in the late 1960s, I remember hearing for the first time about the scourge of Dutch elm disease; the disease quickly took its toll. In just a few years the elms of my childhood, the elms that are such a feature of Constable's paintings, the elms across the whole of Britain, were all gone. A magnificent tree that had been part of the English landscape for thousands of years was almost obliterated. A bald spot appeared in the Kew arboretum.[6]

One morning in early October 2006, soon after I arrived back in Chicago from our years at Kew, I drove down to the University of Chicago with my daughter, who was attending school on campus. There had been a big storm overnight. Where we live in Oak Park, to the west of the city, there was just lots of water. As we approached the South Side, the scene was different; everywhere there were trees with big limbs on the ground, and others had been completely uprooted. On campus, in the Quadrangle outside the Divinity School, a big oak, perhaps older than the university itself, was down on its side and already being dismantled by staff from the grounds department. More than nine hundred large mature trees came down that night on the South Side of Chicago; a thousand more were badly damaged. In a matter of hours trees were lost that had always been there, trees that were there before my daughter's grandparents were born. We had taken them for granted, and we hadn't known what we had until they were gone. Trees take time to grow, but their loss can be quick and easy. It takes only a moment to dynamite an acre of rich Appalachian forest to plunder the coal beneath. It takes just a few hours to fell giant dipterocarps in the rain forest of Borneo, or old-growth hemlocks in the Pacific Northwest of the United States. Trees and forests that have stood firm for centuries in the face of repeated natural assaults have no power to resist our fleeting, but often devastating, attention.

These timescales—hours, days, or even a few years—contrast with the timescales of ginkgo's life story. Hundreds, thousands, millions, tens of millions, hundreds of millions of years: these spans are not easy for us to grasp, but they are perhaps more relevant to the way we should think about ourselves, and our true place in the world. They should make us pause. Trees help calibrate the speed of current environmental change: they provide a context more in tune with the tempo of the Earth. They slow us down, they teach us the virtue of patience, and they remind us to think about all that has gone before and what is to come; the legacy we received and the legacy that we will leave. The modern-day mantra of more, better, faster is all very well; but followed unthinkingly it is a recipe for disaster. Trees, especially trees like ginkgo, which connect us to the deep history of our planet, ask us to reflect more often and think more carefully about all we lose when the short view rules our world and everything in it.

Appendix

List of Common Plant Names Used

in the Text and Latin Equivalents

(fossils indicated by †)

African oil palm = *Elaeis guineensis*

Alder = *Alnus* spp.

†*Allicospermum xystum* = fossil seed related to living ginkgo

Almond = *Prunus dulcis*

American beech = *Fagus americana*

American elm = *Ulmus americana*

apple = *Malus pumila*

apricot = *Prunus armeniaca*

†*Archaefructus* = early fossil flowering plant

†*Archaeopteris* = fossil "progymnosperm" related to living and fossil seed plants

argan = *Argania spinosa*

aroids = several genera in the family Araceae

aromatic ginger = *Kaempferia galanga*

ash = *Fraxinus* spp.

aspen = *Populus* spp.

†*Asteroxylon* = fossil plant from Rhynie Chert related to living fir clubmoss

Atlantic cedar = *Cedrus atlantica*

†*Avatia* = fossil seed-bearing structure related to living ginkgo

azalea = *Rhododendron* spp.

†*Baiera furcata* = fossil leaf related to living ginkgo

†*Baiera gracilis* = fossil leaf related to living ginkgo

†*Baiera hallei* = fossil leaf related to living ginkgo

bald cypress = *Taxodium* spp.

banyan = *Ficus benghalensis*

baobab = *Adansonia digitata*

bay tree = *Laurus* spp.

bearded fig = *Ficus citrifolius*

beech = *Fagus* spp.

betel = *Piper betle*

bigleaf hydrangea = *Hydrangea macrophylla* (syn. *Hydrangea otaksa*)

birch = *Betula* spp.

blackberry = *Rubus fruticosus*

black locust = *Robinia pseudoacacia*

black tupelo = *Nyssa sylvatica*

bright-green cave-moss = *Cyclodictyon laete-virens* (syn. *Hookeria laete-virens*)

bristlecone pine = *Pinus longaeva* (other closely related, long-lived species, which are sometimes also called bristlecone pine, are *Pinus aristata* and *Pinus balfouriana*)

butterfly bush = *Buddleja* spp.

cabbage palmetto = *Sabal palmetto*

Café Marron = *Ramosmania rodriguesii*

camellia = *Camellia* spp.

Carolina mahogany = *Persea borbonia*

castor oil = *Ricinus communis*

Cathay silver fir [yin shan] = *Cathaya argyrophylla*

Caucasian wingnut = *Pterocarya fraxinifolia*

cedar of Lebanon = *Cedrus libani*

celery = *Apium graveolens*

cherry = *Prunus avium*

chestnut = *Castanea sativa*

chili = *Capsicum annuum*

†China fir = *Cunninghamia cheneyi* fossil shoots of living China fir in the Clarkia fossil flora

China fir = *Cunninghamia lanceolata*, *Cunninghamia konishii* (two closely related forms that may just be a single species)

Chinese date = *Ziziphus jujuba*

Chinese swamp cypress = *Glyptostrobus pensilis*

cinnamon = *Cinnamomum aromaticum*

Clanwilliam cedar = *Widdringtonia cedarburgensis*

clubmoss = several genera in the family Lycopodiaceae

coast redwood = *Sequoia sempervirens*

cork oak = *Quercus suber*

crepe myrtle = *Lagerstroemia indica*

crocus = *Crocus* spp.

croton = *Croton tiglium*

currant = *Ribes* spp.

cycads = about eleven genera in several families of the order Cycadales

†*Cyclocarya brownii* = fossil fruits of the living wheel wingnut in the Almont fossil flora

cypress = several genera in the family Cupressaceae

dawn redwood = *Metasequoia glyptostroboides*

dense logwood = *Croton congestus*

dipterocarp = several genera in the family Dipterocarpaceae

divi-divi = *Caesalpinia coriaria*

dove tree (handkerchief tree) = *Davidia involucrata*

dragon tree = *Dracaena draco*

drautabua = *Acmopyle sahniana*

elm = *Ulmus* spp.

English oak = *Quercus robur*

†*Eoasteria* = fossil, probable pollen-producing structures related to living ginkgo

ephedra [ma huang] = *Ephedra* spp.

eucalypt = *Eucalyptus* spp.

fig tree = *Ficus* spp.

fir = *Abies* spp.

fir clubmoss = *Huperzia selago*

fishtail palm = *Caryota* spp.

Franklinia = *Franklinia alatamaha*

garlic = *Allium sativum*

giant sequoia = *Sequoiadendron giganteum*

ginger = *Zingiber officinalis*

ginkgo = *Ginkgo biloba*

†*Ginkgo adiantoides* = fossil leaf related to living ginkgo

†*Ginkgo apodes* = fossil leaf related to living ginkgo

†*Ginkgo australis* = fossil leaf related to living ginkgo

†*Ginkgo cordilobata* = fossil leaf related to living ginkgo

†*Ginkgo cranei* = fossil leaf related to living ginkgo

†*Ginkgo florinii* = fossil leaf related to living ginkgo

†*Ginkgo huttoni* = fossil leaf related to living ginkgo

†*Ginkgo orientalis* = fossil leaf related to living ginkgo

†*Ginkgo rajmahalensis* = fossil leaf related to living ginkgo

†*Ginkgo yimaensis* = fossil leaf related to living ginkgo

†*Ginkgoites matatiensis* = fossil leaf related to living ginkgo

†*Ginkgoites muriselmata* = fossil leaf related to living ginkgo

†*Ginkgoites taeniata* = fossil leaf related to living ginkgo

†*Ginkgoites telemachus* = fossil leaf related to living ginkgo

†*Ginkgoites ticoensis* = fossil leaf related to living ginkgo

†*Ginkgoites tigrensis* = fossil leaf related to living ginkgo

gnetum = *Gnetum* spp.

grape = *Vitus* spp.

grape family = Vitaceae

guanacaste = *Enterlobium cyclocarpum*

handkerchief tree (dove tree) = *Davidia involucrata*

hardy rubber tree = *Eucommia ulmoides*

Hawaiian cotton tree = *Kokia drynarioides*

hazel = *Corylus* spp.

heath = *Erica* spp.

hemlock = *Tsuga* spp.

Henry honeysuckle = *Lonicera henryi*

Henry lily = *Lilium henryi*

hickory = *Carya* spp.

hornbeam = *Carpinus* spp.

hydrangea = *Hydrangea* spp.

Japanese cedar = *Cryptomeria japonica*

Japanese elm = *Zelkova serrata*

Japanese fig = *Ficus erecta*

Japanese maple = *Acer japonica*

†*Kannaskoppia* = fossil seed-bearing structure possibly related to living ginkgo

†*Kannaskoppianthus* = fossil pollen-producing structure possibly related to living ginkgo

†*Kannaskoppifolia* = fossil leaves possibly related to living ginkgo

†*Karkenia asiatica* = fossil seed-bearing structure related to living ginkgo

†*Karkenia incurva* = fossil seed-bearing structure related to living ginkgo

katsura = *Cercidiphyllum japonicum*

†*Kerpia* = fossil leaf possibly related to living ginkgo

koompassia = *Koompassia* spp.

laurel = *Laurus* spp.

licorice = *Glycyrrhiza glabra*

lilac = *Syringa* spp.

lily bulbs = *Lilium* spp.

linden (lime) = *Tilia* spp.

live oak = *Quercus virginiana*

Lombardy poplar = *Populus nigra*

London plane = *Platanus* x *acerifolia*

lotus = *Nelumbo nucifera*

lychee = *Litchi chinensis*

magnolia = *Magnolia* spp.

mahogany = *Swietenia* spp.

ma huang (Ephedra) = *Ephedra* spp.

maidenhair fern = *Adiantum* spp.

maize = *Zea mays*

maple = *Acer* spp.

Mediterranean fig = *Ficus carica*

mimosa = *Mimosa* spp.

mistletoe = *Viscum album*

monkey puzzle = *Araucaria araucana*

Monterey pine = *Pinus radiata*

moonseed family = Menispermaceae

Mormon tea = *Ephedra* spp.

mountain cedar = *Widdringtonia nodiflora*

Mulanji cypress = *Widdringtonia whytei*

mulberry = *Morus* spp.

Norway maple = *Acer platanoides*

nypa = *Nypa fruticans*

oak = *Quercus* spp.

oil palm [African] = *Elaeis guineensis*

onion = *Allium cepa*

pagoda tree = *Styphnolobium japonicum*

†*Palaeocarpinus* = fossil fruit in the birch family

palmetto = *Sabal palmetto*

palm family = Arecaceae

Panama hat palm = *Carludovica palmata*

Pau Brasil = *Caesalpinia echinata*

pea = *Pisum sativum*

peach = *Prunus persica*

peanut = *Arachis hypogaea*

peepal (pipal) = *Ficus religiosa*

†*Petriellaea* = fossil seed-bearing structure potentially related to fossil *Kannaskoppia*

pine = *Pinus* spp.

pinyon pine = *Pinus* subsection *Cembroides* and *Nelsonianae*

pistachio = *Pistachio vera*

plane (sycamore) = *Platanus* spp.

plum = *Prunus* spp.

plum yew = *Cephalotaxus* spp.

poison ivy = *Toxicodendron radicans*

poison oak = *Toxicodendron diversilobum*

pomegranate = *Punica granatum*

Portuguese laurel = *Prunus lusitanica*

potato = *Solanum tuberosum*

prickly ash = *Zanthoxylum* spp.

†*Psygmophyllum* = fossil leaves of uncertain relationship

pumpkin = *Cucurbita pepo*

quaking aspen = *Populus tremuloides*

red maple = *Acer rubrum*

rhododendron = *Rhododendron* spp.

rice = *Oryza sativa*

Robinson Crusoe cabbage tree = *Dendroseris litoralis*

rockrose = *Cistus* spp.

rose = *Rosa* spp.

rubber tree = *Hevea brasiliensis*

Sago cycad = *Cycas revoluta*

Saint Helena ebony = *Trochetiopsis ebenus*

sandalwood = *Santalum* spp.

sassafras = *Sassafras* spp.

sesame = *Sesamum indicum*

sessile oak = *Quercus petraea*

she-oak = *Casuarina* spp.

silk tree = *Albizia* spp.

silverberry = *Elaeagnus* spp.

silver fir = *Abies alba*

snowdrop = *Galanthus* spp.

Spanish dagger = *Yucca gloriosa*

†*Sphenobaiera umaltenis* = fossil leaves related to living ginkgo

spruce = *Picea* spp.

†*Stalagma samara* = enigmatic fossil plant perhaps related to living conifers

stonecrops = *Chara* spp.

stone pine = *Pinus pinea*

strawberry = *Fragaria* x *ananassa*

strawberry tree = *Arbutus unedo*

sugar maple = *Acer saccharum*

swamp cypress = *Taxodium distichum*

swamp gum = *Eucalyptus regnans*

sweet gum = *Liquidambar styraciflua*

sweetsop family = Annonaceae

sycamore = *Platanus* spp.

tamarix = *Tamarisk* spp.

teak = *Tectona* spp.

titan arum = *Amorphophallus titanum*

Toromiro tree = *Sophora toromiro*

tree of heaven = *Ailanthus altissima*

†*Trichopitys heteromorpha* = fossil seed plant of uncertain relationship

umbrella pine = *Sciadopitys verticillata*

valley oak = *Quercus lobata*

Venus flytrap = *Dionaea muscipula*

walnut = *Juglans* spp.

watercress = *Nasturtium officinale*

water lily = several genera of aquatic plants in the family Nymphaeaceae

watermelon = *Citrullus lanatus*

welwitschia = *Welwitschia mirabilis*

wheel wingnut = *Cyclocarya* and Asian *Pterocarya* spp.

willow = *Salix* spp.

willowmore cedar = *Widdringtonia schwarzii*

witch hazel = *Hamamelis* spp.

Wollemi pine = *Wollemia nobilis*

yam = *Dioscorea* spp.

yevaro = *Eperua purpurea*

yew = *Taxus* spp.

†*Yimaia capituliformis* = fossil seed-bearing structure related to living ginkgo

†*Yimaia qinghaiensis* = fossil seed-bearing structure related to living ginkgo

†*Yimaia recurva* = fossil seed-bearing structure related to living ginkgo

yin shan (Cathay silver fir) = *Cathaya argyrophylla*

Notes

1. Time

1. Epigraph: Conan Doyle (1912, 63). Note that Conan Doyle's original spelling transposed the *g* and the *k*, a common error in the spelling of ginkgo.

2. In March 2011, the Old Lion ginkgo at Kew stood about sixty-three feet tall and had a trunk diameter of a little more than five feet just below the level where the trunk splits into its two main leaders (Tony Kirkham, Royal Botanic Gardens, Kew, personal communication).

3. Darwin referred to another botanical oddity, *Welwitschia,* as a platypus for the vegetable kingdom. He wrote to J. D. Hooker on December 18, 1861: "Your African plant seems to be a vegetable *Ornithorhynchus,* and indeed much more than that"; see Darwin and Seward (1903, 281).

4. The new kinds of plants that rose to prominence around 100 million years ago, and that now dominate almost all terrestrial environments with about 350,000 living species, are the angiosperms (flowering plants). The geologic timescale used in this book is that of Gradstein et al. (2004).

5. Earlier records of modern humans in China, for example from 100,000 years ago (Liu et al., 2010), are controversial and could potentially be interpreted as gracile forms of *Homo erectus* (Dennell, 2010).

2. Trees

1. The Marquess of Blandford who planted the arboretum on the Whiteknights estate later became the fifth Duke of Marlborough.

2. Bill Burger, a colleague at the Field Museum in Chicago, first introduced me to Warren Woods.

He describes the significance of this special woodland in *Chicago Wilderness* magazine; see Trigg (2008).

3. Nearly eighty thousand people every year visit the coast redwoods at Muir Woods, just north of San Francisco.

4. The photographs in *Meetings with Remarkable Trees* and *Remarkable Trees of the World*, by Thomas Pakenham (2002, 2003), come closer than most to revealing the true power of the world's great trees.

5. Eiseley (1958) argues that the prehensile hands and forward-directed vision that evolved in humankind's ancestors developed as adaptations to life in the trees and facilitated movement through a three-dimensional maze of branches.

6. My climb up the koompassia took place at the Danum Valley Field Centre in Sabah, Malaysia, probably the leading center for rain forest research in the Old World tropics. The center is adjacent to the Danum Valley Conservation Area, one of the largest remaining, most important, and best-protected areas of pristine lowland rain forest in southeastern Asia. Koompassia trees are often left standing in previously logged areas because their timber is poor and silica deposits in the wood make them hard to fell.

7. The story of Adam and Eve is from Genesis 2:9; mention of the tree of knowledge is from Genesis 2:17.

8. The giant peepal at the Mahabodhi Temple in Bodh Gaya, northern India, is known commonly as the Bodhi tree. The Buddha is said to have achieved Enlightenment under this tree. *Bodhi* means "awakening" or "enlightenment" in Sanskrit. The name is also used to refer to other trees propagated from the Sri Maha Bodhi. Ongoing studies by John Rashford are elucidating the significance of fig trees in Candomblé.

9. The significance of the palmetto, sometimes called the cabbage palmetto, to South Carolina dates to the Revolutionary War, when the British fleet retreated from its attack on Sullivan's Island on June 28, 1776, after the palmetto-log fort withstood its cannon fire. In 1950 Mrs. John Raymond Carson incorporated the story into the pledge to the state flag: "I salute the flag of South Carolina and pledge to the Palmetto State love, loyalty and faith." For the significance of the bearded fig in Barbados and its presence on that country's coat of arms, see Rashford (2007). In China, in 1942, the scientist, poet, and historian Guo Moruo proposed ginkgo as the national tree. This also became the goal of Jiangsu Congress Deputy Ju Zhangwang, who from 2003 through 2008 proposed such adoption to the National People's Congress. In a national poll sponsored by the State Forestry Administration in 2005, ginkgo won more than 1.7 million of the total 1.8 million votes cast. Ginkgo remains to be officially designated as China's national tree.

10. The Council Oaks and the Emancipation Oak are live oaks. The Hooker Oak is a valley oak. Joseph Dalton Hooker (1817–1911) was the second director of the Royal Botanic Gardens, Kew. He succeeded his father, William Jackson Hooker (1785–1865), in 1865 and retired in 1885. See Allen (1967) for more on these first two directors of Kew, and Desmond (1999) and Endersby (2008) for more on Joseph Dalton Hooker.

11. According to some estimates the Yongmunsa, "Dragon Gate," Temple was first built in 913 and expanded in 1392, although it has been rebuilt several times. It burned down in 1592, and many of the old structures were burned in 1907 by the Japanese after a military uprising. The temple was also badly damaged during the Korean War. Reconstruction of the current temple was completed in the 1980s; see *Sky News* (2007).

12. The Silla Dynasty began in 57 B.C. and ended in A.D. 935. If the Yongmunsa Ginkgo was planted at the end of this period, it has survived for almost eleven centuries. The Yongmunsa Ginkgo reaches an impressive 200 feet. It remains imposing and among the tallest ginkgo trees on the planet; see Invitation ForestOn, "Story of forest: Old gigantic trees in Korea."

13. Between 1983 and 1996, the National Forest Seed Centre in Burkina Faso distributed seventeen tons of seeds of about sixty tree species to help restore degraded woody vegetation and stock local plantations and nurseries. This project was expanded and enhanced through the work of the Millennium Seed Bank and also has yielded important data on storing and conserving seeds from the semiarid tropics. (Moctar Sacande, Royal Botanic Gardens, Kew, personal communication); see also Sanon et al. (2004).

14. Estimates of U.S. paper consumption are from the World Resource Institute's 2005 Statistics; see Kahl (2009) and Nadkarni (2008).

15. Olson et al. (2001) document the "OneTree Project."

16. Bill Vaughan (1915–1977) wrote a regular feature, "Starbeams," for the *Kansas City Star*.

17. The Angel Oak is a massive live oak on John's Island just outside Charleston, South Carolina. The "tree sit" on the University of California, Berkeley, campus is recounted by Burress (2008).

18. President Clinton's 1993 "Forest Plan for a Sustainable Economy and Sustainable Environment," known as the Northwest Forest Plan, aimed to integrate management and conservation and balance different demands on forest resources; see Tuchmann et al. (1996). Recognizing that the cutting of forests contributes perhaps between 6 and 17 percent of annual global carbon dioxide emissions, REDD (Reduced Emissions from Deforestation and Forest Degradation) has been a recurring issue in the global negotiations about climate change since it was first introduced in the United Nations Framework Convention on Climate Change, Conference of the Parties 13, in Bali in 2007. For an estimate of CO_2 emissions from forest loss, see Van der Werf et al. (2009).

19. The Millennium Ecosystem Assessment, carried out between 2001 and 2005, concluded that over the past fifty years, humans have modified ecosystems "more rapidly and extensively than in any comparable period of time in human history," resulting in "a substantial and largely irreversible loss in the diversity of life on Earth." Cultivated systems now cover one-quarter of our planet's terrestrial land. "Global forest cover loss" is estimated at around 628,206 square miles between 2000 and 2005; see Hansen et al. (2010) and Millennium Ecosystem Assessment (2005).

3. Identity

1. Epigraph: inscription on plaque by the ginkgo trees near the LuEsther Mertz Library, New York Botanical Garden.

2. The descriptions of William Hooker's character are from William Henry Harvey, an Irish botanist who specialized in algae and named the bright-green cave-moss (*Hookeria laete-virens*) for his lifelong friend; see Allen (1967).

3. For the history of the Royal Botanic Gardens, Kew, see Desmond (2007). An outline timeline of the development of the Royal Botanic Gardens, Kew, can be found at www.kew.org/heritage/index. html; see also Blomfield (1994, 2000) for a broader history of Kew village and the Kew community.

4. In leading the revitalization of Kew, William Hooker helped to realize the vision of one of his mentors, Sir Joseph Banks. Banks wished to see Kew present interesting, beautiful, and useful plants from all over the world to the public, and distribute plants of economic importance throughout the British Empire; see Desmond (2007) and Allen (1967).

5. See Drayton (2000) for a view of how Kew and other botanic gardens became instruments of broader colonial objectives in Victorian and earlier times. See Griggs et al. (2000) for an introductory overview of the Kew Economic Botany collection.

6. See Quin (1882, 199) and also Prendergast et al. (2001) for more on John Quin and his collection at Kew.

7. The Qin (or Ch'in) Dynasty began in 221 B.C. and ended in 206 B.C.

8. The ginkgo leaf in the Kew collection is of no great value except for the unusual way that it combines an archetypical Chinese image with one of the most potent natural symbols of the East. Ancient ginkgo trees occur in Shizilin, the Lion Grove Garden, and also in Liu Yuan, the Lingering Garden, in Suzhou, China. The ginkgo growing by a small pavilion in the Lingering Garden is 110 feet tall.

9. Some have speculated that the resemblance of the ginkgo leaf to a fan, a revered symbol, may have been important in the tree's adoption into Buddhism and Shintoism.

10. The Nongso, Seonsan, Ginkgo is in the city of Seonsan in the South Chungcheong Province, South Korea. Information on this and other ginkgo trees designated as natural monuments in South Korea can be found at http://english.cha.go.kr/.

11. For a more complete list of Japanese names that incorporate *icho*, "ginkgo," see Hori and Hori (1997).

12. Larry Kirkland's mural can be viewed in the Keck Center Lobby of the National Academy of Sciences building in Washington, D.C., at Sixth and E Streets.

13. Frank Lloyd Wright may not have removed the ginkgo tree, but he did complain about its smelly seeds in the autumn, according to my former colleague Laurel Ross (personal communication). Her father apprenticed with Frank Lloyd Wright in Oak Park and often recounted Wright's comments about the female ginkgo. The Ginkgo Dreams Web site features a wide variety of products and art that use the ginkgo leaf as part of the design; see www.ginkgodreams.com. See also, for

example, the ginkgo collection of Michael Aram; www.michaelaram.com and Schmid and Schmoll (1994) for the influence of ginkgo on diverse decorative arts.

14. For the use of ginkgo in the Art Nouveau architecture of Nancy and Prague see Kwant, "*Ginkgo biloba* and Art Nouveau in l'Ecole de Nancy," and Kwant, "*Ginkgo biloba* and Art Nouveau in Prague."

15. Gilbert and George cite their first experience smelling ginkgo in New York City as their inspiration for the exhibit. They first attributed the smell to a dog, then concluded, erroneously, that "the female ginkgo leaf simply smells like that." For a review of the Gilbert and George retrospective see Vogel (2005), Wyman (2008); see also the catalogue (Birnbaum and Bracewell, 2005).

16. The ginkgo tree under which Confucius is reputed to have sat and taught is often depicted as an apricot, a mistake stemming from use of the Chinese term *hsing* for "silver apricot." For more on Confucianism, ginkgo, and Wofo Si, the Temple of the Reclining Buddha, see Taylor and Choy (2005) and Porter and Johnson (1993).

17. For a summary of ancient ginkgo trees in Japan see Hori and Hori (1997, 395). For a more comprehensive list, with many photographs of individual trees, see Hori and Hori, (2005). Yasukuni Jinja is a Shinto shrine devoted to the spirits of those who died fighting in the service of the Emperor of Japan since 1853.

18. See the Ginkgo Pages Web site for lists of many individual ginkgo trees in different parts of the world, as well as much other fascinating information. See Zukowski, "In Hoboken, Trees for 9/11," for more on the Hoboken Memorial. Yoko Ono's living sculpture in Detroit's Times Square comprises a ginkgo tree, a block of granite, and a bronze plaque with the inscription "WISH TREE for Detroit. Whisper your wish, To the bark of the tree. Yoko Ono 2000 spring." It is one of a series of Wish Tree installations; others can be found in Brazil, California, Italy, and Japan. Ono explained that she was moved by ginkgo trees in the temple gardens of Japan where she grew up, with tiny rolls of paper bearing hopes and prayers adorning the branches like blossoms; see Nawrocki and Clements (2008, 49).

19. Six ginkgos survived the explosion of the atomic bomb at Hiroshima, one less than a mile from its epicenter; see Kwant, "A-bombed ginkgo trees in Hiroshima, Japan." On walks during his later years, Harry Truman often paid tribute to the old ginkgo tree near his home in Independence, Missouri, with a friendly pat and a few words. According to the pastor Thomas Melton, who accompanied Truman on occasion, he told the tree, "You're doing a good job" (McCullough, 1992). The ginkgo is marked by a plaque as a stop on the Truman Historic Walking Tour; see Fischer (2010).

20. The Morton Arboretum is the second-oldest arboretum in the United States, after Harvard's Arnold Arboretum. Its seventeen hundred acres include more than forty-one hundred species of trees and shrubs from around the globe. In addition to its role in conserving plant species and engaging the public, the arboretum is committed to creating greener communities by planting and protecting trees in urban areas: see Ballowe and Klonowski (2003) and www.mortonarb.org.

21. For more on Goethe and Marianne Willemer see Unseld (2003). Goethe's book *Versuch die Metamorphose der Pflanzen zu erklären*—"An attempt to explain the metamorphosis of plants" (Goethe, 1790)—is widely regarded as the first scientific study of plant form.

22. "From top to bottom a plant is all leaf" is a translation from part of a letter that Goethe wrote

to Johann Gottfried van Herder, May 17, 1787. For more on Goethe's search for a "Bauplan" and a "fundamental organizational theme" in plant form, see Kaplan (2001).

4. Energy

1. Epigraph: Coolidge (1919, 13). According to Arnott (1959), some rejoining of the veins occurs in 13.4 percent of ginkgo long-shoot leaves and 8.2 percent of short-shoot leaves, with at most five junctions on a single leaf. On average, leaf vein unions are found on fewer than one in ten leaves. For other studies of leaf venation in ginkgo see Florin (1936b) and Arnold (1947).

2. Creating artificial photosynthesis, and improving on its efficiency, which is two to three times less than the best photovoltaic devices under optimal conditions, has long been a goal in solar energy research: see Hohmann-Marriott and Blankenship (2011) and Blankenship et al. (2011). Gary Brudvig's research group has created "artificial leaves" that split water into oxygen gas and hydrogen gas, and that can be used to power a new kind of fuel cell. While solar panels provide electricity only during the day, these new fuel cells potentially provide a constant energy supply.

3. It is possible to examine the stomata on well-preserved fossil ginkgo leaves. Measurements of stomatal density from fossil ginkgo leaves have been used to trace changing levels of atmospheric carbon dioxide in the distant past. When concentrations of atmospheric carbon dioxide are high, the density of stomata needed to ensure an adequate supply of carbon dioxide into the leaves is low, and vice versa; see Retallack (2001), Royer et al. (2001), Beerling and Royer (2002), and Royer (2003).

4. In almost all plants the leaves are the main site of photosynthesis, but in ginkgo the seed also photosynthesizes. The nutritive tissues in the developing seed contain chlorophyll and provide a significant contribution to the energy required for its own growth and ultimately growth of the embryo that it nourishes. Light penetrates the fleshy coat and hard seed wall at levels equivalent to full shade on a clear day; see Friedman and Goliber (1986).

5. Among the several kinds of photosynthesizing bacteria, the cyanobacteria (blue-green algae) first exhibited the kind of photosynthesis seen in plants.

6. A single ginkgo leaf of average size may contain perhaps fifty million cells, based on rough estimates from scanning electron microscope images of upper and lower ginkgo leaf surfaces and several leaf cross sections. The so-called thylakoid membranes, in which chlorophyll molecules are embedded, are approximately five nanometers thick (five millionths of a millimeter).

7. Another important class of light-gathering pigments, the carotenoids, is responsible for the yellow and orange shades in the autumn colors of many deciduous trees.

8. As they branch and diverge from the leaf base, the veins in a ginkgo leaf gradually become finer and contain fewer and fewer water-conducting cells (tracheids). A more simple but comparable system is seen in the needles of many pines: see Zwieniecki et al. (2006).

9. The mucilage-like resin in ginkgo leaves is deposited in a single file in each interveinal area. Late leaves contain much more resin than leaves produced early in the season; see Critchfield (1970). This number of ginkgo trees on the streets of Japan reflects an increase of about twenty thousand

from the 1992 statistics cited by Handa et al. (1997); (Toshiyuki Nagata, Hosei University, personal communication, 2011).

10. For more on factors that cause trees to shed their leaves, see Treshow (1970).

11. For more on the recovery of nutrients from senescing leaves, see Andersson et al. (2004), Buchanan-Wollaston (1997), and Killingbeck (1996). "The Two Ginkgo Leaves," by Otto Crusius, former president of the Bavarian Academy of Sciences, cited in the translation of Unseld by North-cott (Unseld, 2003): "Ginkgo leaves tired of summer, brilliant as a brimstone butterfly. Flutter down onto the bench, Flutter, whisper, 'do you remember?'"

12. The Meiji-Jingu Park in the Kasumigaoka district of Tokyo commemorates the Meiji Emperor who died 1912. Completed in 1926, it contains sports and cultural facilities the centerpiece of which is Seitoku Memorial Museum (Meiji Memorial Picture Gallery). The gallery is approached along an avenue about a quarter of a mile long with ginkgo trees on either side: see Handa et al. (1997, 272). Similarly, at the Memorial Showa Garden, constructed on the site of a former American military base, the long central canal that is the centerpiece of the landscape design is flanked by a double row of ginkgos planted in 1983; see Handa et al. (1997).

13. The ginkgo of the Dorsch Public Library in Monroe is the partial namesake for the *Lotus Ginkgo Show,* the self-described longest-running program on Michigan's Monroe Public Access Cable Television.

14. "The Consent," Nemerov (1977, 476). A small group of ginkgo trees that grew outside Neme-rov's office on the campus of Washington University in Saint Louis was the inspiration for this poem (Peter Raven, Missouri Botanical Garden, Saint Louis, personal communication). This poem has a different tone from "Ginkgoes in Fall," appearing one page before: "Their fallen yellow fruit mimics the scent / Of human vomit"; the same leaves described as "fluttering fans of light" are "filtering a urinary yellow light" in the other; Nemerov (1977, 475).

5. Growth

1. The comment on artistic portrayals of trees is from Hargraves (2010). Ginkgos are easy to spot from a distance, especially in winter, from their distinctive pattern of branching. When I have occasionally been confused, the other tree has generally turned out to be a conifer, most often a pine.

2. The tradition of learning how to identify trees from their twigs in winter comes from Germany, and for more than a century, German botany students have benefited from Camillo Karl Schneider's classic *Dendrologische Winterstudien,* which includes photographs and illustrations of bark and buds from 434 species of trees and shrubs; see Schneider (1903). A friend at Berkeley tells the story of a colleague who grabbed a naked ginkgo twig hoping to bamboozle a prospective Ph.D. student. He left chastened when the candidate quickly identified the specimen as ginkgo (Bruce Baldwin, University of California, Berkeley, personal communication).

3. During the April 2007 cold snap, gardens all across the midwestern United States suffered as spring flowering trees and shrubs were damaged by wind chills that fell to minus 30° Fahrenheit

(roughly minus 34° Celsius). Because ginkgo is pollinated at the same time that the leaves emerge (see Chapter 7), the cold snap also reduced significantly the production of ginkgo seeds that year.

4. For more on the expression and development of short and long shoots in ginkgo, see Gunckel et al. (1949).

5. See Chapter 3 and Unseld (2003) for more on Goethe and the significance of the poem he sent to Marianne Willemer. See Chapter 28 for more on Linnaeus and the origin of the name of ginkgo.

6. Studies by Leigh et al. (2010) on the structural and physiological differences between long- and short-shoot leaves document an interesting difference in vein density. Even though long-shoot leaves have a lower density of veins than short-shoot leaves, they are significantly more effective at conducting water. It is not completely clear how this is accomplished, which highlights the fact that there is still much research to be done before we fully understand how the strange leaves of ginkgo actually work. One possibility, which needs to be tested by future measurements and experiments, is that the veins in long-shoot leaves are better connected to the upper and lower leaf epidermis and that the epidermal cells play an important role in leaking water out of the veins and into the surrounding leaf tissues.

7. The cells in the outer part of the woody cylinder through which the water passes in most trees are dead. They form from the cylinder of actively dividing living cells (cells of the cambium), but once formed they quickly undergo preprogrammed cell death; see Chapter 6.

8. Early in the season in many trees, as the sap begins to rise, sugars stored in the roots and the living tissues in the lower part of the stem, including the ray cells in the wood, are mobilized and taken into solution within the cells. This draws in water from the soil, which expands the volume of fluid in the lower parts of the plant and causes the sap to rise. This is the process that is exploited by tapping a sugar maple, and in this special circumstance, at this particular time of the year, there is pressure from below.

9. The volume of water taken up by the yevaro tree (260 gallons a day) was the maximum recorded in a review of fifty-two studies conducted over thirty years on sixty-seven species. In this review, 90 percent of trees around sixty-five feet in height take up only between 2 and 44 gallons of water a day. For more details and a breakdown of water requirements by species, see Wullschleger et al. (1998).

10. The same minute pores are also present in conifers. For more on the structure of ginkgo wood, see Dute (1994).

6. Stature

1. Epigraph: from Eknath Easwaran's translation of the classic Buddhist text (Easwaran, 2007, 126). For a biography of Philipp Franz von Siebold (1796–1866) see Thiede et al. (2000), Kouwenhoven and Forrer (2000), and the exhibits and publications of the Siebold Museum in Leiden (www.sieboldhuis.org) and Nagasaki (www.city.nagasaki.nagasaki.jp/siebold). Glover's house still stands above Nagasaki harbor, with statues of Puccini and Cho-Cho San looking out to sea from the garden. For more on the origins of Madame Butterfly, see Van Rij (2001).

2. The Siebold Museum in Nagasaki, modeled on Siebold's former house in Leiden, is located next to the former site of Siebold's medical school, Narutaki Juku.

3. In *Flora Japonica,* Siebold named *Hydrangea otaksa* after his nickname for Sonogi. The accepted name for this species now is *Hydrangea macrophylla* (Thunb.) Ser. Today, bigleaf hydrangea, as it is commonly known, is the most popular species of hydrangea planted in home gardens (Kouwenhoven and Forrer, 2000).

4. The map supplied to Siebold had been completed in 1818 and engraved only in 1823. It was the result of exhaustive surveying of the coasts and islands of Japan by a team of fourteen and was the most detailed account for northern Japan then available; see Murdoch (2004, 555–558).

5. An alternate version of the incident from Murdoch's *A History of Japan,* based on Siebold's diary entry for December 16, 1828, describes his betrayal by Yoshio Tsujiro, an interpreter who was helping him translate from Japanese books; see Franz (2005, 37); Totman (1993, 510). Siebold transcribed the maps for the archives at Deshima before handing them over.

6. Oine was an inquisitive child and at the age of nineteen was directed by Ninomiya Keisaku to study obstetrics. She was appointed to the post of imperial obstetrician in 1877.

7. For more on the life of Oine see Kouwenhoven and Forrer (2000, 25).

8. All three works were published in multiple parts over the next quarter century. *Nippon* (Siebold, 1832–1852) describes the ethnography and geography of Japan, including an account of his journey to Edo. *Fauna Japonica* (Siebold et al., 1833–1850) was a series of monographs based on the collections of Siebold and his Deshima successor, Heinrich Bürger. *Flora Japonica* built on the work of Siebold's predecessors Kaempfer and Thunberg and was undertaken in collaboration with the German botanist Joseph Gerhard Zuccarini (Siebold and Zuccarini, 1835–1870). It was begun in 1835 and came to a halt after the death of Zuccarini in 1848, but ultimately, after Siebold's death in 1866, F. A. W. Miquel from the National Herbarium in Leiden published some additional parts. After thirty-five years, *Flora Japonica* was completed in 1870. The first volume contained twenty parts and the second volume ten. Plate 136, a superb illustration of *Ginkgo biloba,* was published in the second volume (see Chapter 5).

9. After William Jackson Hooker's death, his library, a collection of around 4,000 volumes, was purchased by the British Government for £1,000 in 1866; an additional £1,000 bought his correspondence, manuscripts, portraits, and other miscellanea. The collections of botanical art at Kew comprise more than 200,000 items, including astonishing work from the eighteenth, nineteenth, twentieth, and twenty-first centuries. Most of these paintings are housed in an annex to the Kew Library, with selections on display in the Shirley Sherwood Gallery of Botanical Art and the Marianne North Gallery.

10. The Berlin collection was acquired in 1911 from Paul Kuegler, a senior staff physician in the German navy. He probably obtained the collection in Japan, toward the end of the nineteenth century (Lack, 1999). The Koishikawa Botanic Garden holds a set of twenty-five boards in the same style. Harvard University Museum also has a small collection of boards brought to New England by Edward Sylvester Morse, who was the first professor of zoology at Tokyo University between 1877 and 1879.

There is also a small private collection in London. The Koishikawa Botanic Garden, or Koishikawa Shokubutsuen, was founded as the Koishikawa Medicinal Herb Garden in 1684 by the Tokugawa Shogunate, and was the birthplace of modern Japanese research in plant science following the Meiji Restoration. It contains one of the most famous of all ginkgo trees: see Chapter 9.

11. Chikusai Kato prepared sketches of ginkgo for *Koishikawa-shokubutsuen-somoku-zusetsu* (Illustrations of the trees and herbs in the Koishikawa Botanical Gardens), edited by Keisuke Ito (1803–1901) and Hika Kaku (1786–1884); see Ito and Kaku (1881–1883). Kaku's elder brother studied under Siebold. Keisuke Ito gave Siebold fourteen collections of dried plants, including a specimen of ginkgo, to bring back from Japan.

12. The way that rays are cut in preparing a piece of timber is often responsible for the distinctive "grain" of certain high-quality woods. In ginkgo the rays are not sufficiently thick to impart a distinctive grain. For more on ray cells and their development in ginkgo wood, see Barghoorn (1940, 321).

13. Extreme cork production occurs in the cork oak, the main source of corks for wine bottles. The high suberin content of cork oak bark also makes it more pliant and water resistant than the bark of other trees.

14. The dead, water-conducting cells, the so-called tracheids, reach up to two to three thousandths of an inch wide and a tenth to three-tenths of an inch long in the trunk and nearly four-tenths of an inch long in roots. The specialized water-conducting cells (vessel elements) joined together in the wood of many flowering plants make tubes an order of magnitude longer, up to about fourteen inches (Sperry et al. 2006; Wilson and Knoll, 2010). The tiny valves in the water-conducting cells of ginkgo, the so-called torus-margo structures associated with the pit membranes, improve water conductivity by minimizing hydraulic resistance while also preventing the formation of embolisms; see Hacke et al. (2004) and Pitterman et al. (2005).

15. Eleven annual rings are visible in the small branch that makes up the top left-hand corner of the ginkgo board in the Kew xylotheque. The rings on the other branches are obscured by tool marks and varnish (Mark Nesbitt, Royal Botanical Gardens, Kew, personal communication).

16. The different uses of ginkgo wood are listed by Hori and Hori (1997).

17. Ojiya is a small town known as an important center for the breeding of koi carp. It gained notoriety as the epicenter of a fatal, magnitude 7.2 earthquake in October 2004. Li Shizhen recorded the use of ginkgo wood by Daoist shamans in his A.D.1596 herbal *Bencao Gangmu,* writing, "The wood of the tree is white with fine texture and lasts a long time. Alchemists carve chops with the wood saying that such thing is good to summon the spirits" (Shizhen and Xiwen, 2003).

18. The legacy of carved wooden Buddhas, *mokujikibutsu,* that Mokujiki Shonin left across Japan, deeply moved Yanagi Sōetsu, the philosopher and founder of the *mingei* (folk art) movement in Japan. Yanagi retraced Mokujiki's original route, cataloguing the mokujikibutsu, which he described as "simple," "natural," and "ego-less" in their beauty and tradition; see Kibuchi (1997). See also the drawing that introduces Part VII.

7. Sex

1. Epigraph: Dawkins (1976, xxi).

2. Camerarius (1665–1721) reported his results on reproduction in plants in his 1694 publication *De sexu plantarum epistola*. The botanists John Ray (1627–1705) and Nehemiah Grew (1641–1712) were among the earliest fellows of the Royal Society, the U.K. national academy of science.

3. According to *Science and Civilization in China,* the earliest reference to ginkgo comes from the *Ko Wu Tshu Than* (Simple discourses on the investigation of things), written by Tsan-Ning, a "learned monk." This is a series of short statements about natural phenomena, written about A.D. 980; see Needham (1986, 491) and Chapter 24.

4. The evolutionary transition from a species composed entirely of hermaphrodite individuals to species with separate male and female individuals (dioecy) is best documented among flowering plants. In many cases remnant or vestigial organs from the other sex remain in the flower, suggesting that the separation into male and female plants occurred relatively recently in evolutionary history. Based on observations like these, Darwin (1876, 1877) developed his initial ideas on the evolutionary advantages of separate male and female plants, which have been further elaborated by later evolutionary biologists (e.g., Charnov et al., 1976; Lloyd, 1982). Dioecy, monoecy (separate male and female flowers on the same plant), and dichogamy (separation in the timing of maturation of male and female parts of the same flower) in flowering plants are all effective in promoting cross-pollination between individuals, which increases the genetic variation in the next generation upon which natural selection can act.

5. For description of the development of ginkgo pollen cones, see Liu et al. (2006). See Christianson and Jernstedt (2009) for details on the position of pollen cones and seed-bearing structures in ginkgo.

6. Ginkgo pollen grains probably do not remain viable for long after they are shed, but under laboratory conditions they can remain alive for up to sixteen months (Newcomer, 1939). Under sterile conditions Tulecke (1954) achieved a germination rate of 35 to 45 percent for ginkgo pollen after storage for two years.

7. Andrew Leslie's estimate is based on observations of trees growing outside the Hinds Geophysical Laboratory at the University of Chicago with the following estimated parameters: a single pollen sac contains about twenty thousand pollen grains; each pollen cone contains about seventy-seven side branches, each with two pollen sacs, or about three million grains per cone; each short shoot produces about seven cones, and there might be around 17,500 short shoots on a forty-foot tree.

8. Often in trees with separate sexes, males begin to produce pollen slightly before females produce ovules, which is consistent with theoretical predictions: see Lloyd and Webb (1977).

9. After pollination, the dried mucilaginous residue from the pollination drop seals the micropyle, and pollen development and ultimately germination proceed within the sealed cavity (Lee, 1955). For more on how pollination drops function in conifers, see Takaso (1990), and for video imagery of pollination and fertilization in ginkgo and cycads, see the Tokyo Cinema film *The Sea in*

the Seed. A pollen tube is formed once the pollen has been taken into the micropyle, and in ginkgo and cycads the tube appears to be modified for nutrient uptake. It penetrates the nutritive tissues in the central part of the ovule, forming extensive branching networks of fine haustoria among the cells; see Friedman and Gifford (1997) and Chapter 9.

10. Occasionally seed-bearing stalks have only a single seed at their tip or several, but two is the norm. The development of more than one mature embryo in a single seed occurs in about 2 percent of ginkgo seeds and also occurs occasionally in conifers: see Cook (1902, 1903), Buchholz (1920), and Berlyn (1962) for additional information. For a photograph of two seedlings emerging from a single ginkgo seed see Stuppy et al. (2009, 24).

11. For more on the development of the embryo of ginkgo, see Lyon (1904). The fleshy outer layer appears to inhibit germination if it is not removed (Rothwell and Holt, 1997). Ginkgo differs from most conifers in having hypogeal germination: the cotyledons remain embedded in the nutritive tissue of the seed, often underground. Only the lower parts of the cotyledons project from the seed shell; see Seward and Gowan (1900, 116), Chick (1903).

8. Gender

1. Epigraph: Angelou (1990).

2. An important question is why there is apparently so little change in the evolutionary history of ginkgo, as well as in other classic examples of so-called stasis. Standard explanations suggest that either ginkgo has tracked the same environmental conditions for more than 200 million years or the same form (perhaps in spite of genetic variation) has been maintained by strong stabilizing selection. However, a third possibility is that there is some kind of strong, inbuilt constraint during development that has kept ginkgo more or less unchanged for a long period of time. Potentially, all three factors may be at work.

3. Joseph Franz von Jacquin (1766–1839) succeeded his father, Nikolaus Joseph von Jacquin (1727–1817), as professor of botany and director of the botanical garden at the University of Vienna. The family lived in a house on the Rennweg, near the modern Institute of Systematic Botany of the university, close to the Belvedere Palace. Along with his brother and sister, Joseph von Jacquin was taught the piano by Mozart. The composer, part of his father's aristocratic circle, regularly visited the Jacquin home. Mozart's "Kegelstatt Trio," dedicated to the Jacquin family, was first performed at their home in August of 1786 by Franziska, Joseph Jacquin's sister.

4. Nikolaus von Jacquin had connections to the palace at Schönbrunn, for which he worked as a plant collector from 1755 to 1759. An account of early ginkgos in Europe is provided by Loudon (1838); see also Jacquin (1819) and Chapter 30.

5. August Pyramus De Candolle was the first to recognize ovule-bearing shoots on a ginkgo in Europe at Bourdigny, a village about six miles outside Geneva, in 1814. See also Chapter 30. The Bourdigny tree was cut down in 1866 by a new owner of the estate; see Wilson (1920, 56). Jacquin was the first botanist to use grafting for a scientific purpose. He published the results from his grafting experiment in the same year that Goethe first published his ginkgo poem (Jacquin, 1819).

6. The slightly delayed development of the female branch is consistent with theoretical predictions: see Lloyd and Webb (1977).

7. The experiments carried out by Hugo de Vries at the University of Amsterdam (commemorated by a plaque in the garden) were some of the most important ever done in a botanical garden. The term *genes* was introduced by the Danish botanist Wilhelm Johannsen in 1909. Johannsen also was the first to use the terms *phenotype* and *genotype*.

8. Theodor Boveri (1862–1915) worked in Germany while Walter Sutton (1877–1916) worked in the United States. Thomas Hunt Morgan (1866–1945) received the Nobel Prize in 1933 for his role in linking chromosomes with heredity. The research of Nettie Stevens (1861–1912) at Bryn Mawr College and of Edmund Beecher Wilson (1856–1939) at Columbia was conducted on the mealworm, the larvae of a species of beetle (*Tenebrio molitor*), which has the advantage of having relatively large chromosomes that are easily observed.

9. Human males and females both have twenty-two pairs of matching chromosomes in all their cells, but in the twenty-third pair, the sex chromosomes, the two chromosomes differ in the male. Females have two matching copies of chromosome twenty-three, which are designated XX. Males have one long chromosome, the X chromosome, and one short chromosome, the Y chromosome.

10. The first chromosome count for ginkgo was reported by the Japanese scientist Mitsuharu Ishikawa (1910).

11. The suggestion of an XY-type sex determination system in ginkgo was based on several early studies that reported a satellite on the arm of chromosome eleven (Tanaka et al., 1952; Newcomer, 1954). Later research found various satellites associated with various chromosomes in male and female plants (e.g., Ho, 1963; Chen et al., 1987). For more on the chromosomes of ginkgo see Hizume (1997).

12. A similar situation occurs in certain animals where the sex of offspring is determined after the embryo has begun to develop and depends on environmental conditions. For example, in many lizards and turtles higher temperatures during a critical phase in the development of the embryo favor the production of females, while in alligators they favor the production of males.

13. The story of the graft on the Old Lion at Kew is recounted in Bean (1973). See Crane (2006) for the report of seeds on the same tree. The example in the Jena botanical garden is cited by Melzheimer and Lichius (2000).

14. "Witches' brooms," dense masses of shoots growing from a single point often high up in the canopy of a tree, can be caused by damage of various kinds—for example, by pests or diseases, as well as by human interference. The result is a kind of cancer, an abnormal proliferation of growth, probably caused by a local loss of control of normal development in that part of the plant. In the Cave Hill tree the result has been to change the sex of that part of the tree. The Japanese botanist Seiichiro Ikeno, who had earlier been the first to observe swimming sperm in cycads, also noted that male ginkgos can occasionally produce seeds; see Ikeno (1901), Miyoshi (1931).

15. Observations at the Blandy Experimental Farm are reported by Santamour et al. (1983a).

16. Lloyd and Webb (1977).

9. Seeding

1. Epigraph: Shakespeare (1623c), act 1, scene 3. The Great Kanto Earthquake struck on September 1, 1923, and the most destructive air raid on Tokyo took place on March 10, 1945. Information about the Koishikawa Ginkgo was provided by Tetsuo Ohi-Toma, Koishikawa Botanical Gardens, University of Tokyo (personal communication). The ginkgo at the Koishikawa Botanical Garden was visited by His Majesty the Emperor of Japan and Her Majesty the Empress of Japan in 2006 (His Majesty the Emperor of Japan, 2007).

2. The Koishikawa Ginkgo was planted around 1680, and an attempt was made to fell it in August 1868. The scars from the axe cuts made at the time were visible until about fifty to seventy years ago. Part of the motivation in trying to fell the tree may have been profit (Tetsuo Ohi-Toma, University of Tokyo, and Toshiyuki Nagata, Hosei University, personal communications; see also Primack and Ohkubo, 2008).

3. The University of Tokyo has undergone various name changes over the years, from Tokyo University (1877–1886), to Imperial University (1886–1896), to Tokyo Imperial University (1896–1948), and finally to the University of Tokyo (1948–present). Ryokichi Yatabe, the first Japanese graduate of Cornell University, became the first professor of botany and curator of botanic gardens at Tokyo University. Jinzo Matsumura was the second professor and curator. Keisuke Ito, a student of Siebold, oversaw the garden in the late nineteenth century and retired from the university in 1886. Appointed at the age of seventy-five, he was given a special title and did not have to teach (Toshiyuki Nagata, Hosei University, personal communication).

4. Stopes studied at Tokyo Imperial University for a year and a half during the period when the Department of Botany was based at the Koishikawa Garden (1897–1935). The garden is now part of the Department of Biological Sciences, which is located on the main campus (Toshiyuki Nagata, Hosei University, personal communication). Marie Stopes's visit to Japan was funded by a fellowship from the Royal Society. The studies of Cretaceous petrified fossil plants from Hokkaido, initiated by Stopes and Fujii (1910) in the course of that fellowship, continue today through Harufumi Nishida at Chuo University, Tokyo; see Nishida (1991) for a review. Stopes and Fujii became involved in an ill-fated love affair. Fujii, who was married, feigned leprosy to break it off. Stopes later published their letters under the pseudonym of G. N. Mortlake as *Love Letters of a Japanese;* see Mortlake (1921) and Hall (1977).

5. This list of significant books was compiled in 1935 by asking a number of American academics to name the twenty-five most influential books of the previous fifty years (Hall, 1977). More on Marie Stopes can be found in several biographies; see Hall (1977), Briant (1962); for selections of Stopes's writing, see Stopes (1918), Garrett (2007).

6. According to some sources, the University of Manchester attempted to retract its offer when administrators realized that Stopes was a woman, but she nevertheless was appointed. Marie Stopes was as flamboyant as she was brilliant. Bill Chaloner, one of my paleobotanical mentors, recalls meeting her in 1952 at the Geological Society of London. After he had explained some of his research to her, she exclaimed loudly so all could hear, "Ah dear boy, that is wonderful! Of course, paleobotany was my *first* love!" (Chaloner, 2005).

7. Details of sexual reproduction in ginkgo, based on the material sent from Vienna, were published by Strasburger (1892). German students of plant science still study from a revised version of the massive textbook that Strasburger wrote. Hirase worked at the botanical laboratory of the College of Science of the Tokyo Imperial University. Soon after his discovery of motile sperm in ginkgo, the Department of Botany was moved to the Botanical Garden at Koishikawa, where it remained from 1897 until 1935 (Toshiyuki Nagata, Hosei University, personal communication).

8. Hirase had been appointed by Ryokichi Yatabe as the first professor of botany at the Imperial University, Tokyo. Initial observations on fertilization in ginkgo were published by Hirase (1895a, b), but the key paper recognizing the motile sperm appeared in 1896 (Hirase, 1896). According to Singh (2006, 236), Hirase began his career as a technical illustrator, but taught himself botanical techniques and began to study ginkgo fertilization and embryo formation in 1893. He made his discoveries while preparing microscope slides of ginkgo ovules, when he observed a peculiar ellipsoid body with an attached coiled band in the pollen tube. He correctly noted that it might be a spermatozoid in a lecture on April 25, 1896, and continued cutting and examining ovules until at last finding motile sperm a few months later on September 9. An excellent film of the swimming sperm, and many other aspects of the biology of ginkgo, is available in *The Sea in the Seed*.

9. Not long after Hirase made his discovery, his most important mentors left the Imperial University. Hirase left soon after. Because he lacked formal botanical training, the strictly hierarchical system may have made his position awkward. For the rest of his career Hirase worked as a schoolteacher; see Ikeno and Hirase (1897) and Nagata (1997) for more on Hirase and his discovery.

10. Marie Stopes's observations on ginkgo sperm came exactly twelve years after Hirase's observation: see Stopes (1910, 218). Using the dates recorded in Stopes's journal, Professor Toshiyuki Nagata, a former director of the Koishikawa Garden, collected ginkgo seeds from the same and other trees on September 9, 1997. He found that fertilization within each tree was broadly synchronized, but that there were differences in timing among trees.

11. The work by Oliver and Scott was published in 1903 and 1904; for more on its significance see Andrews (1980). Since Oliver and Scott's work many different kinds of fossil seed plants have been grouped together as seed ferns, but studies of the relationships among living and fossil seed plants show that seed plants are a very heterogeneous and unnatural group of diverse relationships (e.g., Crane, 1985; Doyle and Donoghue, 1986; Hilton and Bateman, 2006).

12. Aberrant ovule-producing leaves in ginkgo were first described by Mitsuharo Shirai (1891) and then by Kenjiro Fujii (1896). When fertilized, the ovules developed into seeds, although these remained smaller than those of normal trees. The leaves bearing ovules and pollen sacs were smaller than normal leaves, and those leaves bearing viable seeds were even smaller still. The significance of such aberrations is still debated.

13. See Favre-Ducharte (1958) and Eames (1955). At EWHA University, Seoul, South Korea, I collected nearly mature seeds that fell from several trees well before the normal time for fertilization. The seeds were left outside for several months, and a few produced viable embryos. More detailed work would be needed to securely establish that fertilization occurred after the seeds were shed.

14. Liz Jaeger, councilor for Whitton (personal communication). Masamichi Takahashi, Niigata

University (personal communication; see also Kochibe, 1997). Ginkgolic acid may also be called ginkgoic acid.

10. Resilience

1. Epigraph: Gandhi (1961, 133).

2. In Utah, a single clonal colony of quaking aspen called Pando is considered to be the world's heaviest, and by some measures, oldest organism. With more than forty-seven thousand stems growing over 107 acres, the plant weighs an estimated six thousand tons. Though the average stem age is 130 years old, according to some estimates, genetically identical plants—parts of the same clone—may have existed ten thousand or more years ago. See also Chapter 24.

3. *Chi-chi* has also been translated as "nipples" in Japanese. Fujii (1896) investigated the internal anatomy of ginkgo chi-chi and showed that close to the point of attachment to the parent shoot each contains an embedded short shoot, the buds of which keep growing so as to maintain their position on the surface of the downward growth. These buds have the potential to burst into life when they reach the ground, or even before, and produce new upward-growing shoots. For more on the development of chi-chi see Barlow and Kurczyńska (2007).

4. For more on the development of ginkgo lignotubers see Del Tredici (1992a).

5. See the drawing that introduces Part II for an illustration of the Kitakanegasawa Ginkgo.

11. Origins

1. The founding collections of the Swedish Museum of Natural History were the collections of the Royal Swedish Academy.

2. Alfred Nathorst (1850–1920) retired in 1919; see Seward (1921) and Andrews (1980) for brief biographies.

3. Nathorst's studies of the Spitsbergen fossils are revised and discussed by Kvaček et al. (1994). See Schweitzer and Kirchner (1995) for additional information on *Ginkgo cordilobata* and the other fossil plants with which it occurs. The Stockholm collections include about thirty of Nathorst's ginkgoalean specimens (Else Marie Friis, Swedish Museum of Natural History, personal communication).

4. In hindsight, Mackie recalled discovering a piece of plant-bearing chert around 1880, more than thirty-five years before he first announced his discovery; see Mackie (1913, 225) and Trewin (2004).

5. The Rhynie Chert formed in the Early Devonian (Kenrick and Crane, 1997).

6. Clubmosses are still native to Scotland, and *Asteroxylon* is especially similar to the fir clubmoss, a plant that is common in the Scottish Highlands.

7. See Kenrick and Crane (1997) for an overview and analysis of the early fossil record of plants on land.

8. The closest living relatives of land plants are the freshwater "charophycean green algae," which

include the stonecrops. This provides additional support for the idea that the land was colonized by plants from freshwater rather than directly from the sea.

12. Ancestry

1. Epigraph: According to a letter written by Huxley to his friend Dr. Dryster about two months after the debate, his response was: "If then, said I, the question is put to me would I rather have a miserable ape for a grandfather or a man highly endowed by nature and possessed of great means of influence and yet who employs these faculties and that influence for the mere purpose of introducing ridicule into a grave scientific discussion, I unhesitatingly affirm my preference for the ape." For a complete account of the Wilberforce-Huxley encounter and the associated controversies, see Jensen (1988).

2. The Southern African paleoflora from the Devonian to the Cretaceous is described by Anderson and Anderson (1985). See Anderson and Anderson (1983, 1989, 2003, 2008) and Anderson et al. (2007) for descriptions of and context for fossil plants from the Molteno Formation.

3. For a review of ancient ginkgolike fossils from North America see Ash (2010). For an early occurrence of ginkgo species from the early Middle Triassic of Australia see Holmes and Anderson (2007).

4. The two most informative of the three specimens of *Trichopitys heteromorpha* studied by Florin (1949) from Lodève, France, were formerly in the collection of the École Nationale Supérieure des Mines de Paris but are now at the University of Lyon. The other is in the Natural History Museum in London. However, Florin did not examine Saporta's original material, which is in the Natural History Museum in Paris. The specimen of *Trichopitys* illustrated by Taylor et al. (2009, 745) is from another locality at Montpellier (Hans Kerp, University of Münster, personal communication) and is preserved in a different way. Even though shoots with multiple ovules are occasionally produced in living ginkgo, the ovules rarely develop to maturity. The largest number of mature seeds I have seen on a single shoot is three.

5. See Meyen (1988, 344–346) for his interpretation of Saporta's original material of *Trichopitys*.

6. For treatment of Permian fossils from Argentina similar to *Trichopitys* see Archangelsky and Cúneo (1990). More likely candidates for Permian members of the ginkgo lineage are plants that produced Permian *Sphenobaiera*-like leaves, some of which have leaf cuticles and resin similar to ginkgo. Many of these *Sphenobaiera* leaves have been described under different names such as *Ginkgophyton* and *Ginkgophytopsis* (Zhou Zhiyan, Nanjing Institute of Geology and Palaeontology, personal communication; see also Zhou, 2009).

7. See Meyen (1984).

8. *Kerpia* has some similarities to pinnate leaves assigned to *Psygmophyllum*. The fossil seed clusters were assigned the genus *Karkenia*; see Naugolnykh (1995, 2007) and Chapter 15 for more on *Karkenia*.

9. In ginkgo, fusion among the veins on the leaf, which creates reticulations, does occur but is relatively rare; see Chapter 4.

10. Glossopterid leaves (*Glossopteris*) have been discovered in Antarctica, Australia, India, South Africa, and South America. Continental drift is often referred to as the theory of plate tectonics.

11. The remarkable occurrence of probable sperm in glossopterid fossils was reported by Nishida et al. (2004), based on pollen grains preserved inside a fossil ovule. The illustrations show what appear to be the bases of flagellae arranged in a spiral as in living ginkgo.

12. Anderson and Anderson (2003) and Anderson et al. (2007) suggest that the seed-bearing structures *Kannaskoppia* may be similar to fossils previously described as *Petriellaea* (Taylor et al. 1994). Although fossils of *Petriellaea* are much more fragmentary than fossils of *Kannaskoppia*, some details of their structure are better understood, and there are two to six small ovules inside each recurved cuplike structure. Each ovule is triangular in cross section. *Petriellaea* is quite different from the seed-bearing structures of ginkgo but the differences may be possible to reconcile (see also Meyen, 1984); in this regard the recurved ovules of *Karkenia* may be significant.

13. Relationships

1. A synopsis of Hennig's ideas appeared in English in 1965 (Hennig, 1965). Hennig's book was translated into English mainly by Rainer Zangerl, a specialist in fossil fishes and one of my predecessors as chairman of the Department of Geology at the Field Museum in Chicago; see Hennig (1966). Something of the flavor of the vigorous and often vituperative debates around the development of cladistics, and the personalities involved, can be gleaned from Hull (1988).

2. The group of plants defined by the production of wood (secondary xylem) is termed *lignophytes*.

3. These numbers are the number of potential rooted phylogenetic trees, based upon the most recent common ancestor of all the associated entities. The formula to calculate the number of rooted trees for n different organisms is: $(2n-3)!/(2^{(n-2)} \times (n-2)!)$. For further explanation see http://www.scientific-web.com/en/Biology/Evolutionary/PhylogeneticTree.html. For three taxa the number of rooted trees is 3, for four taxa it is 15, for five taxa it is 105, for six taxa it is 945, for seven taxa it is 10,395, for eight taxa it is 135,135, for nine taxa it is 2,027,025.

4. Cladistic analysis is a method used to build these so-called phylogenetic diagrams, or cladograms (sometimes called phylogenetic trees) and to test which provides the simplest explanation of the data.

5. A classic early paper that used molecular data to develop a new understanding of the relationships among a wide range of flowering plants was coordinated by Mark Chase at the Royal Botanical Gardens, Kew; see Chase et al. (1993). For an overview of our current understanding of relationships among major groups of angiosperms, based mainly on DNA data, see Stevens (2008); for more on the angiosperm fossil record see Friis et al. (2011). For a formal classification of angiosperms based on recent phylogenetic discoveries see APG (2009).

6. The Gnetales include three superficially different but fundamentally similar kinds of plants; the genus *Ephedra*, sometimes referred to in North America as Mormon tea, the genus *Welwitschia*, a bizarre plant that is well known in Namibia and produces only two leaves in its entire life; and the

genus *Gnetum,* a tree or climber of tropical forests with leaves that look very like those of angiosperms (flowering plants).

7. The first paper to apply cladistic analysis to the study of relationships among the major groups of living plants was Parenti (1980). This was followed by a more detailed treatment of living seed plants by Hill and Crane (1982) and later by studies that included living and fossil seed plants in the same cladistic analysis (Crane, 1985; Doyle and Donoghue, 1986). There have been many cladistic analyses since, but the real need is to understand a broader range of fossil plants in greater detail. For example, there remains much potential for paleobotanical collecting in the cherty layers of the Molteno Formation and for comparison of the Molteno seed plants with increasingly well-known fossils of similar age from Antarctica (John Anderson, personal communication); see also Taylor and Taylor (2009) for fossil seed plants from Antarctica.

14. Recognition

1. Epigraph: Wieland (1768, canto II).

2. The work carried out by Zhou Zhiyan at Reading under Harris's supervision resulted in the description of a strange conifer, *Stalagma samara* (Zhou, 1983). The details of the life and work of Thomas M. Harris (1903–1983) in this chapter lean heavily on the biography written by his student and one of my own paleobotanical mentors William G. Chaloner (1985).

3. Harris's work on fossil plants from Yorkshire covered mosses, liverworts, clubmosses, and ferns (Harris, 1961); cycads and various extinct seed plants (Harris, 1964); an important extinct group of seed plants, Bennettitales (Harris, 1969); ginkgo and its possible relatives (Harris et al., 1974); and conifers (Harris, 1979).

4. Harris's B.Sc. was from the University of London. At that time the University of Nottingham did not have its own degree-awarding authority. For more on H. S. Holden, including his work in forensic science, see Andrews (1980). Paleobotany first took root in Manchester when William C. Williamson was appointed to the Chair of Natural History at Owens College in 1851, which eventually became the Victoria University of Manchester in 1880 (one of the two institutions that later merged to form the modern University of Manchester). In addition to Williamson other influential paleobotanists who have been associated with the University of Manchester over more than 150 years have included Marie Stopes, Ernest Weiss, William Lang, John Walton, Isabel Cookson, and Joan Watson. For a full account of the history of paleobotany in Manchester, see Watson (2005).

5. Sir Albert Charles Seward (1863–1941) served as vice chancellor of Cambridge University in 1924 and 1925; see Andrews (1980, chapter 6) for an engaging account of Seward's life and work. Francis Darwin (1848–1925) was also a Cambridge don and a specialist in plant physiology.

6. The overused "abominable mystery" quotation is from a letter that Darwin wrote to Joseph Dalton Hooker on July 22, 1879. Hooker was then director of the Royal Botanic Gardens, Kew (Darwin and Seward, 1903). See Friedman (2009) for a modern discussion of the background to this famous quotation and its meaning.

7. See Chaloner (1985) and Andrews (1980) for the background to Harris's work on fossil plants from East Greenland.

8. Thor G. Halle (1884–1964) was appointed as an assistant in the Swedish Museum of Natural History in Stockholm in 1913. He succeeded Alfred Nathorst and became professor and director of the Department of Palaeobotany in 1918. Harris's comment on his first encounter with Lauge Koch is from Chaloner (1985).

9. Harris described fourteen species of fossil ginkgolike leaves from East Greenland. The seeds that he believed to be associated with *Ginkgoites taeniata* he named *Allicospermum xystum;* Harris (1935).

10. The coastline of northeastern England around Whitby provided a suitably foreboding backdrop for Bram Stoker's *Dracula.*

11. The possible ginkgo pollen cone from Scalby Ness was collected and first described by Han van Konijnenburg–van Cittert (1971); see also Harris et al. (1974).

15. Proliferation

1. I was an undergraduate in the Department of Botany at the University of Reading from 1972 to 1975, a Ph.D. student from 1975 to 1978, and then a temporary lecturer there from 1978 to 1981.

2. The first publication on the ginkgolike fossils from Yima is by Zhou and Zhang (1988).

3. See Zhou and Zhang (1989) for the first detailed account of these fossils. Zhou and Zhang noted that in living ginkgo there are occasionally aberrant ovule-bearing stalks in which as many as ten young ovules are borne on distinct side branches; see also Florin (1949).

4. For additional information on the leaves of *Baiera hallei* and the associated seed-bearing structures, *Yimaia recurva,* see Zhou and Zhang (1992).

5. *Yimaia qinghaiensis* is preserved in a paper coal from the Lucaoshan coal mine, Qinghai Province, northwestern China. It is known from deeply divided leaves with very narrow segments, and also from seeds and seed-bearing structures. The seeds are a little smaller than those of *Yimaia recurva,* and there are fewer seeds at the tips of the seed-bearing structures, but they are otherwise very similar; for more on this fossil plant see Wu et al. (2006). Zhou has also described a third *Yimaia* species, *Yimaia capituliformis,* from Daohugou, Inner Mongolia; see Zhou et al. (2007). With hindsight we now recognize also that fossil plants very similar to *Yimaia* have been described from Europe. Black (1929) described seeds and seed-bearing axes associated with the leaf he described as *Baiera gracilis.* Harris et al. (1974) reassigned *Baiera gracilis* to *Baiera furcata.* Similar material from Germany was described by Schenk (1867) and Kirchner (1992).

6. Archangelsky (1965), Del Fueyo and Archangelsky (2001), Zhou et al. (2002).

7. Since Archangelsky's original description, there have been discoveries of *Karkenia* seed-bearing structures in the Northern Hemisphere. For example, *Karkenia asiatica,* from the Upper Jurassic of Bureya, Russia (Krassilov, 1970), is very similar and is associated with leaves very like those of *Ginkgoites tigrensis* (named *Sphenobaiera umaltensis*). About six reasonably well-understood species

of *Karkenia* are now known across Europe and Asia. They differ mainly in the size and number of seeds that they bear and in the details of the leaves with which they are associated, but all are fundamentally similar. For further treatment of *Karkenia*-like plants and a review, see Krassilov (1970) and Zhou (2009).

8. A case in point is the abundant leaves of *Ginkgo australis* in the Koonwarra Fossil Bed in Victoria, Australia, which date from about 125 million years ago (Drinnan and Chambers, 1986). They are always deeply divided in two but are variable in the degree of dissection; some leaves have only four leaf segments, while others are divided into as many as sixteen. *Ginkgo australis* is very similar to *Ginkgo rajmahalensis* from the Jurassic of northeast India and *Ginkgoites ticoensis* from the Early Cretaceous of Tico, Argentina. No ovulate organs other than *Karkenia* have been linked to the ginkgo-like leaves found in the southern continents, including at the Koonwarra locality; see Drinnan and Chambers (1986). Several species of *Sphenobaiera* and *Ginkgo* leaves are associated with male catkins, although the associated ovulate organs are quite different from those of ginkgo; see Anderson and Anderson (1989, 2003); Holmes and Anderson (2007); Anderson et al. (2007); Zhou (2009).

9. *Archaefructus,* an intriguing early flowering plant from the Jehol Biota, is especially fascinating and controversial. The U.S. television documentary series *Nova* built a whole program around the origin of flowers, and *Archaefructus* was a central part of that story; see Lewis (2007). The Yixian Formation is in western Laoning Province, China.

10. For more on the significance of *Ginkgo apodes* see Zhou and Zheng (2003). For a complete description and illustration of *Ginkgo apodes* and its associated seed-bearing shoots, see Zheng and Zhou (2004).

16. Winnowing

1. Epigraph: Murdoch (1970, 170). A ginkgo leaf is included in the recent portrait of Iris Murdoch by Tom Phillips. He recalls wanting "a bit of nature to be present," and he soon discovered they were both enthusiasts for "the world's oldest tree"; see Phillips, "Portraits: Dame Iris Murdoch."

2. Zhou and Wu (2006).

3. The twenty-two species of fossil ginkgo leaves recorded by Zhou in the Early Cretaceous of China include species assigned to the two genera *Ginkgo* and *Ginkgoites.*

4. For more on the diversification and vegetational expansion of flowering plants in the Cretaceous, see Friis et al. (2011). See Kvaček et al. (2005) for a description of *Nehvizdyella,* a strange putative ginkgo relative from the Late Cretaceous of the Czech Republic.

5. Peter Del Tredici cites the well-developed olfactory lobes and jaws adapted to opening hard seeds as evidence that multituberculates were more likely candidates for ginkgo dispersers than the dinosaurs or early birds of the time: see Del Tredici (1989); but see also Van der Pijl (1982), Janzen and Martin (1982), and Tiffney (1984). Biomechanical analysis of multituberculate jaws suggests that they were unlikely to have been effective dispersers of ginkgo seeds (Wall and Krause, 1992).

6. Angiosperm fossils are relatively rare in the Horseshoe Canyon Formation, but seeds, wood,

leaf impressions, and pollen indicate the presence of early relatives of sassafras, katsura, plane ("syca-more"), alder, sweet gum, and other trees. For more on plant fossils from the Horseshoe Canyon Formation, Alberta, see Aulenback (2009).

7. K-T boundary (Cretaceous-Tertiary boundary = Cretaceous-Paleogene boundary). Of 130 Cretaceous leaf species appearing at multiple levels in the Hell Creek and Fort Union Formations, only 29 are also present in the Paleocene; an estimated 30 to 57 percent of the flora present in the final five million years of the Cretaceous became extinct at the boundary; see Wilf and Johnson (2004). The presence of ginkgo leaves below the K-T boundary is clearly documented based on fossils collected from seven different localities in the Hell Creek Formation of North Dakota; see Johnson (2002). See Zhou and Wu (2006) for more on the decline of the ginkgo in the mid-Cretaceous.

8. Following rapid diversification, the diversity of grazing horses reached its acme (about sixteen species) between about 15 million and 18 million years ago. Today there is only one, or under some interpretations two, extant species; see MacFadden and Hulbert (1988).

17. Persistence

1. Epigraph: Carlyle (1858, 286). I had corresponded with David Dilcher while I was at the University of Reading, and he had brought me to the United States to work with him on fossil plants as a postdoc in his laboratory.

2. David Dilcher and Steven Manchester were the first to publish on Almont fossil plants (Manchester and Dilcher, 1982). They described fossil fruits of *Cyclocarya,* a special kind of Asian wingnut, the wheel wingnut, and found only minor differences from the fruits of the single living species, which is native to the rich deciduous forests of central China. The fossils I described from southern England that have a counterpart at Almont were assigned to the fossil genus *Palaeocarpinus;* see Crane (1981).

3. For our preliminary account of the Almont flora, see Crane et al. (1990).

4. The fossil ginkgo from Almont has now been studied in detail by Zhou and his colleagues (Zhou et al., 2012) and formally named *Ginkgo cranei.*

5. For more on living fossils, see Eldredge and Stanley (1984) and Schopf (1984). *Lingula,* the clamlike organism, is a brachiopod, not a mollusk.

6. See Crane et al. (1990, Fig. 28a) for the possible ginkgo pollen catkin. A fragment of a ginkgo pollen catkin, very like that of the living species, is known from the Horseshoe Canyon Formation (Rothwell and Holt, 1997). Zhou et al. (2012) note minor differences between the cuticles of *Ginkgo cranei* and living ginkgo.

7. See Royer et al. (2003). Taken together, the distribution of ginkgo fossils through time supports the idea that the genus has, since the Cretaceous, preferred growing in warm temperate climates with moist hot summers and cold winters; see Del Tredici (2000), Tralau (1968), and Uemura (1997).

18. Prosperity

1. Epigraph: Attributed to Goethe, but the specific source is not confirmed. For details on the paleobotanical career of John Starkie Gardner (1844–1930), see Andrews (1980, 372). See *British Eocene Flora,* vol. 1, for the early collaborative work of Gardner and Ettingshausen on fossil ferns (Gardner and Ettingshausen, 1879–1882); vol. 2, on fossil gymnosperms, was published by Gardner alone (Gardner, 1883–1885).

2. See Andrews (1980) for commentary on the paleobotanical work of Constantin von Ettingshausen (1826–1897).

3. See Andrews (1980) for commentary on the paleobotanical work of James Scott Bowerbank (1797–1877), Eleanor Reid (1860–1953), and Marjorie Chandler (1897–1983).

4. In a series of publications during the 1960s, Marjorie Chandler expanded and revised the original *London Clay Flora* from 1933 and placed it in the context of other Eocene fossil floras from Southern England; see Chandler (1961, 1962, 1963, 1964). Additional revisions have since been made by Professor Margaret Collinson of Royal Holloway, University of London, and Dr. Hazel Wilkinson of the Jodrell Laboratory at Kew. For more on the importance of work done on the London Clay flora, see Crane and Carvell (2007). The seed described by Bowerbank (1840) and assigned to *Ginkgo* (see also Gardner and Ettingshausen, 1879–1882) was reexamined by Reid and Chandler (1933), and reassigned to the flowering plant family Icacinaceae.

5. The London Clay is known today to contain more than 500 kinds of plants and about 350 named species. For additional information on the London Clay and the fossil plants that have been described from it, see the classic works of Reid and Chandler (1933) and later reviews and updates by Chandler (1961) and Collinson (1983).

6. The exposed columnar basalts that create Fingal's Cave were part of the inspiration for the music of the Norwegian Romantic composer Edvard Grieg.

7. For the geological observations of the Duke of Argyll see Duke of Argyll and Forbes (1851).

8. For further details on the Paleocene fossil plants from the Isle of Mull, see Boulter and Kvaček (1989).

9. In 1995, the Messel Pit Fossil Site became a UNESCO World Heritage Site. For details on the fossil Panama hat palm, see Smith et al. (2008); for more on the Messel flora see Collinson et al. (2012). *Ginkgo orientalis* appears in several Paleocene localities from Eastern Europe; see Samylina (1967).

10. Well-preserved fossil fish are particularly common, earning one especially rich bed the name "split fish layer." Millions of fish fossils have been collected from these lake beds, which have been a major source of fossils for commercial collectors; see Grande (1984, 2013).

11. For more on the fossil flora of the Clarno Formation see Manchester (1981) and Wheeler and Manchester (2002).

12. Wes Wehr was a gifted artist and poet as well as a dedicated paleobotanist and collector of fossil plants. Fossil floras similar to those from the Pacific Northwest, which also contain ginkgo leaves, are known on the other side of the Pacific, for example, from Fushun (Liaoning Province), Huachian

(Jilin Province), and Yilan (Heilongjiang Province) in northeastern China; see Endo (1942), Manchester et al. (2005), He and Tao (1997). Ginkgo is also known from fossils of Late Cretaceous age in this region (Sun et al., 2007).

13. Ginkgo is reasonably common in the Middle Eocene formations of northern Washington State and British Columbia at Driftwood Creek, Quesnel, Horsefly, Tranquilo, McAbee, Quilchena, Princeton, and Republic localities; see Mustoe (2002).

14. Ginkgo was among the fossil plants encountered by Alfred Nathorst, the paleobotanist on several of Otto Nordenskiold's expeditions. Nathorst collected ginkgo at several locations, including most notably from Spitsbergen at 80° North Latitude. Nathorst also traveled on Nordenskiold's remarkable Vega Expedition that sailed from Sweden across the northern coast of Asia and down to Japan through the Bering Strait in 1883. For more on the Eocene fossil flora of Ellesmere Island, see McIver and Basinger (1999).

19. Constraint

1. Epigraph: Nietzsche (1896, 98).

2. Minus 20° Fahrenheit is equivalent to minus 29° Celsius, and minus 45–50° F corresponds to minus 43–46° C. The coldest temperature on record in Chicago is minus 27° F (minus 33° C), with a wind chill of minus 83° F (minus 64° C) on January 20, 1985.

3. For more on freezing resistance in North American trees, see Sakai and Weiser (1973). At minus 40° the Fahrenheit and Celsius scales converge.

4. For more on the effect of climate on plant growth and productivity see Skre (1990), Dahl (1990), and Melillo et al. (1993).

5. The fate of the ginkgo seedlings was relayed by Wolfgang Stuppy (Royal Botanic Gardens, Kew, personal communication).

6. The Arnold Arboretum, one of the world's great collections of living trees, is located in Jamaica Plain, just outside Boston. It has a long history of sending expeditions all over the world to bring trees back for cultivation. The collections are especially rich in material from China, and strong collaborations with Chinese botanists gave Peter Del Tredici the opportunity to work on the timing of sexual reproduction in ginkgo trees growing in a near wild situation; see Del Tredici (2007).

7. While there is no natural period of dormancy, retention of the outer fleshy layer on the seed does inhibit germination; see Rothwell and Holt (1997). Cold stratification, although not required, improves the evenness of germination and may improve the overall germination percentage; see Holt and Rothwell (1997), Rothwell and Holt (1997), and Del Tredici (2007).

8. This account of the process from fertilization to germination, and the factors that cause variations in this process, is based on Del Tredici (2007).

9. Other manifestations of the temperature sensitivity of ginkgo include earlier bud burst and leaf drop in response to warming temperatures and a difference of forty days in bud burst and leaf drop in southern versus northern Japan (Matsumoto et al., 2003).

10. The existence of warm winters at high latitudes during the Eocene is also consistent with the presence of crocodiles among the high-latitude fauna at that time.

11. Forty-three degrees F is equivalent to 6° C.

12. Recent research found that the number of hours of chilling at temperatures below 38° F (3.5° C) was an important factor in determining the tree's height and shoot development; see Wilson et al. (2004).

13. Forty-one degrees F is equivalent to 5° C.

14. It seems to me that ginkgo never looks quite as robust, luxuriant, and fresh in northwest Europe as it does in Seoul or Chicago. Ginkgo seems to like more warmth during the growing season, perhaps even needs great warmth to really flourish, but it also needs a cold winter, just so long as it is not overly tough. Similarly, ginkgo in southern California and other dry places rarely seems as robust as it is in places with a more pronounced winter.

20. Retreat

1. Epigraph: Attributed to Groucho Marx, but the specific source has not been confirmed. For a description of the Paleogene ginkgolike leaf fossils from Tasmania, see Hill and Carpenter (1999).

2. For a description of the Paleocene ginkgo leaves from the Isle of Mull, see Boulter and Kvaček (1989, 34–39). Ginkgo is missing, for example, from the well-studied fossil flora from Kreuzau, and also from the exceptionally preserved Pliocene assemblage of fossil plants recovered from the fill of an ancient sinkhole at Willershausen near Göttingen, in Germany. The Willershausen flora is especially rich, "more than thirty thousand specimens have been collected, representing at least 130 species of plants in 77 genera," but so far those plants do not include ginkgo; see Ferguson and Knobloch (1998). The occurrence of ginkgo in the Miocene flora from Frankfurt, Germany, is unusual in the context of fossil floras of the time. The Frankfurt specimen was described as *Ginkgo adiantoides* by Florin (1936a) but renamed *Ginkgo florinii* by Samylina (1967).

3. For more on the Selárdalur flora in Iceland, see Denk et al. (2011); see also Akhmetiev et al. (1978). For more on ginkgo fossils from southeastern Europe see Kovar-Eder et al. (1994, 2006) and Denk and Velitzelos (2002). For the work of Royer et al. (2003) see Chapter 17.

4. Only one ginkgo leaf has been recovered among the many thousand leaf fossils collected from the Florissant fossil flora (Bret Buskirk and Herb Meyer, Florissant Fossil Beds National Monument, personal communication). The occasional occurrence of ginkgo leaves in the Oligocene Ruby River fossil flora from southwestern Montana (see Becker, 1961) also indicates that the geographic distribution of ginkgo at this time was complex. See also Chaney and Axelrod (1959), Schorn et al. (2007), and Wolf (1987).

5. At the same time, much farther south, Miocene ginkgo is present in the Cedarville flora of northwestern Nevada and neighboring California, about two hundred miles inland from the present-day coastline.

6. See Scott et al. (1962) and Wheeler and Dilhoff (2009) for additional information on the Middle Miocene woods from Vantage, Washington. There is a hint of a pattern of increasing restriction in eastern Asia similar to that seen in western North America. There seems to be a decline of ginkgo in China through the Cenozoic. Ginkgo is present at several different Eocene sites and continues into the Oligocene in far eastern sites along the Russian border. Ginkgo is missing from the fossil record of China from the Early Miocene, about twenty million years ago, but persists in southeastern Russia along the Sea of Okhotsk. Ginkgo is also known from several fossil occurrences in Japan, during the Pliocene and Pleistocene, between about two million and five million years ago, for example in the fossil floras from Hoshiwara, Hiradoguchi, and Daiwa in southern Japan. For a thorough account of the Cenozoic distribution of ginkgo in East Asia, see Uemura (1997). Ginkgo, however, is far from ubiquitous at this time. It is missing, for example, from the classic Mogi flora that Nathorst collected in Honshu when Nordenskiold's Vega expedition spent the summer there in 1879.

21. Extinction

1. Epigraph: From Jimmy Cliff's album *Jimmy Cliff*, 1969, Trojan Records. For more information on the Early Pliocene fossil vegetation in southern Europe, see Kovar-Eder et al. (2006).

2. The Willershausen flora is especially rich; more than 130 species have been collected, representing more than one hundred different kinds of plants. See Straus (1967); Ferguson (1967); Ferguson and Knobloch (1998).

3. Like ginkgo, the Caucasian wingnut has been reintroduced by people into many of the places where it once grew; there are large specimens of the Caucasian wingnut at Kew, for example, that date from the late nineteenth century. The nearest native populations are in the Caucasus, with its closest relative, a similar species native to China.

4. The gomphotheres may have persisted until as recently as six thousand years ago in present-day Colombia; see Rodríguez-Flórez et al. (2009). For a complete list of now-extinct large herbivores of Central America, see Janzen and Martin (1982, 21).

5. For a popular discussion and elaboration of Janzen and Martin's idea, see Barlow (2002).

6. Known foragers of the seeds of living ginkgo include the catlike *Paguma larvata* in China and the Japanese badger *Nyctereutes procyonoides*. Rothwell and Holt (1997) note the improved germination rates of seeds scarified by passing through the digestive tract of badgers.

7. By the end of the Pliocene, ginkgo had disappeared from the fossil record everywhere except perhaps for a small area of southern Japan; see Uemura (1997).

8. Castiglioni had visited North America between 1785 and 1787, and also had consulted *Flora Virginica*, published by Gronovius in 1739 and 1743, and Thunberg's *Flora Japonica* published in 1784: see Spongberg (1993). Asa Gray at Harvard was a frequent correspondent and staunch supporter of Darwin in North America. Darwin's letter to him on "botanical geography" was written on October 12, 1856. New information that Gray had at his disposal included Siebold's *Flora Japonica* as well as specimens brought back from the Rodgers-Ringgold Expedition (1853–1856), also known as the

North Pacific Exploring and Surveying Expedition, a United States scientific and exploring project with the broader purpose of finding shorter trade routes for merchant ships in the Pacific; see Cole (1947). Also available were specimens brought back from Japan by Charles Wright.

9. Gray (1859, 422) listed about 580 Japanese species "which have particular relatives in other and distant parts of the northern temperate zone," along with the corresponding plants in the floras of Europe; central and northern Asia; western North America; and eastern North America. To explain the widely separated but highly similar floras of eastern Asia and eastern North America, Gray suggested that before the glacial epoch, the flora of the North Temperate Zone had been relatively homogeneous and that regional extinction during the Ice Ages resulted in greater losses from western North America and Europe. In some cases the impact of regional extinction was less pronounced. The sweet gum, for example, has widely separated remnants not only in eastern Asia and eastern North America, but also in southeastern Europe.

22. Endurance

1. Epigraph: Thoreau (1862, 517).

2. See Chapter 29 for additional information on initial Western encounters with ginkgo.

3. The Royal Horticultural Society estate where Robert Fortune worked as a gardener is now Chiswick House. The first Opium War lasted from 1832 to 1842. Fortune's four trips to China—in 1843–1846, 1848–1851, 1853–1856, and 1858–1859—are recounted in Fortune (1847, 1852, 1857, 1863). Fortune's trip to Japan from 1860 to 1862 was one of the first by a Western botanist following the Treaty of Amity and Commerce (the Harris Treaty), which was signed on July 29, 1858, and expanded trade to five major Japanese ports and allowed for diplomatic exchange.

4. Fortune (1847, 118). See Chapter 30 for more information on bonsai ginkgo.

5. Père Jean Marie Delavay (1834–1895); Père Paul Guillaume Farges (1844–1912); Père Jean Pierre Armand David (1826–1900). Père David arrived in Canton in 1862. His journeys into the interior were as a missionary priest, but he devoted much energy to studies of native plants and animals. His success was emulated by his contemporaries Père Delavay and Père Farges. Their collective contributions resulted in the introduction of hundreds of new plants into Europe, including the butterfly bush, many species of rhododendron, and the spectacular dove tree. They also sent thousands of botanical specimens back to the Muséum national d'Histoire naturelle in Paris. Their work often carried a significant price in personal hardships; Delavay contracted bubonic plague, which eventually cost him his life. David at various times suffered from typhus, smallpox, leprosy, rabies, cholera, plague, dysentery, and malaria. For further information on the contributions of the French missionary botanists see PlantExplorers.com (1999–2012). The dove tree is sometimes more descriptively called the handkerchief tree.

6. Augustine Henry writes about the classification, fossil history, varieties, distribution, uses, and history of ginkgo, also providing a list of important ginkgo trees in the British Isles. Of the seeds he wrote, "The nuts are sometimes eaten boiled or roasted, but are not much thought of," and plates

21–23 show photographs of old ginkgos; see Elwes and Henry (1906, 55–62). For more on the career and contributions of Augustine Henry (1857–1930), see Nelson (1983). Wilson received guidance from Henry in his efforts to re-collect the much sought-after dove tree, which had originally been described by and named for Père David.

7. Ernest Henry Wilson (1876–1930) was appointed at the Arnold Arboretum in 1927. Tragically, after surviving many adventures in China, including having his leg crushed by boulders in an avalanche, he was killed in an automobile accident in Worcester, Massachusetts, at the age of fifty-four.

8. Wilson began his career as an apprentice gardener at a local nursery near his home in Chipping Campden, but after working briefly at the Birmingham Botanic Gardens, and also at Kew, he accepted a position with the horticultural firm of James Veitch and Sons as their collector of Chinese plants. Wilson traveled to China via the United States, stopping in Boston to visit Charles Sargent at the Arnold Arboretum, and arrived in Hong Kong on June 3, 1899. On his first visit Wilson gathered thirty-five cases of living plants, which he sent home as bulbs, corms, rhizomes, and tubers as well as seed. There were also herbarium specimens of nearly a thousand plant species, many of which are in the Herbarium at Kew. Wilson returned to China as a collector for the Arnold Arboretum in 1907, 1908, and 1910. Between 1911 and 1915 he was in Japan, and in 1917 and 1918 in Korea and Formosa. He also traveled widely in the Southern Hemisphere, collecting plants in Australia, New Zealand, India, South America, and Africa. Wilson ended his career directing the Arnold Arboretum at Harvard University. His remarks on ginkgo are recorded in Wilson (1913, 45). Wilson's claims regarding the link between ginkgo and Buddhism (Wilson, 1920) were later challenged by Li (1956) among others; see Chapter 26.

9. For an overview and discussion on the significance of the ginkgos growing on Tianmu Mountain, see Del Tredici (1990, 1992b) and Del Tredici et al. (1992).

23. Relic

1. Nikolai Vavilov is the father of modern studies of crop diversity. He accumulated more than 200,000 collections of crop plant seed from around the world and identified the areas in which the domestication of many crops took place. These collections were protected during the twenty-eight-month Siege of Leningrad by Vavilov's colleagues, twelve of whom starved to death while surrounded by the edible seeds in the collection. Vavilov was arrested in 1940 after disputing the pseudoscience encouraged by Stalin's regime and died of malnutrition in prison in 1943. For additional information on centers of crop diversity and crop origins, see Vavilov (1992). For a biography of Vavilov, his career, and his persecution see Pringle (2008).

2. Most of our modern studies of DNA depend on being able to replicate large quantities of DNA from small samples using the polymerase chain reaction technique (PCR). Kary Mullis received the Nobel Prize for this discovery in 1993, and the process is well described in "The Polymerase Chain Reaction"; see Mullis et al. (1994).

3. Fan et al. (2004).

4. For the RAPD study see Fan et al. (2004).

5. The DNA for the second kind of study was obtained from the chloroplasts in the leaves. For more on this so-called restriction fragment length polymorphism (RFLP) study see Shen et al. (2005).

6. More precise characterization of the DNA fragments would certainly be possible using the sophisticated techniques of modern molecular biology but is beyond the resolution of these two approaches.

7. See Gong et al. (2008a, b) and Zhao et al. (2010).

8. Based on low genetic variability among forty ginkgo trees from the population on Tianmu Mountain, Wu et al. (1992) concluded that the trees had probably originated from those planted near the old temple. Subsequent work has revealed greater genetic variation supporting the possibility that both Jinfo and Tianmu Mountain may preserve relic ginkgo populations; see Gong et al. (2008 a, b) and Zhao et al. (2010). It is hard to exclude completely the possibility that some of the genetic variation in particular populations resulted from ancient people bringing together ginkgos from multiple sources, but at Jinfo and Tianmu Mountain this seems unlikely.

9. Ginkgo is not the only "living fossil" to come from China; in the mid-1940s, the dawn redwood was found in a small population of large trees along small streams and slopes of northeastern Sichuan Province, only four years after the genus had been described first from fossils. Like ginkgo, the dawn redwood once had a much more extensive former range, covering parts of North America and Asia, and was on the verge of extinction within its native habitat; see Merrill (1948). The dawn redwood is now a popular street tree, especially in China; see Chapter 35.

24. Antiquity

1. Epigraph: Emerson (1883, 478). The large tree at Jinfo Mountain, Naquan County in Chongqing Municipality, was partially destroyed by fire in the 1960s but survives through natural resprouting (Fan et al., 2004). For more on old ginkgo trees in China, see Li et al. (1999).

2. For a list of old, large ginkgos in different Chinese provinces, see Lin et al. (1995).

3. The small hamlet of Li Jiawan is a little more than sixty miles west of Guiyang, the capital of Guizhou Province; see Xiang et al. (2009). There has been much discussion about the possible fire resistance of ginkgo (e.g., Handa, 2000), which has sometimes been attributed to water retention in its leaves, or even to the fire-retardant properties of its sap. To view what happens when ginkgo and maple leaves are burned side by side, see Japan Probe, "Ginkgo Trees Protect Shrines and Temples from Fire." Best known of the six ginkgo trees that survived within a mile of where the atomic bomb exploded at Hiroshima is the ginkgo on the grounds of the Hosenji Temple. Only about half a mile from the epicenter of the destruction, it survived and leafed out again, after having been stripped by the blast. The new temple built around it celebrates the hope of renewal that the tree embodied amid terrible devastation. For more on the ginkgos that survived the Hiroshima bomb, see Hageneder (2005) and Kwant, "A-bombed ginkgo trees in Hiroshima, Japan."

4. The Li Jiawan Grand Ginkgo King has been described as "five-generations-in-one-tree," based on its successive episodes of resprouting; see Xiang et al. (2009).

5. The estimate for the age of the Li Jiawan Grand Ginkgo King is the maximum calculated by

Xiang et al. (2009). Ages of ginkgo trees that are believed to be more than two thousand years old are cited by He et al. (1997). Trunk diameters (DBH) are based on Lin et al. (1995).

6. Coast redwoods, the tallest of all trees, have trunks up to 370 feet tall and 24 feet in diameter. A slice across the trunk of a giant sequoia in the Natural History Museum in London, which is about 14 feet across, is estimated to be 1,335 years old; the beginnings of Islam, the spread of Buddhism into Japan, the fall of major civilizations, the Black Death of Europe, the birth of Shakespeare, and other historic landmarks are marked against their corresponding rings. That we felled so many of these spectacular trees, while knowing their great antiquity, must rank among the greatest acts of human hubris. The giant sequoia, known as the Mark Twain Tree, was felled in 1891 just to prove the existence of such massive trees to disbelievers in the East. The American Museum of Natural History in New York City and other museums hold slices of its trunk. The greatest age so far recorded from a bristlecone pine is from a specimen nicknamed Prometheus that was mistakenly felled by a graduate student researcher in eastern Nevada in 1964. Based on both radiocarbon dating and its tree rings, it had lived for at least 4,862 years and possibly more than 5,000; see Ferguson and Graybill (1983). A potentially older, but much less spectacular, tree is a Norway spruce in the Dalarna region of Sweden. Though the tree is only sixteen feet tall, its rooting system has been dated at around 9,550 years old; see Kullman (2005).

7. Humboldt alludes to the baobab in describing another impressive tree, the Dragon Tree of Tenerife in the Canary Islands: "Among organic creations, this tree is undoubtedly, together with the Adansonia or baobab of Senegal, one of the oldest inhabitants of our globe"; see Humboldt and Bonpland (1852, 62). The provocative age suggested by Adanson approached that calculated by Bishop Ussher for the age of the Earth. David Livingston was among those enraged; see Wickens and Lowe (2008). Rather than indicating especially great age, the massive trunks of baobabs reflect an ability to store water as a buffer against drought in the dry environments where they grow. In the Limpopo Province of South Africa, an enormous hollow baobab with a diameter of thirty-five feet at the ground and a thirteen-foot-tall hollow chamber in its base has been converted into a pub. It is younger than the 6,000 years claimed by the proprietor's website (www.bigbaobab.co.za): radiocarbon dating of samples from one of the internal cavities gave an age of 1,060 years plus or minus 75 (Adrian Patrut, "Babeș-Bolyai" University, personal communication). For the use of radiocarbon dating to determine the age of the ancient baobab from Namibia, see Patrut et al. (2007). Radiocarbon dating relies on the fact that the carbon atom in each molecule of carbon dioxide captured by photosynthesis may be of one of two kinds: Carbon-12 or Carbon-14, which occur at a known ratio in the atmosphere. Carbon-12 is stable, but Carbon-14 begins to decay to Carbon-12 at a known rate as soon as it is incorporated into the tree. Because the ratio of Carbon-12 to Carbon-14 in the atmosphere is known, along with the exact rate of decay, precise measurement of the amount of Carbon-12 and Carbon-14 in a piece of wood can be used to calculate its age. Eventually the amount of Carbon-14 in a sample becomes too small to measure accurately, and this sets a limit of about 60,000 years on how far back radiocarbon dating will provide an accurate age.

8. For beautiful photographs of Jōmon-sugi on Yakushima see Pakenham (2002). When Ernest

Henry Wilson recorded a stump of a Japanese cedar on Yakushima that was fourteen feet in diameter (Sargent, 1913), it caused a sensation.

9. Dendrochronology takes advantage of variation in the thickness of different annual rings in the same trunk to match and cross-correlate among woods from different sources. It serves the needs of archaeologists interested in dating wood fragments, as well as scientists interested in the history of climates over the past few thousand years. For a description of dendrochronology and its application to determining the age of the bristlecone pine, see Ferguson (1968).

10. Minamoto no Yoritomo established the Kamakura bakufu, the first government in Japan controlled by the samurai, twelve years later.

11. In the spring of 2010, to the distress of the temple and many Japanese people, the great Tsuru-gaoka Big Ginkgo blew down in a storm after heavy rains. While it was possible to count the rings from the outer part of the trunk, attempts to arrive at a definitive age were thwarted by decay at the center (Toshiyuki Nagata, Hosei University, personal communication 2011).

12. The Huiji Temple, also known as the Huiji Yuan Monastery and Xiangquan si, or Fragrant Spring Temple, is located just to the north of Tangquan town in the Pukou district of Nanjing. It was founded in the Liu Song Dynasty in the fifth century (420–479). Zhaoming, also known as Xiao Tong, reputedly read there and took baths in the hot spring. About 1821–1850, unusually, a Confucian academy called Yinghua, or Quintessence Flower, was established in the temple. The temple was destroyed during the Taiping Rebellion between 1850 and 1864.

13. See Needham et al. (1996, 581). For most of his life, Joseph Needham worked at Cambridge University, but he also spent much time in China. His early career included distinguished contributions as an embryologist and biochemist, for which he was elected to the Royal Society in 1941. He spent most of the 1940s in China, and on his return he devoted his life to the study of East Asian science and culture. So far, seven volumes of *Science and Civilization in China,* in twenty-four parts (fifteen written or cowritten by Needham), have appeared. A dawn redwood stands in memory of Joseph Needham, his wife Dorothy Needham, and his companion Gwei-Djen Lu-Needham outside the Needham Research Institute in Cambridge, where the work that he began is carried on. For additional information on Needham's life and career see Cullen (1995). For a popular account, see Winchester (2008). As Menzies notes, it may also be significant that the most important early sources of botanical information from China, the *Shih Ching, Ērh Ya,* and *Nan Fang Tshao Mu Chuang* (Records of the plants and trees of the southern regions), written by Chi Han around A.D. 300, make no mention of ginkgo.

25. Reprieve

1. Epigraph: Pandit and Nagarjuna (1977, 66). See Chapter 24 for more on the ginkgo at the Fu-Yen Ssu Monastery. Purported representations of ginkgo in early Chinese art from the fourth to eighth centuries (e.g., see Kwant, "Seven Worthies of the Bamboo Grove") are not sufficiently diagnostic to be fully reliable and require more detailed study. Tso Ssu (253–307) was a prominent poet of

the Jin Dynasty (265–420), though only fourteen of his poems survive; see Mair (2001) for more on T'ai-k'ang poetry.

2. *Simple Discourses on the Investigation of Things,* a series of short statements about natural phenomena that was written about 980, includes the statement "Let male and female ginkgo trees grow near one another, then fruit will form"; see Chapter 8 for more on this and on gender in ginkgo. The poems about ginkgo were exchanged between Ouyang Hsiu (1007–1072) and the poet Mei Yao-Chēn (also known as Sheng-Yü; 1002–1062); see Needham et al. (1996, 581).

3. See Needham et al. (1996) for a history of Chinese agriculture, which remains an important area of research.

4. The Yuan Dynasty spanned from 1271 to 1368.

5. A comparison of height and girth of old ginkgo trees in different Chinese provinces supports the idea that the old ginkgos of Guizhou are on average both taller and wider than those of Anhui and Zhejiang. Sichuan Province, bordering Guizhou to the north, has even larger ginkgos by both measures; data compiled from Lin et al. (1995). The Anhui origin of ginkgo is cited by Menzies from the *Pen Ts'a Kang Mu,* or the Great Pharmacopoeia (1596, p. 1801), quoting the *Jih Yung Pēn Tshao.* The quote from *Shihuazonggui* is from He et al. (1997, 374). Beginning in the Yuan Dynasty other names have also been used; *pei yen,* "white eye"; *pei kuo,* "white fruit"; *ling yen,* "spirited eye"; *jen hsing,* "nut apricot"; others such as *kung sun shu,* "grandfather-grandchild tree," appeared later. These records, from the Song Dynasty (960–1279) and Yuan Dynasty are reported in He et al. (1997, 374) and Li (1963, 92).

6. The oldest ginkgo trees in South Korea have been designated as natural monuments by the Cultural Heritage Administration; see Invitation ForestOn "Story of forest: Old gigantic trees in Korea."

7. The villagers of Yeongwol believe that a giant snake resides inside the tree, keeping animals and insects away, and that young children who fall from the tree will not be injured. For descriptions and images of the Yeongwol Ginkgo, Duseo-myeon Ginkgo, and Yongmunsa Ginkgo, see Invitation ForestOn, "Story of forest: Old gigantic trees in Korea." The Anbulsa Temple in North Korea is in Tonghung-ri, Kumya County, South Ham-Yong Province.

8. A list of old ginkgo trees in Japan and their associated legends is given by Hori and Hori (1997). According to legend, the Yongmunsa and Zenpukuji ginkgo trees in Korea and Tokyo, respectively, were both grown from sticks planted in the ground by priests wishing for the prosperity of their temples. In another legend, an official accompanying Emperor Kao Tsung as he moved from Kaifeng to Hangchow in the south of China in 1127 picked a branch of ginkgo and stuck it into the ground, declaring that if it lived, he would settle there; the branch grew into a huge tree adorned with many chi-chis. The Senbon Ginkgo ("One Thousand Ginkgo Trees") in Japan allegedly all sprouted from the trunk of a grand ginkgo that was struck by lightning; see Primack and Ohkubo (2008), Handa (2000), Li (1963), and Kwant, "An Old Chinese Legend." At Ubagami Shrine in Miyagi Prefecture, the large female tree is called Uba—"wet nurse"—Ginkgo. Emperor Shōmu ruled from 701 to 756. The Heian period began in 794 and ended in 1185.

9. See Chapter 24 for more on the Tsurugaoka Big Ginkgo. The Nara period in Japan is usually regarded as beginning in 710 and ending in 794.

10. For more on the cultural history of ginkgo in Japan, see Hori and Hori (1997).

11. These ancient trees in Japan are reported by Li (1963).

12. See Hori and Hori (1997).

13. The Kamakura period began in 1185 and ended in 1333; and the Muromachi period began in 1336 and ended in 1573.

26. Voyages

1. Epigraph: Fuller (1732, #1850). The Korean National Maritime Museum is the responsibility of the Ministry of Culture and Tourism; see www.seamuse.go.kr.

2. In 1994 the Marine Antiques Preservation Center became the National Maritime Museum of South Korea. The remains of the Shinan Ship, and much of its cargo, can be seen at the National Maritime Museum. Since the excavation of the Shinan Ship the museum has led the excavation of other wrecks discovered around the Korean coast.

3. For a description and diagrams of the ship's structure and construction, see Green (1983); for descriptions and photographs of the ship, its cargo, and the underwater excavation see Kim (2006a, b, and c).

4. The ginkgo nut recovered from the Shinan Ship is on display at the National Museum of Korea in Seoul, along with ceramics and other materials recovered from the wreck.

5. Among the objects from the Shinan Ship were more than twenty inkstones used in calligraphy and day-to-day writing; a bone die, a little smaller but otherwise identical to dice of today; and several examples of lacquerware. Glass objects included delicate hairpins, buttons, beads, and pieces of small glass bottles. There were also metal bottles, mirrors, incense burners, balance weights, cups, lamp oil containers, scoops, gongs, small cymbals, plates, wine cups, boxes, spoons, chopsticks, shovels, locks, wine dippers, candlesticks, bells, kitchenware, containers, coal basins, cauldrons, pots, instruments, soup bowls, astronomical instruments, Buddhist statues, figurines, acupuncture needles, and rings. The more than three hundred pieces of raw metal recovered, including tin and white copper sheet, would have been destined for manufacturing of all kinds and for making metal alloys. Stored as ballast in the hold of the boat were also about eight million Chinese coins, the heaviest component of the cargo. They had been held together by strong strings threaded through the holes in their center. Together these coins weighed nearly twenty-seven tons and included about seventy different kinds, all from China, that were in circulation from the first to the fourteenth century. Imported Chinese coins were widely used as currency in medieval Japan. Japanese coins were not minted at this time. Poor-quality coins may have been destined to be melted down and made into Buddhist statues and other luxury items.

6. See Seyock (2008). The temples in Fukuoka and Kyoto still have several ancient ginkgos nearby. The trade between China and Japan documented by the Shinan Ship continued through

Japan's period of self-imposed isolation. Through the seventeenth century previously scattered Chinese communities were consolidated at Nagasaki, which then became the main center through which foreign trade was conducted and controlled. Many aspects of Chinese culture were introduced and became assimilated into Japan through this bridgehead. For example, in 1654 Yin Yuan, a high-ranking Buddhist priest, arrived at the Kofukuji Temple in Nagasaki and introduced the teachings of Huan Bo, a particular sect of Zen Buddhism, which later was taken up by the Shogun and many powerful Japanese lords.

7. That ginkgo was present in Japan by the 1300s is also consistent with the discovery of ginkgo leaves pressed inside the pages of books from the fourteenth century at the Kanagawa Provincial Museum. For reasons to do with the history of these collections, the leaves are unlikely to have been placed there recently. The practice of pressing ginkgo leaves between the pages of books continues today. The leaves are thought to have insect repellent properties. See Chapter 25 for a brief review of early written records of ginkgo in Japan. For more details see Hori and Hori (1997).

8. For a list of other artifacts and animals that take their name from *ichou* in Japan see Hori and Hori (1997). The Azuchi-Momoyama period began in 1573 and ended in 1603.

27. Renewal

1. Epigraph: Proust (1919). Engelbert Kaempfer (1651–1716) has been called "the first Interpreter of Japan" (Brown, 1992).

2. For Kaempfer's full quote about the occurrence of ginkgo in Japan see Chapter 28.

3. See Evelyn (1664, 194); Campbell-Culver (2006) raises the possibility that the large trees in China described by John Evelyn may be ginkgo. Given that early contact with China was mainly in the subtropical south of the country, the possibility that they are figs seems more likely.

4. Early reports of Portuguese traders in the 1540s attracted the interest of Jesuit missionary entrepreneurs, who saw opportunities to advance both commercial and spiritual interests. In a letter to attract the interest of private shipowners and traders, the missionary Francis Xavier wrote: "Get someone of them, and that you may tempt his palate with a foretaste of the gains to be gathered in Japan—which happen now to be so serviceable to religion . . ."; see Newitt (2005, 135). A key early convert to Christianity in Japan was the feudal lord Omura Sumitada. With the help of the Portuguese, he established the trading port at Nagasaki in 1571 and by the time other European traders followed, Nagasaki was already well established. The European influence is still evident in Nagasaki's modern-day churches and specialty shops that sell the popular Portuguese sponge cake *castella*. The Japanese word *tempura* may have been derived from the Portuguese *tempero,* but alternatively it may be derived from the Portuguese words for temple—*templo* and *têmpora*—referring to the days when fish and vegetables are eaten instead of meat.

5. Clavell (1975).

6. The Dutch East India Company is often referred to as the voc (Vereenigde Oost-Indische Compagnie). The Shimabara Rebellion began not far to the east of Nagasaki when tens of thousands of peasants and their allies, many of them Christians, rose up against the local authorities. The rebel-

lion was finally put down at the fall of Hara Castle in April 1638, but in the aftermath the existing ban on Christianity was strictly enforced and the Portuguese were finally expelled. The Dutch, who had assisted the victorious Tokugawa Shogunate, were moved from Hirado and began trading from Deshima in 1641. As part of the self-imposed policy of national isolation (*sakoku*), the only other foreign ships allowed to come to Japan were from China.

7. See Chapter 6 for more on Siebold and his time in Japan.

8. Quotations from Kaempfer (1690–1692). Kaempfer, like many early European explorers in Japan, refers to the Shogun as emperor.

9. See Stearn (1948) for an account of Kaempfer's travels, and also the thorough Kaempfer forum; see Michel (2009). On his return to Europe, Kaempfer first gained formal medical qualifications from the University in Leiden. In 1694 he returned to Lemgo to start his medical practice, and in December 1698 Friedrich Adolf, Count of Lippe-Detmold, appointed him as his personal physician. Kaempfer married in 1700, when he was almost fifty, and two daughters and a son were born soon after. All three died in infancy. The full title of his book is *Amoenitatum exoticarum politico- physico- medicarum fasciculi V: quibus continentur variae relationes, observationes & descriptiones Rerum Persicarum & Ulterioris Asiae, multa attentione, in peregrinationibus per universum Orientem, collectae / ab auctore Engelberto Kaempfero*. It was published by a printer in his hometown of Lemgo in northern Germany.

10. Thunberg's botanical collections from Japan and elsewhere are preserved at the University of Uppsala. At the end of the seventeenth century, the Dutch East India Company went bankrupt and was dissolved. Trading continued with the oversight of the Dutch government. Only a few ships visited each year, but the Dutch at Deshima amused themselves by playing badminton and billiards and brewing beer. Their presence had a lasting impact on both Europe and Japan. In the eighteenth century, Deshima became less isolated and found itself at the mercy of political conflict on the other side of the world. During the Napoleonic Wars, when the Netherlands came under control of the French, the tiny colony became vulnerable to the British, who began to prey on Dutch shipping. In 1808 the British warship *Phaeton* entered Nagasaki harbor under a Dutch flag, but on learning that no Dutch ships would be arriving, it left before Japanese reinforcements arrived. A few years later, Deshima was completely cut off from European contact when the British occupied the Dutch base at Batavia in 1811. However, with the help of its Japanese partners and the steady leadership of the Dutch *Opperhoofd* Hedrik Doeff, Deshima survived and trading resumed in 1814.

11. The porcelain ginkgo dish is illustrated as Figure 184 in the Ohashi catalogue, RD30.6, H8.0, BD15.2; see Ōhashi (2006); also the drawing on p. 175.

28. Naming

1. Epigraph: Linnaeus (1751).

2. For more on the etymology of *ginkgo* see Hori and Hori (1997). The Song Dynasty began in 960 and ended in 1279. The Yuan Dynasty began about 1271 and ended in 1368.

3. Kaempfer owned two copies of the *Kinmo Zui*. One is a first edition from 1666, the other is a

slightly different later edition published in 1686. Both are now part of Kaempfer's Japanese Library in the British Library.

4. That the second *g* in *ginkgo* comes from Germany rather than from the East is an intriguing possibility, but the Kaempfer scholar Wolfgang Michel points out that Kaempfer transcribed with a *y* other Japanese words, including other plant names, that contain the Japanese syllable "kyo" or "kyō." Michel thinks it more likely that Kaempfer made a mistake: see his research notes (Michel, 2009). An alternative hypothesis suggests that Kaempfer accurately followed the pronunciation of his interpreter, Imamura Genemon Eisei (Van der Velde, 1995), and transcribed "ginkyo" as it would have sounded in the regional dialect spoken in Nagasaki at that time. The word for *strawberry,* for example, *ichigo,* is still pronounced "itzingo" in the Nagasaki dialect, which is the way that Kaempfer spelled it in his *Amoenitatum Exoticarum.* (Toshiyuki Nagata, Hosei University, personal communication).

5. For more on Kaempfer's collections of manuscripts, drawings, notes, maps, books, and other materials, including forty-nine woodcut books, see Hüls and Hoppe (1982).

6. Sloane was born at Killyleagh, County Down, on April 16, 1660. He arrived in London at the age of eighteen and took up the study of medicine a year later. In 1683 he traveled to Paris with the British physician-botanist Tancred Robinson and attended the botanical lectures of Tournefort. Later, he traveled to Montpellier and took his medical degree in Orange, in southern France, before returning to London in 1684. For more on Sloane's life, acquaintances, and collections, see Dandy (1958) and MacGregor (1994). The so-called Glorious Revolution, by which King James II was deposed by his daughter Mary and her husband William of Orange, was facilitated by many shared elements of British and Dutch culture. Dutch craftsmen, scholars, artists, and merchants had long integrated into Britain, and their influence is apparent in many Dutch-derived English terms, such as *yacht, sketch,* and *landscape.* For more on the Glorious Revolution and links between the Dutch and British at this time see Jardine (2008).

7. Sloane was in Jamaica from 1687 to 1689 (Sloane, 1696; 1707–1725) and recorded the uses there of chocolate, which was mixed with water into a drink. Sloane found it more palatable when mixed with milk, and his recipe was taken over by Cadbury Brothers as "Sir Hans Sloane's Milk Chocolate." The Royal Society, the U.K. national academy of science, had been founded in 1660, during the time of King Charles II. In its second Royal Charter, granted by King Charles II in 1663, the society is named the Royal Society of London for Improving Natural Knowledge. Among its founders were the architect Sir Christopher Wren, the chemist Robert Boyle, the physician and philosopher John Locke, the experimentalist, microscopist, and astronomer Robert Hooke, and the early anatomist and plant physiologist Nehemiah Grew, one of the first to use the microscope to observe structural details of animals and plants. Also among the founding fellows were the botanists John Evelyn and John Ray. Ray, in particular, was a major influence on Sloane. Sloane was president of the Royal Society from 1727 to 1741 and became one of the most influential patrons of science of the time.

8. Sloane's botanical collections also include thirteen thousand samples of seeds, fruits, gums, waxes, bark, and curiosities stored in small sealed boxes with a glass top and bottom to allow study of the contents (Sloane's collection of "Vegetables and Vegetable Substances"). Kaempfer's material

from Japan is in Volume 211 in the Sloane Herbarium, between a collection of plants from Italy made by the German botanist Schröter and a collection of plants from the Carolinas and other parts of eastern North America made by Mark Catesby, author of the *Natural History of Carolina, Florida and the Bahama Islands*. Specimens of ginkgo collected by Kaempfer are on two pages. On folio 91 are specimens from three kinds of plants, among them small sprigs of ginkgo long shoots and a single leaf. On folio 103 are nine different plants, one of them a single ginkgo leaf, nicely laid out, perhaps from a seedling. Beneath it, in ink, in Kaempfer's spindly hand in Latin, is written "Folium Itsjo arboris nucifera, folio adiantino." For more on the Sloane collections in the Natural History Museum in London, see Dandy (1958) and Trustees of the British Museum (1904).

9. For Salisbury's study of Sloane's collections see Salisbury (1817). A detailed catalogue of Kaempfer's plants in the Sloane Herbarium is provided by Hinz (2001).

10. John Gaspar Scheuchzer (1702–1729) received his doctor of medicine degree at Cambridge in 1728. His father, an acquaintance of Sloane's, was well known for his work in natural history. His uncle Johann Scheuchzer (1684–1738) was professor of mathematics and chairman of physics at Zürich, and was also knowledgeable about botany.

11. See Kaempfer (1690–1692, p. 181). By the time *Amoenitatum Exoticarum* was published in 1712, Kaempfer had changed his spelling to "*Gín an.*"

12. For more on William Adams see Chapter 27. The account of his life in Japan that he wrote in two letters between 1611 and 1617 was reprinted from the papers of the Hakluyt Society; see Adams (1896). Swift's *Gulliver's Travels* was published in 1726, just one year before *The History of Japan*. Sloane died on January 10, 1753, and his will led directly to the establishment of the British Museum by an act of Parliament on June 7, 1753. He is also remembered in the name of Sloane Square, which stands at the end of the King's Road in Chelsea, not far from the Chelsea Physic Garden, where Sloane studied, and which was established by the Worshipful Company of Apothecaries in 1673; see Minter (2000). As the size and diversity of the British Museum collections expanded, in the 1880s Sloane's natural history objects (including those of Kaempfer) were transferred to the newly created British Museum (Natural History), now the Natural History Museum, in South Kensington. Kaempfer's books eventually became part of the founding collections of the newly created British Library in 1972.

29. Resurgence

1. Epigraph: Shakespeare (1623b, act 3, scene 2).

2. See also Jarvis (2007) for more on Linnaeus and the significance of his work. Early in his life perhaps the closest Linnaeus came to encountering ginkgo was in 1736, when at the age of twenty-nine he visited Sloane in London. It was a brief social call; Sloane was almost fifty years his senior, and Linnaeus had no time to consult Sloane's vast collection in detail. In retrospect, Linnaeus's encounter with Sloane was a passing of the torch from the great plant collector of one generation to the great collector and classifier of the next.

3. Linnaeus named a genus of ginger *Kaempferia*.

4. Ellis (1770).

5. Ellis's correspondence with Linnaeus is cited in Loudon (1838, 1: 77), who inserts the botanical names in parentheses.

6. The quotation from Ellis to Linnaeus appears in Smith (1821, 207). *Siren lacertina,* the greater siren, is an eel-like amphibian named by Linnaeus in 1766; it occurs along the Atlantic and Gulf coastal plains of eastern North America, south to southern Florida.

7. The ginkgo at the Mile End Nursery was described as the oldest and "most handsome" in England; see Lyman (1885) and Loudon (1838). Dr. John Hope lived from 1725 to 1786. For discussion of Hope's observations in London see Harvey (1981).

8. See Kaigi (1930). Linnaeus's collections and texts are now scattered among various institutions, including the Institut de France in Paris, the British Library in London, the Botanical Museum in Uppsala, and the Swedish Museum of Natural History in Stockholm. The bulk of his collections and the most important were sold in 1784 to James E. Smith and are the founding collections of the Linnean Society of London; see Blunt and Stearn (2001) for further details. The specimen preserved from the Uppsala plant is filed as Specimen 1292.2 in the Linnaean Herbarium, which is preserved in the vaults of the Linnean Society of London beneath Burlington House on Piccadilly.

9. A possible source of Gordon's living material of ginkgo is the tree that still survives in the botanical garden at Utrecht, which was probably in cultivation by then. See Chapter 30 for more on the Utrecht tree.

10. The "Chinese boy" portrait is of Wang-y-Tong, page to the Duchess of Dorset. Sir Joshua Reynolds painted the portrait in about 1776. The painting is now in a private collection. Another version of the portrait is in the Reynolds Room at Knole House in Sevenoaks, Kent (Martin Postle, Paul Mellon Centre for Studies in British Art, personal communication).

11. Bradby Blake's intentions "to procure the seeds of all trees, shrubs, roots, fruits, flowers, &c. &c. which that great empire produces, and are used either for promoting commerce, or useful to mankind" are described in the letters of Henry Laurens (Laurens, 1980), who succeeded John Hancock as the fifth president of the Continental Congress. Laurens was imprisoned in the Tower of London during the Revolutionary War. He was the owner of Mepkin Plantation (now Mepkin Abbey), one of the largest plantations in South Carolina. Bradby Blake died in Canton on November 16, 1773. News of his death arrived in England in May 1774, and after his death, Blake's name was removed from candidacy by the Royal Society, where John Ellis had introduced it just a few years before. For details on the porcelain sent from China by Blake, see Corbeiller (1974). The papers of the British East India Company are housed at the British Library in the India Office Records.

12. Ellis may have had Bradby Blake in mind when in 1770 he wrote his *Directions for Bringing over Seeds and Plants from the East Indies and other Distant Countries in a State of Vegetation,* which was written for "Captains of Ships, Surgeons and other Curious Persons"; see Ellis (1770). This was written nearly sixty years before the 1829 invention of Wardian cases revolutionized the transport of live plant material. Ellis also urged plant collectors to carefully examine seeds before buying them to be sure they were fresh and not "decayed."

30. Gardens

1. Epigraph: Pollan (1991, 64). In Mahayana Buddhism, "Buddha nature" is the intrinsic potential for reaching enlightenment that exists in every sentient being. The Lotus Sutra, one of the most widely revered scriptures of the Mahayana Buddhism that permeates the Buddhist traditions of China, Korea, and Japan, declares "*bodhisattvas* each of these, I call the large trees." For more on "Buddhist ecology," see Swearer (2001).

2. For more on early Jesuit missionaries in China see Hsia (2009). Dimensions of the Geetbets ginkgo tree are from Kwant, The Ginkgo Pages. It is possible that this old female tree was overlooked because, with no male trees nearby, seed was not being produced and it would not have been obviously female. The first female ginkgo to be recognized in Europe failed to produce seed until cuttings were grafted onto male trees at other gardens; see Loudon (1838, 2096).

3. After the French Revolution, the Jardin du Roi became known as the Jardin des Plantes. Ehrhart (1787) also notes ginkgo growing at the Royal Gardens of Herrenhausen in Hannover, Germany.

4. This quotation is from Loudon (1838, 2099), who also reviews other early ginkgos planted in Europe.

5. For more on the introduction of Japanese plants into European gardens, see Farrer (2001) and Crane and Saltmarsh (2002).

6. Lord Bute was close to Princess Augusta, who founded the gardens at Kew in 1759. He was also a mentor and early prime minister (1762–1763) for the young King George III.

7. For discussion of the provenance of the Old Lions, see Fergusson (2006). See Chapter 29 for more on James Gordon.

8. For more on the early cultivation of ginkgo in Europe, see Loudon (1838). The Hortus Botanicus at Leiden was founded in 1590 to instruct medical students at the University of Leiden on the healing properties of plants.

9. Pétigny's ginkgo at the Jardin des Plantes in Paris was first grown in a pot and overwintered in the greenhouse. It was planted outdoors by André Thouin in 1792, but its growth was stunted. Another ginkgo obtained by layering from one of the four other Pétigny seedlings was planted subsequently. The story behind the *l'arbre aux quarante écus* is from Loudon (1838, 2096).

10. Pierre Marie Auguste Broussonet (1761–1807) was a French naturalist and honorary member of the Royal Society; these plant exchanges are cited by Loudon (1838, 2096) and Wilson (1920, 56).

11. Loudon (1838, 2096).

12. Details of the first ginkgo to produce seed in Europe are given by Loudon (1838) and Wilson (1919, 147). The female tree was apparently brought to the Bourdigny Estate in 1790 by an English plant collector named Blakie who had stayed there while collecting plants in the Alps in 1776. Afterward he often sent surplus plants from his stock to the proprietor of Bourdigny, M. Paul Gaussen de Chapeau-Rouge. The ginkgo was part of a collection of trees and plants that Blakie was raising at Monceau for the Duke of Orleans, and was probably grown from stock imported from England. How the ginkgo obtained by Blakie came to Britain is uncertain (Loudon, 1838, 2097).

13. Hamilton's admiration for English woodlands is quoted by Madsen (1989, 14).

14. Hamilton's prominence as an early North American collector of living plants is reflected in his extensive collections of exotic plants. Among other acquisitions, he had three hundred silver firs and five hundred Portuguese laurels shipped from Europe for Woodlands (Smith, 1905). When his greenhouses were catalogued in 1809, they contained nearly ten thousand plants of five thousand to six thousand species collected at great trouble and expense from all around the world (Oldschool, 1809, 507). For Hamilton's correspondence with his secretary, see Smith (1905, 144).

15. When last measured in 1981, Hamilton's ginkgo was nearly sixty-eight feet tall (Del Tredici, 1981, 155). It was felled in the mid-1980s, along with a nearby female of the same vintage, after the caretaker's dog grew sick from gorging on ginkgo seeds (Madsen, 1989, 23). For more on Collinson see Dillingham and Darlington (1851). John Bartram (1699–1777) is sometimes considered the father of American botany; Linnaeus called him the "greatest natural botanist in the world." Bartram's Garden, established around 1728, is the oldest surviving botanical garden in North America. David Hosack is better known as the doctor who tended Alexander Hamilton's fatal wound following his duel with Vice President Aaron Burr in 1804. The exact date of the ginkgo's planting at the Vanderbilt Mansion Estate is unknown, but is estimated as between 1799 and 1835, by either Hosack or Dr. Samuel Bard (Dave Hayes, Roosevelt-Vanderbilt National Historic Sites, personal communication). Hosack was also the founder of Elgin Botanical Garden, a garden mainly of native plants, which he established in New York City in 1770 on a site now occupied by Rockefeller Center.

16. The elder Michaux was tasked with searching out new species of trees through which France could rebuild its forests. Important for shipbuilding, forests had been decimated by overharvesting through nearly a century of warfare with England. Michaux discovered and described more than three hundred different kinds of plants and shipped more than ninety cases of American stock and seeds back to France. Michaux's reputation was tarnished by work he did delivering messages for Citizen Genêt, the new French minister to the United States. Genêt, to the outrage of President Washington, attempted to rally American citizens in taking arms against Spain; see Williams (2004). Cothran (1995) discusses Michaux's garden in Charleston. François Michaux published an account of his travels, *Michaux's Travels to the West of the Alleghany Mountains*, in 1805. The garden that his father founded in Charleston no longer exists, although several fine, but younger, ginkgo trees can still be found there, for example, outside the Charleston Historical Society and at the nearby Middleton Place Estate.

17. Hawke and Rush (1971) describe the parade celebrating the ratification of the Constitution. For more on the horticultural interests of the founding fathers of the United States, see Wulf (2011).

18. Henry Clay was one of the most influential nineteenth-century American politicians. He played a significant role in leading the nation to war in 1812, but was also dubbed the Great Compromiser for his work during the 1820s and 1830s in negotiating agreements during the nullification crisis and on the issue of slavery; see Heidler and Heidler (2010). Clay reputedly acquired ginkgos as seedlings sent from Japan to Washington, D.C.; see Claxton (1940). The ginkgos on his estate (Ashland in Lexington, Kentucky) were planted roughly at the time of the Civil War; see Ashland (2012).

19. See Falconer (1890) for the quotation from Smith. The article was published in *The Garden*,

a British publication, which explains why the cost is given in British pounds. The value of £1 in 1890 would be the equivalent of about $120 today. About fifty years earlier, Loudon (1838, 2100) gives the cost as follows: "Plants, in the London nurseries, are from 1s 6d to 5s, according to the size; female plants, 5s each. At Bollwyller, plants are 5 francs each; and at New York, 2 dollars." For more on the ginkgo trees at the Kentucky Military Institute see Falconer (1890) and Del Tredici (1981, 157).

20. Patricia Colleran, who worked in Giannini Hall on the Berkeley campus, won first prize in a campus photo and writing contest for her poem about the tree, which is available online: see Kell (2005).

21. The ginkgo haiku by Sem Sutter was published in the *University of Chicago Magazine,* April 30, 2007.

22. Santamour et al. (1983b) list eighty horticultural varieties of ginkgo, but Begović Bego (2011) cites at least 220. For a list of ginkgo cultivars based on nut size and shape see Santamour et al. (1983b, 91). A selection of ginkgo cultivars are planted in the Bamboo Garden at the Royal Botanic Gardens, Kew.

23. The leaves of Saratoga recall the frayed margins of several fossil species from the ginkgo lineage, including *Ginkgoites telemachus* and several species of *Sphenobaiera;* see Anderson and Anderson (2003). The Japanese practice of creating bonsai was first brought to the attention of the West by Engelbert Kaempfer, the first Western botanist to describe ginkgo.

31. Nuts

1. Epigraph: Shakespeare (1623a, act 3, scene 2). To a botanist, the so-called ginkgo nut is strictly a seed and is not to be confused with the true nuts (single-seeded, indehiscent fruits) of almond, hazel, and many other flowering plants.

2. The quotation from Ouyang Hsiu is cited in Li (1963).

3. The *Chun Chu Chi Wen,* from the Song Dynasty, is cited by Foster and Chongxi (1992). Del Tredici (1991) comments on the production nuts from a single large ginkgo. The estimate of dried nut production in China is from He et al. (1997).

4. Excavations of a Neolithic settlement in Jordan (ca. 6,760 B.C.) recovered a large basket of carbonized pistachio nuts that would have weighed around forty pounds when fresh (Henry and Servello, 1974); see Rosengarten (1984) for more on the history of nuts and people. Huntley and Birks (1983) describe the changing postglacial forest composition of Northern Europe. Abrams and Nowacki (2008) summarize the likely impact of Native Americans on the pre-Columbian vegetation of North America.

5. Botanically there is little difference between nuts and other convenient "naturally packaged" plant foods, such as grains. Nuts were hard to come by in the early 1600s, and the word came to signify a source of pleasure, as in the eighteenth-century phrase "to be nutts upon." Originally this meant to be "very fond of," but in mid-nineteenth-century American English it came to mean "crazy." See Online Etymology Dictionary, "nuts," www.etymonline.com/index.php?term=nuts. The edible portion of ginkgo nuts has a starch content of 68 percent, a very low fat content of 3 percent, and an energy

value of only 940 calories per pound. By comparison, nuts of pinyon pine have a fat content of 60 percent and an energy value of 2,800 calories per pound. For every 100 grams dry weight, ginkgo nuts contain approximately 403 calories, 10.2–10.5 percent protein, 3.1–3.5 percent fat, 83 percent carbohydrates, 1.3 g fiber, 3.1–3.8 g ash, 11 mg calcium, 327 mg phosphorus, 2.6 mg iron, 15 mg sodium, 1,139 mg potassium, 392 mg beta-carotene equivalent, 0.52 mg thiamine (B1), 0.26 mg riboflavin (B2), 6.1 mg niacin, and 54.5 mg ascorbic acid. For more on the nutritional value of nuts, see Duke (1989) and Kris-Etherton et al. (1999).

6. The original classification of ginkgo nuts (Tsen, 1935) recognized three varieties: the normal so-called variety *typica,* which includes the Meihe-Yinxing group or plum-stone-shaped ginkgo, with a round seed; the *huana,* or "finger-citron" ginkgo, with an elliptic or oblong seed; and the *apiculata,* the Maling-Yinxing or "horse's bell–shaped" ginkgo, with a small point at the top of the seed.

7. The young embryo developing in the seed is bitter to the taste and is sometimes removed by pushing it out with a toothpick.

8. Occasionally access to ripe ginkgo seeds becomes a source of conflict. In one Oakland neighborhood a novice herbalist planning to harvest seeds from the female tree on her property called the police when a neighbor beat her to the crop (Kemba Shakur, Urban ReLeaf, personal communication).

9. See Crosby (2008). Beware "ginkgo" seeds available on the Internet from Cameroon. These are imposters, most often the similar "meats" of gnetum, a very different but equally interesting plant, a climber from tropical forests. Gnetum seeds are often ground and fried into crackers called emping, which are enjoyed in traditional Indonesian and southeast Asian cuisine.

10. The medicinal uses of raw ginkgo nuts appear in the *Bencao Gangmu,* or *The Compendium of Materia Medica,* written by Li Shizhen (1518–1593) during the Ming Dynasty; see also Hori and Hori (1997), Foster and Chongxi (1992, 256), and Chapter 34. Yoshimura et al. (2006) report on the content of ginkgotoxin in raw and cooked seeds.

11. Many nuts and seeds used traditionally for oil production are finding specialty markets, but ginkgo is not, so far, among them. Watermelon seed oil, traditionally used in West Africa, is a popular emollient in cosmetic lotions. Oil from pumpkin seeds is a culinary delicacy in Austria, Slovenia, and Hungary. Argan oil, produced from a tree endemic to the desert valleys of southwestern Morocco, is an important constituent of high-end cosmetics. On a much larger scale, the African oil palm yields an estimated 117.5 million tons of palm oil and 2.1 million tons of palm kernel oil per year, from more than sixteen million acres of plantations in southeast Asia alone.

12. Hori and Hori (1997). Shogun Ashikaga Yoshiteru reigned from 1536 to 1565. The traditional pairing of ginkgo nuts with sake is described by Hageneder (2005).

13. The Edo period in Japan began in 1603 and ended in 1868. The source of the quotation is Morse (1917, 365).

14. For the putative ability of ginkgo to counteract the effects of alcohol in animals see Duke (1997).

15. For more on the ginkgo at the Tsurugaoka Hachiman-gū shrine in Kamakura, see Chapter 24.

32. Streets

1. Epigraph: Written by S. Kupka and E. Castillo and recorded by Tower of Power on its first album, *East Bay Grease,* 1970 — San Francisco Records/Atlantic Records. For more on the urban forests of the United States see McPherson (2003). For recent statistics on the decline of urban tree cover in the United States see Nowak and Greenfield (2012). For a short perspective on the importance of nature in cities see Crane and Kinzig (2005).

2. For more on Trees for Cities, see www.treesforcities.org. The late Nobel Peace Prize laureate Professor Wangari Maathai, who raised global awareness of the importance of trees through her work as founder of the Greenbelt Movement in Kenya, was the patron of UNEP's Plant for the Planet: The Billion Tree Campaign.

3. Wolf (2003) reports on the economic benefits of street trees in business districts.

4. For an explanation of how i-Tree Streets calculates benefits, see i-Tree Tools (2010).

5. Inventories of urban street trees were conducted by the USDA Forest Service Center for Urban Forest Research. The number of ginkgo trees in New York has increased since the 2007 inventory. The MillionTreesNYC initiative, a campaign to plant and care for one million trees in New York City over a decade, pledged in 2008 to plant 5,190 additional ginkgos in the ten-year period. Of the projected $82.40 in benefits gained from every ginkgo planted, the aggregated values are ranked as follows: aesthetic/other: $38.98; rainfall interception: $20.13; energy savings: $19.40; air pollutants neutralized: $3.39; total CO_2 sequestered: $0.50. In the i-Tree system, aesthetic value accounts for nearly half of the total estimated value of ginkgo as a street tree.

6. For more on the effects of trees on driving speeds see Naderi et al. (2008), and for effects on stress response in drivers, see Parsons et al. (1998); for more on the study of hospital patients, see Ulrich (1984); see also Kuo (2003), Kuo and Faber Taylor (2004), and Kuo and Sullivan (2001).

7. For more on the calculated benefits of urban trees, see USDA (2003). Gerhold (2007) describes the origins and history of urban forestry practice. In five years the Urban Resources Institute is working to plant about five thousand trees in New Haven, including an estimated 150 ginkgos. With an eye to long-term disease resistance in the urban forest as a whole the group plants no more than 5 percent of any one kind of tree. Similarly, in Oakland, California, Urban ReLeaf hires at-risk youths to plant trees in their own communities, putting them to work alongside scientists from the Center for Urban Forest Research to collect data on the many benefits that trees provide. For more on urban forestry as a tool for community development and green job creation, see Walsh (2011) and Pullen (2005).

8. There is also a long tradition of planting plane trees along the roads of France. Often attributed to Napoleon, the practice apparently goes back to Henri IV and has recently drawn the ire of some segments of French society who point to the role of trees in traffic fatalities (*Economist,* "The Killer Trees").

9. For more on the history of garden design, and the history of linking trees with public health in cities, see Gerhold (2007).

10. The quotation is from Thone (1929). Unfortunately, the celebrated ginkgo allée near the De-

partment of Agriculture has long since been uprooted; many turned out to be female as they matured. Only a handful of large ginkgo trees, perhaps of that vintage, remain in that part of the National Mall, for example, near the northwest corner of the Department of Agriculture and not far away near the edge of the Tidal Basin.

11. Chlorpropham is more commonly used to prevent sprouting in potatoes; see US EPA (2002).

12. See Brown (2006) for more on municipal efforts to protect heritage trees, including those on private properties. Information on the New York City urban forest is from Andrew Newman (Project Coordinator, MillionTreesNYC, personal communication).

13. Urban forest statistics for Manhattan are from Peper et al. (2007). Ginkgo is so much part of the streets of New York that Disneyworld uses ginkgo in their re-created New York City streetscape in the Hollywood Studios theme park.

14. For more information on ginkgo in Japanese landscapes, see Handa et al. (1997). The use of ginkgo as a street tree in South Korea is on a similarly massive scale. In central Seoul in particular, ginkgo is by far the most common street tree.

15. "Nasty, brutish and short," see Quammen (1998, 71), who borrows this famous phrase from Hobbes. Gangloff's quotation is from Brown (2006).

16. Studies in Britain show that ginkgo can thrive in hot summers and may benefit from warmer conditions, provided that sufficient water is available (Fieldhouse and Hitchmough, 2004). See Bockheim (1974) and Shober and Toor (2009) for more on the properties of urban soils.

17. Injured ginkgo trees are sometimes said to be slow to isolate (compartmentalize) injured parts of the tree, but the arborist who worked on the Everett ginkgo reported, "The tree had compartmentalized better than any tree I'd ever seen, and I've been in this business for 46 years"; see Mallow (2008).

33. Pharmacy

1. Epigraph: Aristotle, 350 B.C., in Ogle (1912).

2. This folktale is from the information sign at the Myeoncheon Ginkgo. In another folktale, *Lon Po Po,* the Chinese version of Little Red Riding Hood, three children outsmart a wolf that tries to eat them by tricking it into falling to its death from a ginkgo tree; see Cai (1994).

3. According to the doctrine of signatures, a plant "provides clues" to its medicinal efficacy through its resemblance to parts of the human body. It was long believed that the chi-chi of ginkgos were the result of nutritional excess and that nursing mothers would benefit from an infusion made from the shavings. Marks from such shavings are evident on some old trees, for example, at Jounichiji Temple in Japan; see Himi City, "The Jounichiji Ginkgo."

4. The Chinese term for the meat from a single ginkgo nut, *bai guo,* means literally "white fruit." *Ma huang* is the common name applied to several species of *Ephedra* that are native to the western United States. For a description of the many uses of ginkgo nuts in traditional Chinese medicine see Foster and Chongxi (1992, 257). Ginkgo nuts are crushed and applied topically to clear the complex-

ion. Flavonoid-rich extracts from the leaves have shown activity as free radical scavengers, which affect pigmentation. This supports the earliest documented use of the leaf extract in Chinese traditional medicine to treat freckles; today an extract is recommended as an alternative treatment for vitiligo; see Soni et al. (2010), Zhu and Gao (2008).

5. The first chemical investigation of ginkgo (Peschier, 1818) described butyric acid from the seeds. K. Nakanishi, who worked on early studies to characterize the structure of chemical compounds in ginkgo, once quipped that the ginkgolide molecule "is a diterpene with an aesthetically beautiful cage skeleton. . . . Because the Ginkgolides can be obtained in several large polymorphic crystalline forms, and because of their extreme stability, we finally thought they might be nice as pendants. However, they are not to be licked because they will be very bitter tasting" (Nakanishi, 2000). Ginkgolide B was first synthesized by Elias Corey of Harvard University, who received the Nobel Prize for chemistry in 1990 for his novel approach to making complex organic molecules by working backward from the desired product.

6. According to Loudon (1838, 2098), "Thunberg says that even the fleshy part of the fruit is eaten in Japan, though insipid or bitterish; and that, if slightly roasted, skin and all, it is not unpalatable. Some of the fruit which ripened in the botanic garden of Montpelier were tasted by M. Delille and M. M. Bonafous of Turin, who found their flavor very like that of newly roasted maize." Lyman (1885) notes, "The juice of the thick pulp outside the nut is very astringent and is used in making a somewhat waterproof, tough paper, and a preservative black wash for fences and buildings."

7. Naxi recipes provided by Selena Ahmed, Tufts University (personal communication, 2011). Ginkgotoxin (methoxypyridoxine) and its derivatives are made through the vitamin B6 pathway involving genes (*PDX1* and *PDX2*) that are necessary for seedling growth and survival in the model plant *Arabidopsis*. A single piece of raw ginkgo nut contains about eighty micrograms of ginkgotoxin. Ginkgotoxin suppresses the formation of GABA, essential for proper function of the brain and the central nervous system, inducing repetitive seizures, and also interferes with the body's uptake of vitamin B6. In Japan about 27 percent of the cases of ginkgo nut food poisoning result in deaths, probably because of the increased susceptibility of children. The negative effects can be counteracted by supplementing the diet with foods high in B vitamins; see Gengenbacher et al. (2006), Wada et al. (1985), and Wada (2000).

8. For estimates on global use of ginkgo leaf extract, see Pérez (2009). *Ben Cao Pin Hue Jing Yaor* was written by Liu Wen-Tai in 1505.

9. Flavonoids are part of a class of chemical compounds known as polyphenols, and their ability to scavenge free radicals is thought to give them helpful antibacterial, antiviral, antitumor, antiinflammatory, and antiallergenic activity (Robberecht and Caldwell, 1983).

10. Yoshitama (1997) provides a comprehensive overview of the flavonoids of ginkgo.

11. The extract is standardized to 22–27 percent flavonoids and 65–67 percent terpenes. Ginkgolic acids, the allergenic compounds in ginkgo nuts that are also present in the leaves, are kept below 5 parts per million. For more on the extraction process, see Juretzek (1997) and Sticher et al. (2000). For more on the history of commercialization of ginkgo extract, see van Beek (2000, 316).

12. Before the reform of the German health care system in 2004 most herbal medicines were deemed reimbursable by public health insurance. However, the reforms abolished reimbursement for all herbal medicines except for standardized preparations of ginkgo, Saint John's wort, and mistletoe; see Bonakdar (2010, 96). For more on the market for ginkgo leaf extract, see Jensen et al. (2002) and Diamond et al. (2000).

13. Global sales of ginkgo extract exceed $1 billion annually; see van Beek (2000). In 1997 annual sales broken down by country were $280 million in Germany, $200 million in the rest of Europe, $205 million in the United States, and $200 million in Asia (Pérez, 2009). Most of these products use a 50:1 ratio by weight: fifty pounds of leaf are reduced to one pound of the resulting extract. For use as a dietary supplement, the recommended daily dose, according to *The Physician's Desk Reference for Herbal Medicines,* is 120 mg of dry extract, in divided doses; see Diamond et al. (2000) and Chabrier and Roubert (1988).

14. The standardized purified extract from ginkgo leaves is commonly referred to as EGb 761. Side effects are few; in a study of 739 patients, 2.6 percent reported gastrointestinal upset, 0.9 percent experienced headaches, 0.4 percent reported sleep disturbance or dizziness, and 0.3 percent suffered skin eruptions. While there are some reports of antiplatelet activity leading to increased bleeding in those taking anticoagulants, a recent review concluded that there is scant evidence that ginkgo potentiates the effects of such drugs; see Bone (2008). Ginkgo treatments are from Letzel et al. (1996) and Diamond et al. (2000). For potential contraindications, see Medline Plus (2011).

15. For a careful assessment of the likely efficacy of ginkgo extract for enhancing cognitive functions, see Gold et al. (2003) and Gold et al. (2002). See also the Ph.D. thesis by York (2006) for a meta-analysis of more than a thousand clinical studies.

34. Risk

1. Epigraph: Darwin (1859, 392–393). Raup and Stanley's book *Principles of Paleontology* (Raup and Stanley, 1971) was a landmark in the development of paleontology and helped increase recognition of its relevance to evolutionary theory (Sepkoski and Ruse, 2009). The book has now been revised by two of Raup's students (Foote and Miller, 2007). The Chicago School of Paleontology is carried forward by my former colleagues in the Department of the Geophysical Sciences, Kevin Boyce, Michael Foote, Dave Jablonski, Sue Kidwell, Michael LaBarbera, and Mark Webster, as well as by many students of the Chicago program who are now scattered across the United States and the world.

2. Jack Sepkoski's compilation of data continued even as he was already analyzing it. Raup and Sepkoski (1984) analyzed data from 3,500 families. Raup and Sepkoski (1986) analyzed data on nearly 11,800 genera.

3. According to some commentators, we are currently in the middle of a sixth mass extinction, of similar magnitude to those of the past, which is driven by people and their increasing planetary footprint. This may be an apt description of the current biodiversity crisis, but the data we have today

on extinctions relate mainly to land-dwelling vertebrate animals, such as birds and mammals, rather than to marine invertebrates on which the idea of the "big five" extinctions is mainly based. Raup and Sepkoski (1982, 1983) analyzed the long-term effects of the different extinctions, and the nature of the subsequent recoveries. They also noted some regularity in the interval between twelve extinction events, which pointed toward periodic extinction peaks on a roughly twenty-six-million-year cycle (Raup and Sepkoski, 1984). Together with growing evidence in the early 1980s that the K-T extinction had been caused by an asteroid impact, this led to the idea that all twelve extinctions may have had an extraterrestrial basis — for example, periodic showers of comets caused by predictable astrophysical phenomena. Despite initial interest in this idea, the likelihood of the twelve extinctions having a similar cause is now discounted by most paleontologists.

4. See Chapter 23 for more on possible relict living populations of ginkgo in China.

5. Luis Alvarez and his son Walter were the first to suggest that an asteroid impact may account for the extinctions at the K-T boundary; see Alvarez et al. (1980).

6. Gould (1986) posed this question and later pursued a similar theme in book form (Gould, 1989).

7. The IUCN's Red List ranks species under different categories of threat that range from critically endangered through endangered to vulnerable. For additional details on the categories and the formal criteria on which they are based see www.iucnredlist.org.

8. According to the IUCN, Père David's deer (*Elaphurus davidianus*) has been extinct in the wild since 2008; the scimitar oryx (*Oryx dammah*) has been extinct in the wild since 2000. A global assessment of the state of plant diversity is more difficult to accomplish than for vertebrate animals because of the much larger number of species involved. However, an important initial unbiased evaluation was made through a Sampled Red List Index Assessment presented at the 2010 Conference of the Parties to the Convention on Biological Diversity held in Nagoya, Japan. The results show that the world's plants are just as threatened as the world's mammals and more threatened than the world's birds; one in five of the species sampled is threatened with extinction. For a summary of this joint study by the Royal Botanic Gardens, Kew and the Natural History Museum, under the auspices of the IUCN, see Kew (2010). For more on threatened conifers, see the IUCN Conifers Status Survey and Conservation Action Plan.

9. The World Wide Fund for Nature is known as the World Wildlife Fund in the United States; both are generally abbreviated to WWF. For more details on the Living Planet Index see WWF (2010).

35. Insurance

1. Epigraph: From Joni Mitchell's album *Ladies of the Canyon*, "Big Yellow Taxi," words and music by Joni Mitchell, © 1970 (Renewed) Crazy Crow Music. All Rights Administered by SONY/ATV Music Publishing. The Yale School of Forestry and Environmental Studies, the oldest graduate forestry program in the United States, was founded in 1901. From its inception, the school strongly influenced the practice of American forestry, with the first five chiefs of the U.S. Forest Service coming

from its graduates and faculty. Aldo Leopold graduated from the Yale School of Forestry and Environmental Studies in the Class of 1909.

2. In a similar way to ginkgo, the Monterey pine survives today in only three native stands, all of which are threatened by pitch canker infection. As a timber tree it is planted over at least seven million acres worldwide (Conifer Specialist Group, 1998).

3. Despite stringent protection, in 2005 some of the Wollemi pines were found to be infected with *Phytophthera cinnamomi,* a potentially devastating pathogenic fungus probably introduced by hikers who had found their way into the restricted area. For more information see Salleh (2005).

4. The Global Trees Campaign is a partnership between Fauna and Flora International, Botanic Gardens Conservation International, UNEP World Conservation Monitoring Centre, and others: see www.globaltrees.org.

5. For additional information on the drautabua and its conservation status see Farjon and Page (1999).

6. Currently only about one in one hundred seedlings of the Mulanji cypress survive, including those in cultivation. Clement Chilima of the Forestry Research Institute of Malawi described recent management and conservation interventions for this and other plant species from Malawi at the October 2009 conference Plant Conservation for the Next Decade: A Celebration of Kew's 250th Anniversary. The fourth species of *Widdringtonia,* the mountain cedar, is relatively widespread from southern Malawi down to the Cape.

7. See Bartram (1791). The estimate of when the ancestor of the Hawaiian cotton tree arrived in Hawai'i is based on Seelanan et al. (1997). For additional information on some of the plant species no longer found in the wild but part of the collections of the Royal Botanic Gardens, Kew, see Kew, n.d.

8. For a review of *ex situ* plant conservation including effective integration with *in situ* conservation, see Guerrant et al. (2004).

9. The premise is that there is only so much money to go around; funds spent on one kind of conservation will not be available to be spent on other conservation initiatives. However, my experience is that different funders have different priorities; funds available for one purpose are often not available for another. A case in point was the funding provided for the Millennium Seed Bank project by the British government, which resulted in the *ex situ* conservation of twenty-four thousand plant species in ten years. This was possibly the largest single effort ever undertaken to conserve plant diversity. It drew on new funds that were unavailable for *in situ* conservation.

10. Many modern botanic gardens are increasingly moving along a similar trajectory to that followed by zoos over the past fifty years from menageries to proactive conservation organizations.

11. I thank my Yale colleague Michael Donoghue for helpful discussion of extinction as a loss of knowledge.

36. Gift

1. Epigraph: Hyde (1983, 101).

2. The Convention to Combat Desertification was opened for signature in October 1994 and entered into force on December 26, 1996. Official information on the Convention on Biological Diversity can be found at www.cbd.int.

3. Rachel Carson's book *Silent Spring* (Carson, 1962) is credited with helping to launch the environmental movement. The plight of Chico Mendes and the Amazonian rubber tappers was brought to international attention in 1990 with a series of films called *A Decade of Destruction* by the British filmmaker Adrian Cowell, which gave greater visibility to the destruction caused by thousands of forest fires in the Amazon.

4. An assessment of plant diversity in the Atlantic rainforest is provided by Murray-Smith et al. (2009).

5. Biological diversity is treated as national patrimony of particular countries under the CBD, but it is self-evident that the distributions of species do not respect national borders, that the boundaries of countries only rarely have a basis in biological or ecological reality, and that sensible conservation strategies will require integration rather than a country-by-country approach.

6. CBD Article 15 recognizes the sovereign rights of states over their natural resources. On October 29, 2010, the Nagoya Protocol on Access and Benefit-sharing (ABS) was adopted by the Conference of the Parties to the CBD, following eleven meetings in six years by the Ad Hoc Open-ended Working Group on ABS. The Nagoya Protocol aims to provide a transparent legal framework for the fair and equitable sharing of benefits arising out of the utilization of genetic resources: see www.cbd .int/abs.

7. Official information on the International Treaty on Plant Genetic Resources for Food and Agriculture can be found at www.planttreaty.org. Conditions by which amendments to the Treaty may be made are elaborated in "Article 23—Amendments of the Treaty"; Article 23.3 specifies that amendments shall be made only by consensus.

8. Perhaps the kindest thing that can be said about the CBD is that it has helped keep biodiversity on the international policy agenda, and at least for plants, one positive outcome was the adoption of the Global Strategy for Plant Conservation (GSPC) by the Conference of the Parties at their meeting in The Hague in the spring of 2002. This had the helpful effect of focusing the world's botanists, especially those in the botanical garden community, on the common goal of conserving plant diversity. The global strategy emerged from an appeal by Peter Raven, then director of the Missouri Botanical Garden, at the International Botanical Congress held in Saint Louis in 1999, which was followed by a meeting in the Canary Islands that resulted in the Gran Canaria Declaration. Additional information on the global strategy can be found at www.cbd.int/gspc. The Global Strategy for Plant Conservation is now into its second ten-year iteration, the original 2010 deadline for the completion of the first round of GSPC goals having now passed.

9. Since 1996 the Forum on Religion and Ecology, organized by my Yale colleagues Mary Evelyn

Tucker and John Grimm, has highlighted the important roles that religions play in helping guide how people interact with their environment; for additional information see www.religionandecology.org.

10. For additional information on the philosophy of Thomas Berry see Berry (2009). For more on the theory of biophilia, see Kellert and Wilson (1993).

11. See National Academy of Sciences (2011, 1).

37. Legacy

1. Epigraph: Hyde (1983, 182).

2. Tony Kirkham updated the classic *The Pruning of Trees, Shrubs, and Conifers*, first published by George Ernest Brown in 1977 (Brown and Kirkham, 2004).

3. On his 1908 expedition for the Arnold Arboretum, Wilson photographed a large ginkgo at the village of Lengji in the valley of the Tung River. The caption to his photograph, preserved in the archives of the Arnold Arboretum, records the tree at an elevation of three thousand feet. Wilson estimated it as eighty feet tall, and his photograph clearly shows the small shrine placed in the branches near the base of the tree. Tony Kirkham and Mark Flanagan revisited the same ginkgo and photographed it in the summer of 2001; see Flanagan and Kirkham (2010).

4. The fall of the great ginkgo at the Tsurugaoka Hachiman-gū Shrine in Kamakura, Japan, was reported in the *Economist*; see "Japan's favourite tree: An Easter story from Japan" (2010). For more detail on the restoration of the Mizufuki Ginkgo see Handa (2000, 32). The Japanese concern for old ginkgo trees also comes through in the story of the Kubi Kake Ginkgo, the "Head-Stake Ginkgo," sometimes called the "Over My Dead Body Ginkgo," in Hibaya Park. Rather than allowing the tree to be felled for road expansion, Dr. Seiroku Honda, the designer of Hibaya Park, Tokyo, embarked on a project of more than three weeks to move the tree more than a quarter of a mile to safety. He is reputed to have said, "I will have it transplanted even if my head is put on a stake."

5. See Chapter 24 for more on the Huiji Temple.

6. John Montagu (1690–1749) married the daughter of the Duke of Marlborough and served under him at the Battle of Culloden. The elms that he planted were decimated by Dutch elm disease; many of the lindens ("limes"), which were a smaller component of the avenues, still survive.

Bibliography

Abrams, M. D., and G. J. Nowacki. 2008. "Native Americans as active and passive promoters of mast and fruit trees in the eastern USA." *The Holocene* 18: 1123–1137.

Adams, W. 1896. *The Original Letters of the English Pilot, William Adams, Written from Japan Between A.D. 1611 and 1617: Reprinted from the Papers of the Hakluyt Society.* Yokohama: Japan Gazette Office.

Akhmetiev M. A., G. M. Bratoeva, R. E. Giterman, L. V. Golubeva, and A. I. Moiseyeva. 1978. "Late Cenozoic stratigraphy and flora of Iceland." *Transactions (Doklady) of the U.S.S.R. Academy of Sciences.* 316: 188 pp. [In Russian; published in English by the National Research Council, Reykjavik, 1981.]

Allen, M. 1967. *The Hookers of Kew, 1785–1911.* London: M. Joseph.

Alvarez, L. W., W. Alvarez, F. Asaro, and H. V. Michel. 1980. "Extraterrestrial cause for the Cretaceous-Tertiary extinction." *Science* 208: 1095–1108.

Anderson J. M., and H. M. Anderson. 1983. *Palaeoflora of Southern Africa: Molteno Formation (Triassic).* Vol. 1. Part 1. *Introduction.* Part 2. *Dicroidium.* Rotterdam: Balkema.

———. 1985. *Palaeoflora of Southern Africa, Prodromus of South Africa Megafloras, Devonian to Lower Cretaceous.* Rotterdam: Balkema.

———. 1989. *Palaeoflora of Southern Africa, Molteno Formation (Triassic).* Vol. 2, *Gymnosperms (Excluding Dicroidium).* Rotterdam: Balkema.

———. 2003. *Heyday of the Gymnosperms: Systematics and Biodiversity of the Late Triassic Molteno Fructifications.* Strelitzia 15. Pretoria: South African National Botanical Institute.

———. 2008. *Molteno Ferns: Late Triassic Biodiversity in Southern Africa.* Strelitzia 21. Pretoria: South African National Botanical Institute.

Anderson J. M., H. M. Anderson, and C. J. Cleal. 2007. *A Brief History of the Gymnosperms: Classification, Biodiversity, Phytogeography, and Ecology.* Strelitzia 20. Pretoria: South African National Botanical Institute.

Andersson, A., J. Keskitalo, A. Sjödin, R. Bhalerao, F. Sterky, K. Wissel, K. Tandre, et al. 2004. "A transcriptional timetable of autumn senescence." *Genome Biology* 5: R24.

Andrews, H. N. 1980. *The Fossil Hunters: In Search of Ancient Plants.* Ithaca: Cornell University Press.

Angelou, M. 1990. "Address to Centenary College of Louisiana." *New York Times,* March 11, 1990.

APG (Angiosperm Phylogeny Group). 2009. "An update of the Angiosperm Phylogeny Group classification for the orders and families of flowering plants: APG III." *Botanical Journal of the Linnean Society* 161: 105–121.

Archangelsky, S. 1965. "Fossil Ginkgoales from the Tico flora, Santa Cruz Province, Argentina." *Bulletin of the British Museum (Natural History) Geology* 10: 121–137.

Archangelsky, S. A., and R. Cúneo. 1990. "*Polyspermophyllum,* a new Permian gymnosperm from Argentina, with consideration about the Dicranophyllales." *Review of Palaeobotany and Palynology* 63: 117–135.

Arnold, C. A. 1947. *An Introduction to Paleobotany.* New York: McGraw-Hill.

Arnott, H. J. 1959. "Anastomoses in the venation of *Ginkgo biloba.*" *American Journal of Botany* 46: 405–411.

Ash, S. R. 2010. "Late Triassic ginkgoaleans of North America." In *Plants in Mesozoic Time: Morphological Innovations, Phylogeny, Ecosystems.* Ed. C. T. Gee, 172–185. Bloomington: Indiana University Press.

Ashland. 2012. "Flora: Wooded pastures and wilderness remnants." Ashland: The Henry Clay Estate. www.henryclay.org/ashland-estate/the-landscape/flora. Modified 2012.

Aulenback, K. R. 2009. *Identification Guide to the Fossil Plants of the Horseshoe Canyon Formation of Drumheller, Alberta.* Calgary: University of Calgary Press.

Ballowe, J., and M. Klonowski. 2003. *A Great Outdoor Museum: The Story of the Morton Arboretum.* Lisle, Ill.: Morton Arboretum.

Barghoorn, E. S., Jr. 1940. "Origin and development of the uniseriate ray in the Coniferae." *Bulletin of the Torrey Botanical Club* 67: 303–328.

Barlow, C. 2002. *The Ghosts of Evolution: Nonsensical Fruit, Missing Partners, and Other Ecological Anachronisms.* New York: Basic.

Barlow, P. W., and E. U. Kurczyńska. 2007. "The anatomy of the chi-chi of *Ginkgo biloba* suggests a mode of elongation growth that is an alternative to growth driven by an apical meristem." *Journal of Plant Research* 120: 269–280.

Bartram, W. S. 1791. *Travels through North and South Carolina, Georgia, East and West Florida, the Cherokee Country, the Extensive Territories of the Muscogulges, or Creek Confederacy, and the Country of the Chactaws.* Philadelphia: James and Johnson.

Bean, W. J. 1973. *Trees and Shrubs Hardy in the British Isles.* 8th ed., vol. 2. London: John Murray.

Becker, H. F. 1961. "Oligocene plants from the upper Ruby River Basin, Southwestern Montana." *Geological Society of America, Memoir* 82: 1–127.

Beerling, D. J., and D. L. Royer. 2002. "Reading a CO2 signal from fossil stomata." *New Phytologist* 153: 387–397.

Begović Bego, B. M. 2011. *Ginkgo biloba L. 1771.* Vol. 2, *Cultivars and Bonsai Forms.* Croatia: Branko M. Begović Bego.

Berlyn, G. P. 1962. "Developmental patterns in pine polyembryony." *American Journal of Botany* 49: 327–333.

Berry, T. 2009. *The Christian Future and the Fate of Earth.* Ed. M. E. Tucker and J. Grim. Maryknoll: Orbis.

Birnbaum D., and M. Bracewell. 2005. *Gilbert and George: Ginkgo Pictures, Venice Biennale.* London: British Council.

Black, M. 1929. "Drifted plant-beds of the Upper Estuarine Series of Yorkshire." *Quarterly Journal of the Geological Society* 85: 389–439.

Blankenship, R. E., D. M. Tiede, J. Barber, G. W. Brudvig, G. Fleming, M. Ghirardi, M. R. Gunner, et al. 2011. "Comparing photosynthetic and photovoltaic efficiencies and recognizing the potential for improvement." *Science* 332: 805.

Blomfield, D. 1994. *Kew Past.* Chichester: Phillimore.

———. 2000. *The Story of Kew: The Gardens, the Village, the Public Record Office.* Kent: Leyborne.

Blunt, W., and W. T. Stearn. 2001. *Linneaus: The Compleat Naturalist.* Originally published in 1971 as *The Compleat Naturalist: A Life of Linnaeus.* London: Frances Lincoln.

Bockheim, J. G. 1974. *Nature and Properties of Highly Disturbed Urban Soils.* Philadelphia: Soil Science Society of America.

Bonakdar, R. A., ed. 2010. *The H.E.R.B.A.L. Guide: Dietary Supplement Resources for the Clinician.* Philadelphia: Lippincott Williams and Wilkins.

Bone, M. 2008. "Potential interaction of *Ginkgo biloba* leaf with antiplatelet or anticoagulant drugs: What is the evidence?" *Molecular Nutrition and Food Research* 52: 764–771.

Boulter M. C., and Z. Kvaček. 1989. "The Palaeocene flora of the Isle of Mull." *Special Papers in Palaeontology* 42: 1–149.

Bowerbank, J. S. 1840. *A History of the Fossil Fruits and Seeds of the London Clay.* London: John Van Voorst.

Briant, K. 1962. *Marie Stopes: A Biography.* London: Hogarth.

Brown, G. E., and T. Kirkham. 2004. *The Pruning of Trees, Shrubs, and Conifers.* London: Timber.

Brown, P. L. 2006. "New laws crack down on urban Paul Bunyans," *New York Times,* January 30, 2006.

Brown, Y. Y. 1992. *Engelbert Kämpfer: First Interpreter of Japan.* London: British Library Board.

Buchanan-Wollaston, V. 1997. "The molecular biology of leaf senescence." *Journal of Experimental Botany* 48: 181–199.

Buchholz, J. T. 1920. "Embryo development and polyembryony in relation to the phylogeny of conifers." *American Journal of Botany* 7: 125–145.

Burress, C. 2008, "Cal wins big in battle over athletic center," *San Francisco Chronicle,* July 23, 2008.

Cai, M. 1994. "Images of Chinese and Chinese Americans mirrored in picture books." *Children's Literature in Education* 25: 169–191.

Camerarius, R. J. 1694. *Ueber das Geschlecht der Pflanzen (De sexu plantarum epistola)*. Leipzig: Wilhelm Engelmann.

Campbell-Culver, M. 2006. *A Passion for Trees: The Legacy of John Evelyn*. London: Transworld.

Carlyle, T. 1858. *Chartism: Past and Present*. London: Chapman and Hall.

Carson, R. 1962. *Silent Spring*. Cambridge: Houghton Mifflin.

Chabrier P. E., and P. Roubert. 1988. "Effect of *Ginkgo biloba* extract on the blood-brain barrier." In *Rökan, Ginkgo biloba: Recent Results in Pharmacology and Clinic*. Ed. E. W. Fünfgeld, 17–25. Berlin: Springer.

Chaloner, W. G. 1985. "Thomas Maxwell Harris. 8 January 1903–1 May 1983." *Biographical Memoirs of Fellows of the Royal Society* 31: 229–260.

———. 2005. "The palaeobotanical work of Marie Stopes." *Geological Society, London, Special Publications* 241: 127–135.

Chandler, M. E. J. 1961. *The Lower Tertiary Floras of Southern England*. Vol. 1, *Palaeocene Floras. London Clay Flora (Supplement). Text and Atlas*. London: British Museum (Natural History).

———. 1962. *The Lower Tertiary Floras of Southern England*. Vol. 2, *Flora of the Pipe-Clay Series of Dorset (Lower Bagshot)*. London: British Museum (Natural History).

———. 1963. *The Lower Tertiary Floras of Southern England*. Vol. 3, *Flora of the Bournemouth Beds, the Boscombe, and the Highcliff Sands*. London: British Museum (Natural History).

———. 1964. *The Lower Tertiary Floras of Southern England*. Vol. 4, *A Summary and Survey of Findings in the Light of Recent Botanical Observations*. London: British Museum (Natural History).

Chaney, R. W., and D. I. Axelrod. 1959. *Miocene Floras of the Columbia Plateau*. Washington D.C.: Carnegie Institution of Washington Publications.

Charnov, E. L., J. J. Bull, and J. M. Smith. 1976. "Why be an hermaphrodite?" *Nature* 263: 125–126.

Chase, M. W., D. E. Soltis, R. G. Olmstead, D. Morgan, D. H. Les, B. D. Mishler, M. R. Duvall, et al. 1993. "Phylogenetics of seed plants: An analysis of nucleotide sequences from the plastid gene *rbc*L." *Annals of the Missouri Botanical Garden* 80: 528–580.

Chen, R. Y., W. Q. Song, and X. L. Li. 1987. "Study on the sex chromosomes of *Ginkgo biloba*." *Proceedings of Sino-Japanese Symposium on Plant Chromosome Research*: 381–386.

Chick, E. 1903. "The seedling of *Torreya myristica*." *New Phytologist* 2: 83–91.

Christianson, M. L., and J. A. Jernstedt. 2009. "Reproductive short-shoots of *Ginkgo biloba*: A quantitative analysis of the disposition of axillary structures" *American Journal of Botany* 96: 1957–1966.

Clavell, J. 1975. *Shōgun: A Novel of Japan*. Vol. 1. New York: Atheneum.

Claxton, T. B. 1940. "*Ginkgo biloba* in Kentucky." *Trees* 3: 8.

Cole, A. B. 1947. "The Ringgold-Rodgers-Brooke expedition to Japan and the North Pacific, 1853–1856." *Pacific Historical Review* 16: 152–162.

Collinson, M. E. 1983. *Fossil Plants of the London Clay*. London: Palaeontological Association.

Collinson, M. E., S. R. Manchester, and V. Wilde. 2012. "Fossil Fruits and Seeds of the Middle Eocene Messel Biota, Germany." *Abhandlungen der Senckenberg Gesellschaft für Naturforschung* 570: 1–251.

Conan Doyle, A. 1912. *The Lost World.* New York: Hodder and Stoughton.

Conifer Specialist Group. 1998. "*Pinus radiata.*" In IUCN *Red List of Threatened Species.* IUCN, 2011.

Cook, M. T. 1902. "Polyembryony in *Ginkgo.*" *Botanical Gazette* 34: 64–65.

———. 1903. "Polyembryony in *Ginkgo.*" *Botanical Gazette* 36: 142.

Coolidge, C. 1919. *Have Faith in Massachusetts: A Collection of Speeches and Messages.* Boston: Houghton Mifflin.

Corbeiller, C. L. 1974. *China Trade Porcelain.* New York: Metropolitan Museum of Art.

Cothran, J. R. 1995. *Gardens of Historic Charleston.* Columbia: University of South Carolina Press.

Crane, P. R. 1981. "Betulaceous leaves and fruits from the British Upper Palaeocene." *Botanical Journal of the Linnean Society* 83: 103–136.

———. 1985. "Phylogenetic analysis of seed plants and the origin of angiosperms." *Annals of the Missouri Botanical Garden.* 72: 716–793.

———. 2006. "Sex and the single ginkgo." *Kew Magazine* 53: 15.

Crane, P. R., and W. N. Carvell. 2007. "The importance of history." *Curtis's Botanical Magazine* 24: 134–154.

Crane, P. R., and A. Kinzig. 2005. "Nature in the metropolis." *Science* 308: 1225.

Crane, P. R., S. R. Manchester, and D. L. Dilcher. 1990. "A preliminary survey of fossil leaves and well-preserved reproductive structures from the Sentinel Butte Formation (Paleocene) near Almont, North Dakota." *Fieldiana, Geology,* n.s. 20: 1–63.

Crane, P. R., and A. Saltmarsh. 2003. "Lasting connections: Native plants of Japan and the gardens of Europe." *Japan Society* 139: 5–21.

Critchfield, W. B. 1970. "Shoot growth and heterophylly in *Ginkgo biloba.*" *Botanical Gazette* 131: 150–162.

Crosby, S. 2008. "Gathering ginkgo nuts in New York." *Gourmet,* November 3, 2008. http://www.gourmet.com/food/2008/11/gingko-nuts.

Cullen, C. 1995. "Obituary: Joseph Needham (1900–1995)." *Nature* 374: 597.

Dahl, E. 1990. "Probable effects of climatic change due to the greenhouse effect on plant productivity and survival in North Europe." In *Effects of Climate Change on Terrestrial Ecosystems.* Ed. J. I. Holten, NINA Notat 4: 7–17. Trondheim: Norwegian Institute for Nature Research.

Dandy, J. E. 1958. *The Sloane Herbarium: An Annotated List of the Horti Sicci Composing It; with Biographical Accounts of the Principal Contributors.* London: British Museum.

Darwin, C. 1859. *On the Origin of the Species by Means of Natural Selection, or the Preservation of Favoured Races in the Struggle for Life.* London: John Murray.

———. 1876. *The Effects of Cross and Self Fertilization in the Vegetable Kingdom.* London: John Murray.

———. 1877. *The Different Forms of Flowers on Plants of the Same Species.* London: John Murray.

Darwin, F., and A. C. Seward, eds. 1903. *More Letters of Charles Darwin: A Record of His Work in a Series of Hitherto Unpublished Letters.* London: John Murray.

Dawkins, R. 1976. *The Selfish Gene.* Oxford: Oxford University Press.

Del Fueyo, G. M., and S. Archangelsky. 2001. "New studies on *Karkenia incurva* Archang. from the Early Cretaceous of Argentina. Evolution of the seed cone in Ginkgoales." *Palaeontographica Abt. B* 256: 111–121.

Del Tredici, P. 1981. "The ginkgo in America." *Arnoldia* 41: 150–161.

———. 1989. "Ginkgos and multituberculates: Evolutionary interactions in the Tertiary." *Biosystems* 22: 327–339.

———. 1990. "The trees of Tian Mu Shan: A photo essay." *Arnoldia* 50: 16–23.

———. 1991. "Ginkgos and people: A thousand years of interaction." *Arnoldia* 51: 2–15.

———. 1992a. "Natural regeneration of *Ginkgo biloba* from downward growing cotyledonary buds (basal chi-chi)." *American Journal of Botany* 79 (5): 522–530.

———. 1992b. "Where the wild ginkgos grow." *Arnoldia* 52: 2–11.

———. 2000. "The evolution, ecology, and cultivation of *Ginkgo biloba*." In *Ginkgo* biloba. Ed. van Beek, 7–23. Amsterdam: Harwood Academic Publishers.

———. 2007. "The phenology of sexual reproduction in *Ginkgo biloba*: Ecological and evolutionary implications." *Botanical Review* 73: 267–278.

Del Tredici, P., H. Ling, and Y. Guang. 1992. "The *ginkgos* of Tian Mu Shan." *Conservation Biology* 6 (2): 202–210.

Denk, T., F. Grímsson, R. Zetter, and L. A. Símonarson. 2011. *Late Cainozoic Floras of Iceland: 15 Million Years of Vegetation and Climate History in the Northern North Atlantic.* Dordrecht: Springer.

Denk T., and D. Velitzelos. 2002. "First evidence of epidermal structures of *Ginkgo* from the Mediterranean Tertiary." *Review of Palaeobotany and Palynology* 120: 1–15.

Dennell, R. 2010. "Palaeoanthropology: Early *Homo sapiens* in China." *Nature* 468: 512–513.

Desmond, R. 1999. *Sir Joseph Dalton Hooker: Traveller and Plant Collector.* Woodbridge: Antique Collector's Club and Royal Botanic Gardens, Kew.

———. 2007. *The History of the Royal Botanic Gardens Kew.* 2nd ed. London: Royal Botanic Gardens, Kew.

Diamond, B. J., S. C. Shiflett, N. Feiwel, R. J. Matheis, O. Noskin, J. A. Richards, and N. E. Schoenberger. 2000. "*Ginkgo biloba* extract: Mechanisms and clinical indications." *Archives of Physical Medicine and Rehabilitation* 81: 668–678.

Dillingham, W. H., and W. Darlington. 1851. *A Tribute to the Memory of Peter Collinson, with Some Notice of Dr. Darlington's Memorial of John Bartram and Humphry Marshall.* Philadelphia: William H. Mitchell.

Doyle, J. A., and M. J. Donoghue. 1986. "Seed plant phylogeny and the origin of angiosperms: An experimental cladistic approach." *Botanical Review* 52: 321–431.

Drayton, R. H. 2000. *Nature's Government: Science, Imperial Britain, and the "Improvement" of the World.* New Haven: Yale University Press.

Drinnan, A. N., and T. C. Chambers. 1986. "Flora of the Lower Cretaceous Koonwarra fossil bed (Korumburra Group), South Gippsland, Victoria." *Memoirs of the Association of Australian Palaeontologists* 3: 1–77.

Drinnan, A. N., and P. R. Crane. 1989. "Cretaceous paleobotany and its bearing on the biogeography of austral angiosperms." In *Antarctic Paleobiology and Its Role in the Reconstruction of Gondwana*. Ed. T. N. Taylor and E. L. Taylor, 192–219. New York: Springer.

Duke, J. A. 1989. CRC *Handbook of Nuts*. Boca Raton: CRC Press.

———. 1997. *The Green Pharmacy: New Discoveries in Herbal Remedies for Common Diseases and Conditions from the World's Foremost Authority on Healing Herbs*. Emmaus, Pa.: Rodale.

Duke of Argyll and E. Forbes. 1851. "On Tertiary leaf-beds in the Isle of Mull, with a note on the vegetable remains from Ardtun Head." *Quarterly Journal of the Geological Society of London* 7: 89–103.

Dute, R. R. 1994. "Pit membrane structure and development in *Ginkgo biloba*." IAWA *Journal* 15: 75–90.

Dwyer, J. F., D. J. Nowak, M. H. Noble, and S. M. Sisinni. 2000. *Connecting People with Ecosystems in the 21st century: An Assessment of Our Nation's Urban Forests*. General Technical Report—Pacific Northwest Research Station: USDA Forest Service.

Eames, A. J. 1955. "The seed and *Ginkgo*." *Journal of the Arnold Arboretum* 36: 165–170.

Easwaran, E. 2007. *The Dhammapada*. Berkeley, Calif.: Blue Mountain Center of Meditation.

Ehrhart, F. 1787. *Verzeichnifs der Glas- und Treibhauspflanzen, welche sich auf dem Königl. Berggarten zu Herren-haufen bei Hannover befinden*. Hannover: Pockwitz.

Eiseley, L. C. 1958. *Darwin's Century: Evolution and the Men Who Discovered It*. Garden City, N.Y.: Doubleday.

Eldredge, N., and S. M. Stanley, eds. 1984. *Living Fossils*. New York: Springer.

Ellis, J. 1770. *Directions for bringing over seeds and plants from the East-Indies and other distant countries in a state of vegetation: Together with a catalogue of such foreign plants as are worthy of being encouraged in our American colonies, for the purposes of medicine, agriculture and commerce. To which is added the figure and botanical description of a new sensitive plant, called Dionaea muscipula: or, Venus's Fly-Trap*. London: L. Davis.

Elwes, H. J., and A. Henry. 1906. *The Trees of Great Britain and Ireland*. Vol 1. Edinburgh: privately printed.

Emerson, R. W. 1883. *Works*. Boston: Houghton Mifflin.

Endersby, J. 2008. *Imperial Nature: Joseph Hooker and the Practices of Victorian Science*. Chicago: University of Chicago Press.

Endo, S. 1942. "On the fossil flora from the Shulan coal-field, Jilin Province, and the Fushun coal-field, Fengtien Province." *Bulletin of the Central National Museum of Manchou Kuo* 3: 33–47. In Japanese.

Etymology Dictionary, "Nuts." Etymology Dictionary Online. www.etymonline.com/index.php ?term=nuts. Accessed September 5, 2011.

Evelyn, J. 1664. *Silva: or, a Discourse of Forest-Trees, and the Propagation of Timber in His Majesty's Dominions*. London: John Martyn.

Falconer, W. 1890. "Notes from Glen Cove, U.S.A. The ginkgo tree (*G. biloba*)." *The Garden: An Illustrated Weekly Journal of Horticulture in All Its Branches* 38: 602.

Fan, X. X., L. Shen, X. Zhang, X. Y. Chen, and C. X. Fu. 2004. "Assessing genetic diversity of *Ginkgo biloba* L. (Ginkgoaceae) populations from China by RAPD markers." *Biochemical Genetics* 42: 269–278.

Farjon, A., and C. N. Page. 1999. *Conifers: Status Survey and Conservation Action Plan.* Gland, Switzerland: IUCN-SSC Conifer Specialist Group.

Farrer, A., ed. 2001. *A Garden Bequest: Plants from Japan.* London: Japan Society.

Favre-Ducharte, M. 1958. "*Ginkgo:* An oviparous plant." *Phytomorphology* 8: 377–390.

Ferguson, C. W. 1968. "Bristlecone pine: Science and esthetics." *Science* 159: 839–846.

Ferguson, C. W., and D. A. Graybill. 1983. "Dendrochronology of bristlecone pine: A progress report." *Radiocarbon* 25: 287–288.

Ferguson, D. K. 1967. "On the phytogeography of Coniferales in the European Cenozoic." *Palaeogeography, Palaeoclimatology, Palaeoecology* 3: 73–110.

Ferguson, D. K., and E. Knobloch. 1998. "A fresh look at the rich assemblage from the Pliocene sinkhole of Willershausen, Germany." *Review of Palaeobotany and Palynology* 101: 271–286.

Fergusson, K. 2006. "Treasures of Kew." *Kew Magazine* 53: 53.

Fieldhouse, K., and J. Hitchmough. 2004. *Plant User Handbook: A Guide to Effective Specifying.* Oxford: Blackwell.

Fischer, W., Jr. 2010. "Ginkgo tree: Truman historic walking tour stop 9." Historical Marker Database. www.hmdb.org/marker.asp?marker=34740. Accessed July 31, 2010.

Flanagan, M., and T. Kirkham. 2010. *Wilson's China: A Century On.* London: Royal Botanic Gardens, Kew.

Florin, R. 1936a. "Die Fossilien Ginkgophyten von Franz-Joseph-Land nebst Erörterungen über vermeintliche Cordaitales Mesozoischen Alters, II. Allgemeiner Teil." *Palaeontographica, Abt. B* 81: 71–173.

———. 1936b. "Die Fossilien Ginkgophyten von Franz-Joseph-Land nebst Erörterungen über vermeintliche Cordaitales Mesozoischen Alters, II. Allgemeiner Teil." *Palaeontographica Abt. B* 82: 1–72.

———. 1949. "The morphology of *Trichopitys heteromorpha* Saporta, a seed-plant of Palaeozoic age, and the evolution of female flowers in the Ginkgoinae." *Acta Horti Bergiana* 15: 79–109.

Foote, M., and A. I. Miller. 2007. *Principles of Paleontology.* New York: Freeman.

Fortune, R. 1847. *Three Years' Wanderings in the Northern Provinces of China.* London: John Murray.

———. 1852. *A Journey to the Tea Countries of China.* London: John Murray.

———. 1857. *A Residence among the Chinese: Inland, on the Coast, and at Sea.* London: John Murray.

———. 1863. *Yedo and Peking: A Narrative of a Journey to the Capitals of Japan and China.* London: John Murray.

Foster, S., and Y. Chongxi. 1992. *Herbal Emissaries: Bringing Chinese Herbs to the West: A Guide to Gardening, Herbal Wisdom, and Well-Being.* Rochester: Healing Arts.

Franz, E. 2005. *Philipp Franz von Siebold and Russian Policy and Action on Opening Japan to the West in the Middle of the Nineteenth Century.* Munich: IUDICIUM.

Friedman, W. E. 2009. "The meaning of Darwin's 'abominable mystery.'" *American Journal of Botany* 96: 5–21.

Friedman, W. E., and E. M. Gifford. 1997. "Development of the male gametophyte of *Ginkgo biloba*: A window into the reproductive biology of early seed plants." In *Ginkgo biloba—A Global Treasure from Biology to Medicine.* Hori et al. (1997, 29–49).

Friedman, W. E., and T. E. Goliber. 1986. "Photosynthesis in the female gametophyte of *Ginkgo biloba.*" *American Journal of Botany* 73: 1261–1266.

Friis, E. M., P. R. Crane, and K. R. Pedersen. 2011. *Early Flowers and Angiosperm Evolution.* Cambridge: Cambridge University Press.

Fujii, K. 1896. "On the different views hitherto proposed regarding the morphology of the flowers of *Ginkgo biloba* L." *Botanical Magazine Tokyo* 10: 104–110.

Fuller, T. 1732. *Gnomologia: Adagies and Proverbs; Wise Sentences and Witty Sayings, Ancient and Modern, Foreign and British.* London: B. Barker.

Gandhi, M. K. 1961. *Non-violent Resistance (Satyagraha).* New York: Schocken.

Gardner, J. S. 1883–1885. *A Monograph of the British Eocene Flora.* Vol. 2, part 2, *Gymnospermae.* London: Palaeontographical Society.

Gardner, J. S., and C. Ettingshausen. 1879–1882. *A Monograph of the British Eocene Flora.* Vol. 1, *Filices.* London: Palaeontographical Society.

Garrett, W. 2007. *Marie Stopes: Feminist, Eroticist, Eugenicist.* San Francisco: Kenon.

Gengenbacher, M., T. B. Fitzpatrick, T. Raschle, K. Flicker, I. Sinning, S. Müller, P. Macheroux, I. Tews, and B. Kappes. 2006. "Vitamin B6 biosynthesis by the malaria parasite *Plasmodium falciparum*: Biochemical and structural insights." *Journal of Biological Chemistry* 281: 3633–3641.

Gerhold, H. D. 2007. "Origins of Urban Forestry." In *Urban and Community Forestry in the Northeast.* Ed. J. E. Kuser, 1–24. New York: Springer.

Goethe, J. W. v. 1790. *Versuch die Metamorphose der Pflanzen zu erklären.* Gotha: Ettingersche.

Gold, P. E., L. Cahill, and G. L. Wenkin. 2002. "*Ginkgo biloba*: A cognitive enhancer?" *Psychological Science in the Public Interest* 3: 2–11.

———. 2003. "The lowdown on *Ginkgo biloba.*" *Scientific American* 288: 86–91.

Gong, W., C. Chen, C. Dobes, C. X. Fu, and M. A. Koch. 2008a. "Phylogeography of a living fossil: Pleistocene glaciations forced *Ginkgo biloba* L. (Ginkgoaceae) into two refuge areas in China with limited subsequent postglacial expansion." *Molecular Phylogenetics and Evolution* 48: 1095–1105.

Gong, W., Y. X. Qui, C. Chen, Q. Ye, and C. X. Fu. 2008b. "Glacial refugia of *Ginkgo biloba* L. and human impact on its genetic diversity: Evidence from chloroplast DNA." *Journal of Integrative Plant Biology* 50: 368–374.

Gould, S. J. 1986. "Play it again, life." *Natural History* 95: 18–26.

———. 1989. *Wonderful Life: The Burgess Shale and the Nature of History.* New York: Norton.

Gradstein, F. M., J. G. Ogg, A. G. Smith, W. Bleeker, and L. J. Lourens. 2004. "A new geologic time scale, with special reference to Precambrian and Neogene." *Episodes* 27: 83–100.

Grande, L. 1984. *Paleontology of the Green River Formation, with a Review of the Fish Fauna.* Laramie: Geological Survey of Wyoming.

———. 2013. *The Lost World of Fossil Lake: Snapshots from Deep Time.* Chicago: University of Chicago Press.

Gray, A. 1859. "Diagnostic characters of new species of phanerogamous plants collected in Japan by Charles Wright, botanist of the U. S. North Pacific Exploring Expedition. (Published by Request of Captain John Rodgers, Commander of the Expedition.) With Observations upon the Relations of the Japanese Flora to That of North America, and of Other Parts of the Northern Temperate Zone." *Memoirs of the American Academy of Arts and Sciences,* n.s. 6: 377–452.

Green, J. 1983. "The Shinan excavation, Korea: An interim report on the hull structure." *International Journal of Nautical Archaeology* 12: 293–301.

Griggs, P. J., H. D. V. Prendergast, and N. Rumball. 2000. *Plants + People: An Exhibition of Items from the Economic Botany Collections in Museum No. 1.* London: Royal Botanic Gardens, Kew. Centre for Economic Botany.

Guerrant, E. O., K. Havens, and M. Maunder, eds. 2004. *Ex Situ Plant Conservation: Supporting Species Survival in the Wild.* Washington, D.C.: Island.

Gunckel, J. E., K. V. Thimann, and R. H. Wetmore. 1949. "Studies of development in long shoots and short shoots of *Ginkgo biloba* L. IV. Growth habit, shoot expression, and the mechanism of its control." *American Journal of Botany* 36: 309–316.

Hacke, U. G., J. S. Sperry, and J. Pittermann. 2004. "Analysis of circular bordered pit function II. Gymnosperm tracheids with torus-margo pit membranes." *American Journal of Botany* 91: 386–400.

Hageneder, F. 2005. *The Meaning of Trees: Botany, History, Healing, Lore.* San Francisco: Chronicle.

Hall, R. E. 1977. *Marie Stopes: A Biography.* London: Deutsch.

Handa, M. 2000. "*Ginkgo biloba* in Japan." *Arnoldia* 60: 26–33.

Handa, M., Y. Iizuka, and N. Fujiwara. 1997. "*Ginkgo* landscapes." In Hori et al. (1997, 255–283).

Hansen, M. C., S. V. Stehman, P. V. Potapov. 2010. "Quantification of global gross forest cover loss." *Proceedings of the National Academy of Science* 107: 8650–8655.

Hargraves, M. 2010. "Portrait of a tree." *Yale Alumni Magazine,* March–April 2010.

Harris, T. M., 1935. "The fossil flora of Scoresby Sound, East Greenland, 4. Ginkgoales, Coniferales, Lycopodiales and isolated Fructifications." *Meddeleser om Grønland* 112: 1–176.

———. 1961. *The Yorkshire Jurassic Flora I. Thallophyta-Pteridophyta.* London: British Museum (Natural History).

———. 1964. *The Yorkshire Jurassic Flora II. Caytoniales, Cycadales, and Pteridosperms.* London: British Museum (Natural History).

———. 1969. *The Yorkshire Jurassic Flora III. Bennettitales.* London: British Museum (Natural History).

———. 1979. *The Yorkshire Jurassic Flora V. Coniferales.* London: British Museum (Natural History).

Harris, T. M., W. Millington, and J. Miller. 1974. *The Yorkshire Jurassic Flora IV. Ginkgoales and Czekanowskiales.* London: British Museum (Natural History).

Harvey, J. H. 1981. "A Scottish Botanist in London in 1766." *Garden History* 9: 40–75.

Hawke, D. F. 1971. *Benjamin Rush: Revolutionary Gadfly.* Indianapolis: Bobbs-Merrill.

He, C-x., and J-r. Tao. 1997. "A study on the Eocene flora in Yilan County, Heilongjiang." *Acta Phytotaxonomica Sinica* 35: 249–256.

He, S. A., Y. Gu, and Z. J. Pang. 1997. "Resources and prospects of *Ginkgo biloba* in China." In Hori et al. (1997, 373–383).

Heidler D. S., and J. T. Heidler. 2010. *Henry Clay: The Essential American.* New York: Random House.

Hennig, W. 1965. "Phylogenetic systematics." *Annual Review of Entomology* 10: 97–116.

———. 1966. *Phylogenetic Systematics.* Urbana: University of Illinois Press.

Henry, D. O., and A. F. Servello. 1974. "Compendium of Carbon-14 determinations derived from Near Eastern prehistoric deposits." *Paléorient* 2: 19–44.

Hill, C. R., and P. R. Crane. 1982. "Evolutionary cladistics and the origin of angiosperms." In *Problems of Phylogenetic Reconstruction: Proceedings of the Systematics Association Symposium, Cambridge, 1980.* Ed. K. A. Joysey and A. E. Friday, 269–361. New York: Academic Press.

Hill, R. S., and R. J. Carpenter. 1999. "*Ginkgo* leaves from Palaeogene sediments in Tasmania." *Australian Journal of Botany* 47: 717–724.

Hilton, J., and R. M. Bateman. 2006. "Pteridosperms are the backbone of seed-plant phylogeny." *Journal of the Torrey Botanical Society* 133: 119–168.

Himi City. "The Jounichiji Ginkgo: The giant ginkgo tree at Jounichiji Temple." Himi City: Planning and Public Relations Office. http://www.city.himi.toyama.jp/~10000/english/ginkgo.htm.

Hinz, P.-A. 2001. "The Japanese plant collection of Engelbert Kaempfer (1651–1715) in the Sir Hans Sloane Herbarium at the Natural History Museum, London." *Bulletin of the Natural History Museum, Botany Series* 31: 27–34.

Hirase, S. 1895a. "Études sur le *Ginkgo biloba.*" *Botanical Magazine Tokyo,* 9: 240.

———. 1895b. "Études sur la fécondation et l´embryogénie du *Ginkgo biloba.*" *Journal of the College of Science, Imperial University of Tokyo* 8: 307–322.

———. 1896. "On the spermatozoid of *Ginkgo biloba.*" *Botanical Magazine Tokyo* 10: 325–328. In Japanese.

His Majesty the Emperor of Japan. 2007. "Linnaeus and taxonomy in Japan." *Nature* 448: 139–140.

Hizume, M. 1997. "Chromosomes of *Ginkgo biloba.*" In Hori et al. (1997, 109–118).

Ho, T. 1963. "The nucleolar chromosomes of the Maiden-hair tree." *Journal of Heredity* 54: 67–74.

Hohmann-Marriott, M. F., and R. E. Blankenship. 2011. "Evolution of photosynthesis." *Annual Review of Plant Biology* 62: 515–548.

Holmes, W. B. K., and H. M. Anderson. 2007. "The Middle Triassic megafossil flora of the Basin Creek Formation, Nymboida Coal Measures, New South Wales, Australia. Part 6. Ginkgophyta." *Proceedings of the Linnean Society of New South Wales* 128: 155–200.

Holt, B. F., and G. W. Rothwell. 1997. "Is *Ginkgo biloba* (Ginkgoaceae) really an oviparous plant?" *American Journal of Botany* 84: 870–872.

Hori, S., and T. Hori. 1997. "A cultural history of *Ginkgo biloba* in Japan and the generic name *Ginkgo.*" In Hori et al. (1997, 385–412).

Hori, T., and S. Hori. 2005. *Enormous Ginkgo Trees in Japan.* Tokyo: Uchida Rokakuho. In Japanese.

Hori, T., R. W. Ridge, W. Tulecke, P. Del Tredici, J. Trémouillaux-Guiller, and H. Tobe (eds). 1997. *Ginkgo biloba—A Global Treasure from Biology to Medicine.* Tokyo: Springer.

Hsia, F. C. 2009. *Sojourners in a Strange Land: Jesuits and Their Scientific Missions in Late Imperial China.* Chicago: University of Chicago Press.

Hull, D. L. 1988. *Science as a Process: An Evolutionary Account of the Social and Conceptual Development of Science.* Chicago: University of Chicago Press.

Hüls H., and H. Hoppe, eds. 1982. *Engelbert Kaempfer zum 330.* Lemgo, Germany: Geburtstag.

Humboldt, A. v., and A. Bonpland. 1852. *Personal Narrative of Travels to the Equinoctial Regions of America, During the Years 1799–1804.* London: Henry G. Bohn.

Huntley B., and H. J. B. Birks. 1983. *An Atlas of Past and Present Pollen Maps of Europe: 0–13,000 Years Ago.* Cambridge: Cambridge University Press.

Hyde, L. 1983. *The Gift: Creativity and the Artist in the Modern World.* New York: Random House.

Ikeno, S. 1901. "Contribution à l'étude de la fécondation chez le *Ginkgo biloba.*" *Annales des Sciences Naturelles Botanique* 13: 305–318.

Ikeno, S., and S. Hirase. 1897. "Spermatozoids in gymnosperms." *Annals of Botany* 11: 344–345.

Invitation ForestOn. "Story of forest: Old gigantic trees in Korea." Korean Forest Service. http://www.san.go.kr/english/culture/old_4.html.

Ishikawa, M. 1910. "*Ueber die Zahl der Chromosemen von Ginkgo biloba L.*" *Botanical Magazine Tokyo* 24: 225–226.

Ito, K., H. Kaku, and J. Matsumura. 1881–1883. *Koishikawa Shokubutsu-en Somoku Zusetsu; or, Figures and Descriptions of Plants in Koishikawa Botanical Garden.* 2 vols. Tokyo: Maruzen.

i-Tree Tools. 2010. "Reference Cities—The Science Behind i-Tree Streets (STRATUM)." http://www.itreetools.org/streets/resources/Streets_Reference_Cities_Science_Update_Nov2011.pdf. Modified November 2010.

IUCN. "Executive Summary—Conifers: Status Survey and Conservation Action Plan." http://intranet.iucn.org/webfiles/doc/SSC/SSCwebsite/Act_Plans/Executive_Summary_Conifers_Action_Plan.pdf.

Jacquin, J. F. v. 1819. *Ueber den Ginkgo.* Vienna: Carl Gerold.

Janzen, D. H., and P. S. Martin. 1982. "Neotropical anachronisms: The fruits the gomphotheres ate." *Science* n.s. 215: 19–27.

Japan Probe. 2011. "Ginkgo trees protect shrines and temples from fire." http://www.japanprobe.com/2011/01/23/ginkgo-trees-protect-shrines-temples-from-fire/. Modified January 23 2011.

"Japan's favourite tree: An Easter story from Japan." 2010. *Economist,* March 31 http://www.economist.com/node/15826315. Accessed March 18, 2011.

Jardine, L. 2008. *Going Dutch: How England Plundered Holland's Glory.* New York: Harper Collins.

Jarvis, C. E. 2007. *Order Out of Chaos: Linnaean Plant Names and Their Types.* London: Linnean Society of London, and the Natural History Museum.

Jensen, A. G., K. Ndjoko, J. L. Wolfender, K. Hostettmann, F. Camponovo, and F. Soldati. 2002.

"Liquid chromatography-atmospheric pressure chemical ionisation/mass spectrometry: A rapid and selective method for the quantitative determination of ginkgolides and bilobalide in ginkgo leaf extracts and phytopharmaceuticals." *Phytochemical Analysis* 13: 31–38.

Jensen, J. V. 1988. "Return to the Wilberforce-Huxley debate." *British Journal for the History of Science* 21: 161–179.

Johnson, K. R. 2002. "Megaflora of the Hell Creek and Lower Fort Union formations in the western Dakotas: Vegetational response to climate change, the Cretaceous-Tertiary boundary event, and rapid marine transgression." In *The Hell Creek Formation and the Cretaceous-Tertiary Boundary in the Northern Great Plains: An Integrated Continental Record of the End of the Cretaceous.* Ed. J. H. Hartman, K. R. Johnson, and D. J. Nichols, 329–392. Boulder: Geological Society of America.

Juretzek, W. 1997. "Recent advances in *Ginkgo biloba* extract (EGb761)." In Hori et al. (1997, 341–358).

Kaempfer, E. 1690–1692. *The History of Japan, together with a description of the Kingdom of Siam.* Trans. J. C. Scheuchzer, 1727. London: Woodward.

———. 1712. *Amoenitatum Exoticarum Politico-Physico- Medicarum Fasciculi V, Quibus Continentur Variae Relationes, Observationes et Descriptiones Rerum Persicarum et Ulterioris Asiae.* Lemgo, Germany: Meyer.

Kahl, R. 2009. "Population and consumption." World Resources Institute. http://earthtrends.wri.org/updates/node/360. Modified November 13, 2009.

Kaigi, G. K. 1930. *Japanese Journal of Botany: Transactions and Abstracts.* Vol. 4. Tokyo: National Research Council of Japan.

Kaplan, D. R. 2001. "The science of plant morphology: Definition, history, and role in modern biology." *American Journal of Botany* 88: 1711–1741.

Kell, G. 2005. "Liquidambar, tupelo, and ginkgo: Autumn's fire lights up the Berkeley campus." *UC Berkeley News,* November 15, 2005. http://berkeley.edu/news/media/releases/2005/11/15_autumn.shtml.

Kellert, S. R., and E. O. Wilson. 1993. *The Biophilia Hypothesis.* Washington D.C.: Island.

Kenrick P., and P. R. Crane. 1997. *The Origin and Early Diversification of Land Plants: A Cladistic Study.* Washington, D.C.: Smithsonian Institution Press.

Kew. 2010. "New study shows one fifth of the world's plants are under threat of extinction." Royal Botanic Gardens, Kew. September 29, 2010. www.kew.org/news/one-fifth-of-plants-under-threat-of-extinction.htm.

———. N.d. "Island plants: Conserving biological diversity." Royal Botanic Gardens, Kew. http://www.kew.org/plants/islandplants/index.html.

Kibuchi, Y. 1997. "Hybridity and the Oriental Orientalism of *Mingei* theory." *Journal of Design History* 10: 343–354.

"The killer trees: A wrong-headed campaign against roadside trees." *Economist,* February 14, 2004. http://www.economist.com/node/2429069.

Killingbeck, K. T. 1996. "Nutrients in senesced leaves: Keys to the search for potential resorption and resorption proficiency." *Ecology* 77: 1716–1727.

Kim, S. B., ed. 2006a. *The Shinan Wreck*. Vol. 1, *Main*. Mokpo: National Maritime Museum of Korea. In Korean with English abstract.

———, ed. 2006b. *The Shinan Wreck*. Vol. 2, *Celadon/Porcelain*. Mokpo: National Maritime Museum of Korea. In Korean with English abstract.

———, 2006c. *The Shinan Wreck*. Vol. 3, *White Porcelain/Others*. Mokpo: National Maritime Museum of Korea. In Korean with English abstract.

Kirchner, M. 1992. "Untersuchungen an einigen Gymnospermen der fränkischen Rhät-Lias-Grenzschichten." *Palaeontographica Abt. B* 224: 17–61.

Kochibe N. 1997. "Allergic substances of *Ginkgo biloba*." In Hori et al. (1997, 301–308).

Kouwenhoven, A., and M. Forrer. 2000. *Siebold and Japan: His Life and Work*. Leiden: Hotei.

Kovar-Eder, J., L. Givulescu, L. Hably, Z. Kvaček, D. Mihajlovic, Y. Teslenko, H. Walther, et al. 1994. "Floristic changes in the areas surrounding the paratethys during Neogene time." In *Cenozoic Plants and Climate of the Arctic*. Ed. M. C. Boulter, H. C. Fisher, 347–369. Berlin: Springer.

Kovar-Eder, J., Z. Kvaček, E. Martinetto, P. Roiron. 2006. "Late Miocene to Early Pliocene vegetation of southern Europe (7-4 Ma) as reflected in the megafossil plant record." *Palaeogeography, Palaeoclimatology, Palaeoecology* 238: 321–339.

Krassilov, V. A. 1970. "An approach to the classification of Mesozoic "Ginkgoalean" plants from Siberia." *Palaeobotanist* 18: 12–19.

Kris-Etherton P. M., S. Yu-Poth, J. Sabaté, H. E. Ratcliffe, G. Zhao, and T. D. Etherton. 1999. "Nuts and their bioactive constituents: Effects on serum lipids and other factors that affect disease risk." *American Journal of Clinical Nutrition* 70: 504S–511S.

Kullman, L. 2005. "Old and new trees on Mt Fulufjället in Dalarna, central Sweden." *Svensk Botanisk Tidskrift* 6: 315–329.

Kuo, F. E. 2003. "The role of arboriculture in a healthy social ecology." *Journal of Arboriculture* 29: 148–155.

Kuo, F. E., and A. Faber Taylor. 2004. "A potential natural treatment for attention-deficit/hyperactivity disorder: Evidence from a national study." *American Journal of Public Health* 94: 1580–1586.

Kuo, F. E., and W. C. Sullivan. 2001. "Environment and crime in the inner city: Does vegetation reduce crime?" *Environment and Behavior* 33: 343–367.

Kvaček, J., L. Falcon-Lang, and J. Dašková. 2005. "A new Late Cretaceous ginkgoalean reproductive structure *Nehvizdyella* gen. nov. from the Czech Republic and its whole-plant reconstruction." *American Journal of Botany* 92: 1958–1969.

Kvaček, Z., S. B. Manum, and M. C. Boulter. 1994. "Angiosperms from the Palaeogene of Spitsbergen, including an unfinished work by A. G. Nathorst." *Palaeontographica Abt. B* 232: 103–128.

Kwant, C. "A-bombed ginkgo trees in Hiroshima, Japan." Ginkgo Pages. http://kwanten.home.xs4all.nl/hiroshima.htm.

———. "An Old Chinese Legend." Ginkgo Pages. http://kwanten.home.xs4all.nl/bonsai.htm.

———. "*Ginkgo biloba* and Art Nouveau in l'Ecole de Nancy." Ginkgo Pages. http://kwanten.home.xs4all.nl/nancy.htm.

———. "*Ginkgo biloba* and Art Nouveau in Prague." Ginkgo Pages. http://kwanten.home.xs4all.nl/prague.htm.

———.The Ginkgo Pages. http://www.xs4all.nl/~kwanten/.

———. "The Seven Worthies of the Bamboo Grove and Rong Qiqi and ginkgo trees." Ginkgo Pages. http://kwanten.home.xs4all.nl/artgal3.htm.

Lack, H. W. 1999. "Plant illustration on wood blocks—A magnificent Japanese xylotheque of the Early Meiji Period." *Curtis's Botanical Magazine* 16: 124–134.

Laurens, H. 1980. *Papers of Henry Laurens, Vol. 8, October 10, 1771, to April 19, 1773.* Ed. G. C. Rogers, D. R. Chesnutt, and P. J. Clark. Charleston: South Carolina Historical Society.

Lee, C. L. 1955. "Fertilization in *Ginkgo biloba.*" *Botanical Gazette* 117: 79–100.

Leigh, A., M. A. Zwieniecki, F. E. Rockwell, C. K. Boyce, A. B. Nicotra, and N. M. Holbrook. 2010. "Structural and hydraulic correlates of heterophylly in *Ginkgo biloba.*" *New Phytologist* 189: 459–470.

Leopold, A. 1949. *A Sand County Almanac and Sketches Here and There.* Oxford: Oxford University Press.

Letzel, H., J. Haan, and W. B. Feil. 1996. "Nootropics: Efficacy and tolerability of products from three active substance classes." *Journal of Drug Development and Clinical Practice* 8: 77–94.

Lewis, S. K. 2007. "Flowers Modern and Ancient." www.pbs.org/wgbh/nova/nature/flowers-modern-ancient.html. Modified April 17, 2007.

Li, H. L. 1956. "A horticultural and botanical history of *Ginkgo.*" *Bulletin of the Morris Arboretum* 7: 3–12.

———. 1963. *The Origin and Cultivation of Shade and Ornamental Trees.* Philadelphia: University of Pennsylvania Press.

Li, J. W., Z. Y. Liu, Y. M. Tan, and M. B. Ren. 1999. "Studies on the *Ginkgo* at Jinfoshan Mountain." *Forest Research* 12: 197–201. In Chinese, with English abstract.

Li, S., and X. Luo. 2003. *Compendium of Materia Medica (Bencao Gangmu). First Published in Chinese in 1593.* Beijing: Foreign Languages Press.

Lin, J. X., Y-S. Hu, and X-P. Wang. 1995. "Old ginkgo trees in China." *International Dendrology Society Yearbook* 1995: 32–37.

Linnaeus, C. 1751. *Philosophia Botanica.* Stockholm: Kiesewetter.

———. 1753. *Species plantarum: Exhibentes plantas rite cognitas, ad genera relatas, cum differentiis specificis, nominibus trivialibus, synonymis selectis, locis natalibus, secundum systema sexuale digestas.* Stockholm: Impensis Laurentii Salvii.

———. 1771. *Mantissa Plantarum Altera Generum Editionis VI & Specierum Editionis II.* Stockholm: Impensis Laurentii Salvii.

Liu, W., C-Z. Jin, Y-Q Zhang, Y-J Cai, S. Xing, X-J. Wu, H. Cheng,et al. 2010. "Human remains from Zhirendong, South China, and modern human emergence in East Asia." *Proceedings of the National Academy of Sciences* 107: 19201–19206.

Liu, X. Q., C. S. Li, and Y. F. Wang. 2006. "The pollen cones of *Ginkgo* from the Early Cretaceous of China, and their bearing on the evolutionary significance." *Botanical Journal of the Linnean Society* 152: 133–144.

Lloyd, D. G. 1982. "Selection of combined versus separate sexes in seed plants." *American Naturalist* 120: 571–585.

Lloyd D. G., and C. J. Webb. 1977. "Secondary sex characters in plants." *Botanical Review* 43: 177–216.

Loudon, J. C. 1838. *Arboretum et Fruticetum Britannicum; or, The Trees and Shrubs of Britain.* London: A. Spottiswoode.

Lyman, B. S. 1885. "The etymology of 'ginkgo.'" *Science* 6: 84.

Lyon, H. L. 1904. "The embryogeny of *Ginkgo.*" *Minnesota Botanical Studies* 3: 275–290.

MacFadden, B. J., and R. C. Hulbert. 1988. "Explosive speciation at the base of the adaptive radiation of Miocene grazing horses." *Nature* 336: 466–468.

MacGregor, A., ed. 1994. *Sir Hans Sloane: Collector, Scientist, Antiquary, Founding Father of the British Museum.* London: British Museum and Alistair McAlpine.

Mackie, W. 1913. "The rock series of Craigbeg and Ord Hill, Rhynie, Aberdeenshire." *Transactions of the Edinburgh Geological Society* 10: 205–236.

Madsen, K. 1989. "To make his country smile: William Hamilton's woodlands." *Arnoldia* 49: 14–24.

Mair, V. H., ed. 2001. *The Columbia History of Chinese Literature.* New York: Columbia University Press.

Mallow, D. 2008. "Town rallies to save old ginkgo." *Pittsburgh Post-Gazette,* April 13, 2008.

Manchester, S. R. 1981. "Fossil plants of the Eocene Clarno Nut Beds." *Oregon Geology* 43: 75–81.

Manchester, S. R., Z. Chen, B. Geng, and J. Tao. 2005. "Middle Eocene flora of Huadian, Jilin Province, Northeastern China." *Acta Palaeobotanica* 45: 3–26.

Manchester, S. R., and D. L. Dilcher. 1982. "Pterocaryoid fruits (Juglandaceae) in the Paleogene of North America and their evolutionary and biogeographic significance." *American Journal of Botany* 69: 275–286.

Matsumoto, K., O. Takeshi, M. Irasawa, and T. Nakamura. 2003. "Climate change and extension of the *Ginkgo biloba* L. growing season in Japan." *Global Change Biology* 9: 1634–1642.

McCullough, D. G. 1992. *Truman.* New York: Simon and Schuster.

McIver, E. E., and J. F. Basinger. 1999. "Early Tertiary floral evolution in the Canadian High Arctic." *Annals of the Missouri Botanical Garden* 86: 523–545.

McPherson, G. E. 2003. "Urban forestry: The final frontier?" *Journal of Forestry* 101: 20–25.

Medline Plus. 2011, "Ginkgo." U.S. National Library of Medicine of the National Institutes of Health. http://www.nlm.nih.gov/medlineplus/druginfo/natural/333.html. Updated October 11, 2011.

Melillo, J. M., A. D. McGuire, D. W. Kicklighter, B. Moore, C. J. Vorosmarty, and A. L. Schloss. 1993. "Global climate change and terrestrial net primary production." *Nature* 363: 234–240.

Melzheimer, V., and J. J. Lichius. 2000. "*Ginkgo biloba* L. Aspects of the systematical and applied botany." In van Beek (2000, 25–50).

Merrill, E. D. 1948. "Metasequoia, another 'living fossil.'" *Arnoldia* 8: 1–8.

Meyen, S. V. 1984. "Basic features of gymnosperm systematics and phylogeny as evidenced by the fossil record." *Botanical Review* 50: 1–111.

———. 1988. "Gymnosperms of the Angara flora." In *Origin and Evolution of Gymnosperms.* Ed. C. B. Beck, 338–381. New York: Columbia University Press.

Michaux, F. A., 1805. *Michaux's Travels to the West of the Alleghany Mountains.* London: R. Phillips.

Michel, W. 2009. "Engelbert Kaempfer forum." Wolfgang Michel's Research Notes, Kyushu University. http://wolfgangmichel.web.fc2.com/serv/ek/index.html. Modified November 2009.

Millennium Ecosystem Assessment. 2005. *Ecosystems and Human Well-Being: Synthesis.* Washington, D.C.: Island.

Minter, S. 2000. *The Apothecaries' Garden: A New History of Chelsea Physic Garden.* Stroud: Sutton.

Miyoshi, N. 1931. "Merkwürdige *Ginkgo biloba* in Japan." *Mitteilungen der Deutschen Dendrologischen Gesellschaft* 43: 21–22.

Morse, E. S. 1917. *Japan Day by Day 1877, 1878–79, 1882–83.* Vols. 1–2. Boston: Houghton Mifflin.

Mortlake, G. N. 1911. *Love-Letters of a Japanese.* London: Stanley Paul.

Mullis, K. B., F. Ferré, and R. A. Gibbs. 1994. *The Polymerase Chain Reaction.* Boston: Birkhäuser.

Murdoch, I. 1970. *A Fairly Honourable Defeat.* London: Chatto and Windus.

Murdoch, J. 2004. *A History of Japan.* Vol. 3. Hertford: Stephen Austin and Sons.

Murray-Smith, C., N. A. Brummitt, A. T. Oliveira-Filho, S. Bachman, J. Moat, E. M. Nic Lughadha, and E. J. Lucas. 2009. "Plant diversity hotspots in the Atlantic coastal forests of Brazil." *Conservation Biology* 23: 151–163.

Mustoe, G. E. 2002. "Eocene *Ginkgo* leaf fossils from the Pacific Northwest." *Canadian Journal of Botany* 80: 1078–1087.

Naderi, J. R., B. S. Kweon, and P. Maghelal. 2008. "The street tree effect and driver safety." ITE Journal on the Web. http://www.walkable.org/assets/downloads/StreetTreeEffectandDriverSafety_ITEfeb08_.pdf.

Nadkarni, N. M. 2008. *Between Earth and Sky: Our Intimate Connections to Trees.* Berkeley: University of California Press.

Nagata, T. 1997. "Scientific contributions of Sakugoro Hirase." In Hori et al. (1997, 413–416).

Nakanishi, K. 2000. "A personal account of the early ginkgolide structural studies." In van Beek (2000, 165–173).

National Academy of Sciences. 2011. *Twenty-First Century Ecosystems: Managing the Living World Two Centuries after Darwin.* Washington, D.C.: National Academies Press.

Naugolnykh, S. V. 1995. "A new genus of *Ginkgo*-like leaves from the Kungurian of the Urals Region." *Paleontologicheskii Zhurnal* 3: 106–116. In Russian.

———. 2007. "Foliar seed-bearing organs of Paleozoic Ginkgophytes and the early evolution of the Ginkgoales." *Paleontological Journal* 41: 815–859.

Nawrocki, D. A., and D. Clements. 2008. *Art in Detroit Public Places.* Detroit: Wayne State University Press.

Needham, J. 1986. *Science and Civilisation in China.* Vol. 6, *Biology and Biological Technology. Part 1, Botany.* Cambridge: Cambridge University Press.

Needham, J., C. Daniels, and N. K. M. Menzies. 1996. *Science and Civilisation in China: Agro-*

industries: Sugarcane Technology. Vol. 6. *Biology and Biological Technology.* Part 3, *Agro-industries and Forestry.* Cambridge: Cambridge University Press.

Nelson, E. C. 1983. "Augustine Henry and the exploration of the Chinese flora." *Arnoldia* 43: 21–38.

Nemerov, H. 1977. *The Collected Poems of Howard Nemerov.* Chicago: University of Chicago Press.

Newcomer, E. H. 1939. "Pollen longevity of *Ginkgo.*" *Bulletin of the Torrey Botanical Club* 66: 121–123.

———. 1954. "The karyotype and possible sex chromosomes of *Ginkgo biloba.*" *American Journal of Botany* 41: 542–545.

Newitt, M. 2005. *A History of Portuguese Overseas Expansion, 1400–1668.* New York: Routledge.

Nietzsche, F. 1896. *The Works of Friedrich Nietzsche.* Vol. 9, *The Case of Wagner; The Twilight of the Idols; Nietzsche contra Wagner.* Trans. Thomas Common. London: H. Henry.

Nishida, H. 1991. "Diversity and significance of Late Cretaceous permineralized plant remains from Hokkaido, Japan." *Botanical Magazine Tokyo* 104: 253–273.

Nishida, H., K. B. Pigg, K. Kudo, and J. F. Rigby. 2004. "Zooidogamy in the Late Permian genus *Glossopteris.*" *Journal of Plant Research* 117: 323–328.

Nowak, D. J., and E. J. Greenfield. 2012. "Tree and impervious cover change in U.S. cities." *Urban Forestry and Urban Greening* 11: 21–30.

Ogle, W. 1912. *The Works of Aristotle.* Vol. 5, *De Partibus Animalium.* Oxford: Clarendon.

Ōhashi, K. 2006. *Nabeshima: Gifts of Porcelain to the Shogun's Family.* Kyushu: The Kyushu Ceramic Museum. In Japanese.

Oldschool, O. 1809. "American scenery for the Portfolio—the woodlands." *Port Folio* 2: 507.

Oliver, F. W., and D. H. Scott. 1903. "On *Lagenostoma lomaxi,* the seed of *Lyginodendron.*" *Proceedings of the Royal Society London* B 71: 477–481.

———. 1904. "On the structure of the Palaeozoic seed *Lagenostoma lomaxi,* with a statement of the evidence upon which it is referred to *Lyginodendron.*" *Philosophical Transactions of the Royal Society of London* B. 197: 193–247.

Olson, G., P. Toaig, and R. Walker. 2001. *Onetree.* London: Merrell.

Pakenham, T. 2002. *Remarkable Trees of the World.* London: Weidenfeld and Nicolson.

———. 2003. *Meetings with Remarkable Trees.* London: Weidenfeld and Nicolson.

Pandit, S., and Nagarjuna. 1977. *Elegant Sayings.* Berkeley, Calif.: Dharma.

Parenti, L. R. 1980. "A phylogenetic analysis of the land plants." *Biological Journal of the Linnean Society* 13: 225–242.

Parsons, R. L., G. Tassinary, R. S. Ulrich, M. R. Hebl, and M. Grossman-Alexander. 1998. "The view from the road: Implications for stress recovery and immunization." *Journal of Environmental Psychology* 18: 113–140.

Patrut A., K. F. von Reden, D. A. Lowy, A. H. Alberts, J. W. Pohlman, R. Wittmann, D. Gerlach, L. Xu, and C. S. Mitchell. 2007. "Radiocarbon dating of a very large African baobab." *Tree Physiology* 27: 1569–1574.

Peper, P. J., E. G. McPherson, J. R. Simpson, S. L. Gardner, K. E. Vargas, and Q. Xiao. 2007. *New York City, New York Municipal Forest Resource Analysis.* Davis, Calif.: Center for Urban Forest Research.

Pérez, C. M. 2009. "Commentary: Can *Ginkgo biloba* combat diseases?" *Puerto Rican Health Sciences Journal* 28: 66–74.

Peschier, C. G. 1818. "Recherches analytiques sur le fruit du *Ginkgo*." *Bibliothèque Universelle des Sciences, Belles-Lettres et Arts* 7: 29–34.

Philipps, T. "Portraits: Dame Iris Murdoch." Tom Phillips. www.tomphillips.co.uk/works/portraits/item/5456-iris-murdoch.

Pittermann, J., J. S. Sperry, U. G. Hacke, J. K. Wheelers, and E. H. Sikkema. 2005. "Torus-Margo pits help conifers compete with angiosperms." *Science* 310: 1924.

PlantExplorers.com. 1999–2012. "The French Missionary-Botanists." www.plantexplorers.com/explorers/biographies/french-missionaries/index.html.

Pollan, M. 1991. *Second Nature: A Gardener's Education.* New York: Grove.

Porter B., and S. R. Johnson. 1993. *Road to Heaven: Encounters with Chinese Hermits.* San Francisco: Mercury House.

Prendergast, H. D. V., H. F. Jaeschke, and N. Rumball. 2001. *A Lacquer Legacy at Kew: The Japanese Collection of John J. Quin.* London: Royal Botanic Gardens, Kew.

Primack, R. B., and T. Ohkubo. 2008. "Ancient and notable trees of Japan." *Arnoldia* 65: 10–21.

Pringle, P. 2008. *The Murder of Nikolai Vavilov: The Story of Stalin's Persecution of One of the Great Scientists of the Twentieth Century.* New York: Simon and Schuster.

Proust, M. 1919. *À la recherche du temps perdu.* Paris: Gaston Gallimard.

Pullen, S. 2005. "The Jefferson Award: Kemba Shakur, tree planter." sf *Gate,* November 12. http://articles.sfgate.com/2005-11-12/opinion/17397738_1_planting-tree-canopy-fruit-trees.

Quammen, D. 1998. *The Flight of the Iguana: A Sidelong View of Science and Nature.* New York: Scribner.

Quin, J. 1882. "How Japanese Lacquer Ware is made—II." *Furniture Gazette* 18: 199–201.

Rashford, J. H. 2007. "Potential big men, fig trees, and tourist attractions in Barbados." *Society for Applied Anthropology* 29: 31–35.

Raubeson, L. A., and R. K. Jansen. 1992. "Chloroplast DNA evidence on the ancient evolutionary split in vascular land plants." *Science* 255: 1697–1699.

Raup, D. M., and J. J. Sepkoski. 1982. "Mass extinctions in the marine fossil record." *Science* 215: 1501–1503.

———. 1984. "Periodicity of extinctions in the geologic past." *Proceedings of the National Academy of Sciences* 81: 801–805.

———. 1986. "Periodic extinction of families and genera." *Science* 231: 833–836.

Raup, D. M., J. J. Sepkoski, and S. M. Stigler. 1983. "Mass extinctions in the fossil record—Reply." *Science* 219: 1240–1241.

Raup, D. M., and S. M. Stanley. 1971. *Principles of Paleontology.* San Francisco: Freeman.

Reid, E. M., and M. E. J. Chandler. 1933. *The London Clay Flora.* London: British Museum (Natural History).

Retallack, G. J. 2001. "A 300-million-year record of atmospheric carbon dioxide from fossil plant cuticles." *Nature* 411: 287–290.

Robberecht, R., and M. M. Caldwell. 1983. "Protective mechanisms and acclimation to solar ultra-violet-B radiation in *Oenothera stricta*." *Plant, Cell, and Environment* 6: 477–485.

Rodríguez-Flórez, C. D., E. L. Rodríguez-Flórez, and C. A. Rodríguez. 2009. "Revisión de la fauna pleistocénica Gomphotheriidae et Colombia y reporte de un caso para el Valle de Cauca." *Boletín Científico Centro de Museos: Museo de Historia Natural* 13: 78–85.

Rosengarten, F., Jr. 1984. *The Book of Edible Nuts.* New York: Walker.

Rothwell, G. W., and B. F. Holt. 1997. "Fossils and phenology in the evolution of *Ginkgo biloba*. In Hori et al. (1997, 223–230).

Royer, D. L. 2003. "Estimating latest Cretaceous and Tertiary atmospheric CO_2 from stomatal indices." *Geological Society of America Special Paper* 369: 79–93.

Royer, D. L., L. J. Hickey, and S. L. Wing. 2003. "Ecological conservatism in the 'living fossil' *Ginkgo*." *Paleobiology* 29: 84–104.

Royer, D. L., S. L. Wing, D. J. Beerling, D. W. Jolley, P. L. Koch, L. J. Hickey, and R. A. Berner. 2001. "Paleobotanical evidence for near present-day levels of atmospheric CO_2 during part of the Tertiary." *Science* 292: 2310–2313.

Sakai, A., and C. J. Weiser. 1973. "Freezing resistance of trees in North America with reference to tree regions." *Ecology* 54: 118–126.

Salisbury, R. A. 1817. "On the coniferous plants of Kaempfer." *Quarterly Journal of Science, Literature and the Arts* 2: 309–314.

Salleh, A. 2005. "Wollemi pine infected by fungus." ABC *Science: News in Science,* November 4. http://www.abc.net.au/science/articles/2005/11/04/1497961.htm.

Samylina, V. A. 1967. "On the final stage of the history of the genus *Ginkgo* L. in Eurasia." *Botanicheskii Zhurnal* 52: 303–316. In Russian.

Sanon, M. D., C. S. Gaméné, M. Sacandé, and O. Neya. 2004. "Desiccation and storage of *Kigelia africana, Lophira lanceolata, Parinari curatellifolia,* and *Zanthoxylum zanthoxyloides* seeds from Burkina Faso." In *Comparative Storage Biology of Tropical Tree Seeds.* Ed. M. Sacandé, D. Joker, M. E. Dulloo, and K. A. Thompsen, 16–23. Rome: IPGRI.

Santamour, F. S., S-A. He, and T. E. Ewert. 1983a. "Growth, survival, and sex expression in *Ginkgo*." *Journal of Arboriculture* 9: 170–171.

Santamour, F. S., S-A. He, and A. J. McArdle. 1983b. "Checklist of cultivated *Ginkgo*." *Journal of Arboriculture* 9: 88–92.

Sargent, C. S. ed. 1913. *Plantae Wilsonianae. An Enumeration of the Woody Plants Collected in Western China for the Arnold Arboretum of Harvard University during the Years 1907, 1908, and 1910 by E. H. Wilson.* Cambridge: Cambridge University Press.

Schenk, A. 1867. *Die fossile Flora der Grenzschichten des Keuper and Lias Frankens.* Wiesbaden: C. W. Kreidel's.

Schmid, M., and H. Schmoll. 1994. *Ginkgo, Ur-baum und Arzneipfanze, Mythos, Dichtung und Kurst.* Stuttgart: Wissenschafliche.

Schneider, C. K. 1903. *Dendrologische Winterstudien.* Jena: Gustav Fischer.

Schopf, T. J. M. 1984. "Rates of evolution and the notion of 'living fossils.'" *Annual Review of Earth and Planetary Sciences* 12: 245–292.

Schorn, H. E., J. A. Myers, and D. M. Erwin. 2007. "Navigating the Neogene: An updated chronology of Neogene paleofloras from the western United States." *Courier Forschungsinstitut Senckenberg,* 258: 139–146.

Schweitzer, H.-J., and M. Kirchner. 1995. "Die Rhäto-Jurassischen Floren des Iran und Afghanistan: 8. Ginkgophyta." *Palaeontographica Abt. B* 237: 1–58.

Scott, R. A., E. S. Barghoorn, and U. Prakash. 1962. "Wood of *Ginkgo* in the Tertiary of western North America." *American Journal of Botany* 49: 1095–1101.

Secretariat of the Convention on Biological Diversity. 2002. *Bonn Guidelines on Access to Genetic Resources and Fair and Equitable Sharing of the Benefits Arising Out of Their Utilization.* Montreal: Secretariat of the Convention on Biological Diversity.

Seelanan, T., A. Schnabel, and J. F. Wendel. 1997. "Congruence and consensus in the Cotton Tribe (Malvaceae)." *Systematic Botany* 22: 259–290.

Sepkoski, D., and M. Ruse, eds. 2009. *The Paleobiological Revolution: Essays on the Growth of Modern Paleontology.* Chicago: University of Chicago Press.

Seward, A. C. 1921. "Prof. A. G. Nathorst: Obituary." *Nature* 107: 112–113.

Seward, A. C., and J. Gowan. 1900. "The maidenhair tree (*Ginkgo biloba* L.)." *Annals of Botany* 14: 109–164.

Seyock, B. 2008. "Archaeological complexes from Muromachi period Japan as a key to the perception of international maritime trade in East Asia." In *The East Asian Mediterranean: Maritime Crossroads of Culture, Commerce and Human Migration.* Ed. A. Schottenhammer, 179–202. Wiesbaden: Harrassowitz.

Shakespeare, W. 1623a. *As You Like It.* In *Mr. William Shakespeares Comedies, Histories, and Tragedies.* London: Isaac Jaggard and Ed. Blount.

———. 1623b. *Henry IV, Part 2.* In *Mr. William Shakespeares Comedies, Histories, and Tragedies.* London: Isaac Jaggard and Ed. Blount.

———. 1623c. *Macbeth.* In *Mr. William Shakespeares Comedies, Histories, and Tragedies.* London: Isaac Jaggard and Ed. Blount.

Shen, L., X. Y. Chen, X. Zhang, Y. Y. Li, C. X. Fu, and Y. X. Qiu. 2005. "Genetic variation of *Ginkgo biloba* L. (Ginkgoaceae) based on cpDNA PCR-RFLPs: Inference of glacial refugia." *Heredity* 94: 396–401.

Shirai, M. 1891. "Abnormal ginkgo tree." *Botanical Magazine Tokyo* 56: 341–342.

Shober, A. L., and G. S. Toor. 2009. "Soils and fertilizers for master gardeners: Urban soils and their management issues." *University of Florida Institute of Food and Agricultural Sciences* SL 276: 1–3.

Siebold, P. F. v. 1832–1852. *Nippon. Archiv zur Beschreibung von Japan, und dessen Neben- und Schultzländern: Jezo mit den südlichen Kurilen, Sachalin, Korea und den Liukiu-Inseln.* Leiden.

Siebold, P. F. v., C. J. Temminck, H. Schlegel, and W. de Haan. 1833–1850. *Fauna Japonica Sive Descriptio Animalium, Quae in Itinere Per Japoniam, Jussu et Auspiciis Superiorum, qui Summum in India*

Batavia Imperium Tenent, Suscepto, Annis 1823–1830 Collegit, Notis, Observationibus et Adumbrationibus, Illustravit. Leiden: Lugundi-Batavorum.

Siebold, P. F. v., and J. G. Zuccarini. 1835–1870. *Flora Japonica, Sive, Plantae in Imperio Japonico Collegit, Descripset, ex parte in Ipsis Locis Pingendas Curavit.* Leiden: Lugundi-Batavorum.

Singh, V. P. 2006. *Gymnosperm (Naked Seeds Plant): Structure and Development.* New Delhi: Sarup and Son.

Skre, O. 1990. "Consequences of possible climatic temperature changes for plant production and growth in alpine and subalpine areas in Fennoscandia." In *Effects of Climate Change on Terrestrial Ecosystems.* Ed. J. I. Holten, NINA Notat 4: 18–37. Trondheim: Norwegian Institute for Nature Research.

Sky News. 2007. "Heritage guide: Gingko tree with a thousand-year legend." Korean Air. http://www.skynews.co.kr/article_view.asp?mcd=192&ccd=6&scd=9&ano=47. Modified June 10, 2007.

Sloane, H. 1696. *Catalogus Plantarum quae in insula Jamaica sponte proveniunt, vel vulgò coluntur: cum earundem synonymis & locis natalibus; adjectis aliis quibusdam quae in insulis Maderae, Barbados, Nieves, et Sancti Christophori nascuntur. Seu Prodromi historiae naturalis Jamaicae pars prima.* London: D. Brown.

———. 1707, 1725. 2 vol. *A Voyage to the Islands Madera, Barbados, Nieves, S. Christophers and Jamaica, with the Natural History of the Herbs and Trees, Four-footed Beasts, Fishes, Birds, Insects, Reptiles, Etc. of the Last of those Islands.* London: B. M.

Smith, B. H. 1905. "Some letters from William Hamilton to his private secretary." *Pennsylvania Magazine of History and Biography* 29: 70–79.

Smith, J. E. 1821. *A Selection of the Correspondence of Linnaeus and Other Naturalists from the Original Manuscripts.* Vol. 1. London: Longman, Hurst, Rees, Orme and Brown.

Smith, S. Y., M. E. Collinson, and P. J. Rudall. 2008. "Fossil *Cyclanthus* (Cyclanthaceae, Pandanales) from the Eocene of Germany and England." *American Journal of Botany* 6: 688–699.

Soni, P., R. Patidar, V. Soni, and S. Soni. 2010. "A review on traditional and alternative treatment for skin disease 'Vitiligo.'" *International Journal of Pharmaceutical and Biological Archives* 1: 220–227.

Sperry, J. S., U. G. Hacke, and J. Pittermann. 2006. "Size and function in conifer tracheids and angiosperm vessels." *American Journal of Botany* 93: 1490–1500.

Spongberg, S. A. 1993. "Exploration and introduction of ornamental and landscape plants from Eastern Asia." In *New Crops.* Ed. J. Janick and J. E. Simon. New York: Wiley.

Stearn, W. T. 1948. "Kaempfer and the lilies of Japan." *Royal Horticultural Society, Lily Year Book* 12: 65–70.

Stevens, P. F. 2008. Angiosperm Phylogeny Website, Version 9. http://www.mobot.org/MOBOT/research/APweb/. Modified June 29, 2012.

Sticher O., B. Meier, and A. Hasler. 2000. "The analysis of ginkgo flavonoids." In van Beek (2000, 179–202).

Stopes, M. C. 1910. *A Journal from Japan: A Daily Record of Life as Seen by a Scientist.* London: Blackie and Son.

————. 1918. *Married Love; or, Love in Marriage.* New York: Critic and Guide Company.

Stopes, M. C., and K. Fujii. 1910. "Studies on the structure and affinities of Cretaceous plants." *Philosophical Transactions of the Royal Society London* B 201: 1–90.

Strasburger, E. 1892. *Histologische Beiträge.* Vol. 4, *Ueber das Verhalten des Pollens und die Befruchtungsvorgänge bei den Gymnospermen: Schwarmsporen, Gameten, pflanzliche Spermatozoiden.* Jena: Gustav Fischer.

Straus, A. 1967. "Zur Paläontologie des Pliozäns von Willershausen." *Bericht der Naturhistorischen Gesellschaft zu Hannover* 111: 15–24.

Stuppy, W., R. Kesseler, and M. Harley. 2009. *The Bizarre and Incredible World of Plants.* London: Papadakis.

Sun, E., A. Akhmetiev, L. Golovneva, E. Bugdaeva, C. Quan, T. M. Kodrul, H. Nishida, et al. 2007. "Late Cretaceous plants from Jiayin along Heilongjiang River, northeast China." *Courier Forschungsinstitut Senckenberg* 258: 75–83.

Sutter, S. 2007. "Lite of the mind: Champion haiku." *University of Chicago Magazine,* March–April, 80.

Swearer, D. K. 2001. "Principles and poetry, places and stories: The resources of Buddhist ecology." *Daedalus* 130: 225–241.

Takaso, T. 1990. "Drop time at the Arnold Arboretum." *Arnoldia* 50: 2–7.

Tanaka, N., N. Takemasa, and Y. Sinoto. 1952. "Karyotype analysis in Gymnospermae. I. Karyotype and chromosome bridge in the young leaf meristem of *Ginkgo biloba* L." *Cytologia* (Tokyo) 17: 542–545.

Taylor, R. L., and H. Y. F. Choy. 2005. *The Illustrated Encyclopedia of Confucianism.* Vol. 1. New York: Rosen.

Taylor, T. N., G. M. del Fueyo, and E. L. Taylor. 1994. "Permineralized seed fern cupules from the Triassic of Antarctica: Implications for cupule and carpel evolution." *American Journal of Botany* 81: 666–667.

Taylor, T. N., and E. L. Taylor. 2009. "Seed ferns from the late Paleozoic and Mesozoic: Any angiosperm ancestors lurking there?" *American Journal of Botany* 96: 237–251.

Taylor, T. N., E. L. Taylor, and M. Krings. 2009. *Paleobotany: The Biology and Evolution of Fossil Plants.* Burlington: Academic Press, Elsevier.

Thiede, A., Y. Hiki, and G. Keil. 2000. *Philipp Franz von Siebold and His Era: Prerequisites, Development, Consequences, and Perspectives.* Berlin: Springer.

Thone, F. 1929. "Nature ramblings: Ginkgo." *Science Newsletter* 16: 120.

Thoreau, H. D. 1862. "Wild apples." *Atlantic Monthly* 10: 513–526.

Tiffney, B. H. 1984. "Seed size, dispersal syndromes, and the rise of angiosperms: Evidence and hypothesis." *Annals of the Missouri Botanical Garden* 71: 551–576.

Totman, C. D. 1993. *Early Modern Japan.* Berkeley: University of California Press.

Tralau, H. 1967. "The phytogeographic evolution of the genus *Ginkgo* L." *Botaniska Notiser* 120: 409–422.

———. 1968. "Evolutionary trends in the genus *Ginkgo*." *Lethaia* 1: 63–101.

Treshow, M. 1970. *Environment and Plant Response.* New York: McGraw-Hill.

Trewin, N. H. 2004. "History of research on the geology and palaeontology of the Rhynie area, Aberdeenshire, Scotland." *Transactions of the Royal Society of Edinburgh: Earth Sciences* 94: 285–297.

Trigg, R. 2008. "Warren Woods State Park." *Chicago Wilderness Magazine,* Spring. http://www.chicagowildernessmag.org/CW_Archives/issues/spring2008/itw_warrenwoods.html.

Trustees of the British Museum. 1904. *The History of the Collections Contained in the Natural History Departments of the British Museum.* London: Trustees of the British Museum.

Tsen, M. C. 1935. "*Ginkgo* in Zhuji county, Zhejiang province." *Hortus* 1: 157–165. In Chinese.

Tuchmann, T. E., K. P. Connaughton, L. E. Freedman, and C. B. Moriwaki. 1996. *The Northwest Forest Plan: A Report to the President and Congress.* Washington, D.C.: U.S.D.A. Office of Forestry and Economic Assistance.

Tulecke, W. R. 1954. "Preservation and germination of the pollen of *Ginkgo* under sterile conditions." *Bulletin of the Torrey Botanical Club* 81: 509–512.

Uemura, K. 1997. "Cenozoic history of *Ginkgo* in East Asia." In Hori et al. (1997, 207–221).

Ulrich, R. S. 1984. "View through a window may influence recovery from surgery." *Science* 224: 420.

Unseld, S. 2003. *Goethe und der Ginkgo: Ein Baum und ein Gedicht.* Trans. K. J. Northcott. Chicago: University of Chicago Press.

USDA (United States Department of Agriculture). 2003. *A Technical Guide to Urban and Community Forestry: Urban and Community Forestry; Improving Our Quality of Life.* Portland, Ore.: World Forestry Center.

US EPA (United States Environmental Protection Agency). 2002. "Report of FQPA Tolerance Reassessment Progress and Interim Risk Management Decision (TRED) for chlorpropham." http://www.epa.gov/oppsrrd1/REDs/chlorpropham_tred.pdf. Modified July 19, 2002.

Van Beek, T. A. 2000. *Ginkgo Biloba.* Amsterdam: Harwood Academic.

Van der Pijl, L. 1982. *Principles of Dispersal in Higher Plants.* Berlin: Springer.

Van der Velde, P. 1995. "The interpreter interpreted: Kaempfer's Japanese collaborator Imamura Genemon Eisei." In *The Furthest Goal: Engelbert Kaempfer's Encounter with Tokugawa Japan.* Ed. B. M. Bodart-Bailey and Derek Massarella, 44–58. Kent: Japan Library.

Van der Werf, G. R., D. C. Morton, R. S. DeFries, J. G. J. Olivier, P. S. Kasibhatla, R. B. Jackson, G. J. Collatz, and J. T. Randerson. 2009. "CO2 emissions from forest loss." *Nature Geoscience* 2: 737–738.

Van Konijnenburg–van Cittert, J. H. A. 1971. "In situ gymnosperm pollen from the Middle Jurassic flora of Yorkshire." *Acta Botanica Neerlandia* 20: 1–96.

Van Rij, J. 2001. *Madame Butterfly: Japonisme, Puccini, and the Search for the Real Cho-Cho San.* Berkeley, Calif.: Stone Bridge.

Vavilov, N. I. 1992. *Origin and Geography of Cultivated Plants.* Cambridge: Cambridge University Press.

Vogel, C. 2005. "Gilbert and George's artistic mischief." *New York Times,* August 4. http://www.nytimes.com/2005/08/04/arts/04geor.html?pagewanted=1&_r=1.

Wada, K. 2000. "Food Poisoning by *Ginkgo* seeds: the Role of 4-0-methylpyridoxine." In *Ginkgo Biloba*. Edited by T. A. van Beek, 453–465. Amsterdam: Harwood Academic.

Wada, K., and M. Haga. 1997. "Food poisoning by *Ginkgo biloba* seeds." In Hori et al. (1997, 373–383).

Wada K., S. Ishigaki, K. Ueda, M. Sakata, and M. Haga. 1985. "An antivitamin B6, 4′—methoxypyridoxine, from the seed of *Ginkgo biloba* L." *Chemical and Pharmaceutical Bulletin* 33: 3555–3557.

Wall, C. E., and D. W. Krause. 1992. "A biomechanical analysis of the masticatory apparatus of *Ptilodus* (Multituberculata)." *Journal of Vertebrate Paleontology* 12: 172–187.

Walsh, D. 2011. "Prisons, then parks: A therapeutic journey." *New York Times,* August 2. http://green.blogs.nytimes.com/2011/08/02/prisons-then-parks-a-therapeutic-journey/.

Watson, J. 2005. *One Hundred and Fifty Years of Palaeobotany at Manchester University.* In *History of Palaeobotany: Selected Essays.* Ed. A. J. Bowden, C. V. Burek, and R. Wilding, 229–257. London: Geological Society Special Publications.

Wheeler, E. A., and T. A. Dilhoff. 2009. "The Middle Miocene wood flora of Vantage, Washington, U.S.A." *International Association of Wood Anatomists Journal,* Supplement 7: 1–102.

Wheeler, E. A., S. R. Manchester. 2002. "Woods of the Eocene nut beds flora, Clarno Formation, Oregon, USA." *International Association of Wood Anatomists Journal,* Supplement 3: 1–188.

Wickens, G. E., and P. Lowe. 2008. *The Baobabs: Pachycauls of Africa, Madagascar, and Australia.* Berlin: Springer Science.

Wieland, M. 1768. *Musarion oder die Philosophie der Grazien.* Leipzig: Weidmanns Erben und Reich.

Wilf, P., and K. R. Johnson. 2004. "Land plant extinction at the end of the Cretaceous: A quantitative analysis of the North Dakota megafloral record." *Paleobiology* 30: 347–368.

Williams, C. 2004. "Explorer, botanist, courier, or spy? André Michaux and the Genet affair of 1793." *Castanea* 69: 98–106.

Wilson, E. H. 1913. *A Naturalist in Western China with Vasculum, Camera, and Gun.* London: Methuen.

———. 1919. "The romance of our trees—II. The ginkgo." *Garden Magazine* 29–30: 144–149.

———. 1920. *The Romance of Our Trees.* New York: Doubleday, Page.

Wilson, J. C., J. E. Altland, J. L. Sibley, K. M. Tilt, and G. F. Wheeler. 2004. "Effects of chilling and heat on growth of *Ginkgo biloba* L." *Journal of Arboriculture* 30: 45–51.

Wilson, J. C., and A. H. Knoll. 2010. "A physiologically explicit morphospace for tracheid-based water transport in modern and extinct seed plants." *Paleobiology* 36: 335–355.

Winchester, S. 2008. *The Man Who Loved China: The Fantastic Story of the Eccentric Scientist Who Unlocked the Mysteries of the Middle Kingdom.* New York: HarperCollins.

Wolf, K. L. 2003. "Public response to the urban forest in inner-city business districts." *Journal of Arboriculture* 29: 117–126.

"World's oldest living tree 9550 years old—discovered in Sweden." *Science Daily,* April 16, 2008. http://www.sciencedaily.com/releases/2008/04/080416104320.htm.

Wu, J. Y., P. L. Cheng, and S. J. Tang. 1992. "Isozyme analysis of the genetic variation of *Ginkgo biloba* L. population in Tian-Mu Mountain." *Journal of Plant Resources and Environment* 1: 20–23.

Wu, X. W., X. J. Yang, and Z. Y. Zhou. 2006. "Ginkgoalean ovulate organs and seeds associated with *Baiera furcata*–type leaves from the Middle Jurassic of Qinghai Province, China." *Review of Palaeobotany and Palynology* 138: 209–225.

Wulf, A. 2011. *Founding Gardeners: The Revolutionary Generation, Nature, and the Shaping of the American Nation.* New York: Knopf.

Wullschleger, S. D., F. C. Meinzer, and R. A. Vertessy. 1998. "A review of whole-plant water use studies in trees." *Tree Physiology* 18: 499–512.

WWF (World Wide Fund for Nature/World Wildlife Fund). 2010. "The Living Planet Report 2010: Biodiversity, biocapacity, and development." http://awsassets.panda.org/downloads/lpr2010.pdf.

Wyman, A. 2008. "Gilbert and George: A gorgeous retrospective at the de Young." SFStation. www .sfstation.com/gilbert-and-george-a8111. Modified February 26, 2008.

Xiang, Z., Y. Xiang, B. Xiang, and P. Del Tredici. 2009. "The Li Jiawan grand ginkgo king." *Arnoldia* 66: 26–30.

York, K. M. 2006. "A meta-analysis of the effects of *Ginkgo biloba* on cognitive and psychosocial functioning in humans." PhD. diss., University of Southern Mississippi.

Yoshitama, K. 1997. "Flavonoids of *Ginkgo biloba*." In Hori et al. (1997, 287–299).

Yoshimura, T., N. Udaka, J. Morita, Z. Jinyu, K. Sasaki, D. Kobayashi, K. Wada, and Y. Hori. 2006. "High performance liquid chromatographic determination of Ginkgotoxin and Ginkgotoxin-5'-Glucoside in *Ginkgo biloba* seeds." *Journal of Liquid Chromatography and Related Technologies* 29: 605–616.

Zhao, Y., J. Paule, C. Fu, M. A. Koch. 2010. "Out of China: Distribution history of *Ginkgo biloba* L." *Taxon* 59: 495–504.

Zheng, S. L., and Z. Y Zhou. 2004. "A new Mesozoic *Ginkgo* from western Liaoning, China, and its evolutionary significance." *Review of Palaeobotany and Palynology* 131: 91–103.

Zhou, Z. Y. 1983. "*Stalagma samara,* a new podocarpaceous conifer with monocolpate pollen from the Upper Triassic of Hunan, China." *Palaeontographica Abt. B* 185: 56–78.

———. 2009. "An overview of fossil Ginkgoales." *Palaeoworld* 18: 1–22.

Zhou, Z. Y., C. Quan, and Y-S. Liu. 2012. "Tertiary Ginkgo ovulate organs with associated leaves from North Dakota, U.S.A., and their evolutionary significance." *International Journal of Plant Sciences* 173: 67–80.

Zhou, Z. Y., and X. W. Wu. 2006. "The rise of ginkgoalean plants in the early Mesozoic: A data analysis." *Geological Journal* 41: 363–375.

Zhou Z. Y., and B. L. Zhang. 1988. "Two new ginkgoalean female reproductive organs from the Middle Jurassic of Henan Province." *Science Bulletin (Kexue Tongbao)* 33: 1201–1203.

———. 1989. "A Middle Jurassic *Ginkgo* with ovule-bearing organs from Henan, China." *Palaeontographica Abt. B* 211: 113–133.

———. 1992. "*Baiera hallei* Sze and associated ovule-bearing organs from the Middle Jurassic of Henan, China." *Palaeontographica Abt. B* 224: 151–169.

Zhou Z. Y., B. L. Zhang, Y. D. Wang, and G. Guignard. 2002. "A new *Karkenia* (Ginkgoales) from the Jurassic Yima formation, Henan, China, and its megaspore membrane ultrastructure." *Review of Palaeobotany and Palynology* 120: 91–105.

Zhou, Z. Y., and S. L. Zheng. 2003. "The missing link of *Ginkgo* evolution." *Nature* 423: 821–822.

Zhou, Z. Y., S. L. Zheng, and L. J. Zhang. 2007. "Morphology and age of *Yimaia* (Ginkgoales) from Daohugou Village, Ningcheng, Inner Mongolia, China." *Cretaceous Research* 28: 348–362.

Zhu, W., and J. Gao. 2008. "The use of botanical extracts as topical skin-lightening agents for the improvement of skin pigmentation disorders." *Journal of Investigative Dermatology Symposium Proceedings* 13: 20–24.

Zukowski, D. "In Hoboken, trees for 9/11." The Newark Metro. http://www.newarkmetro.rutgers.edu/reports/display.php?id=101. Accessed January 10, 2011.

Zwieniecki, M. A., H. A. Stone, A. Leigh, C. K. Boyce, and N. M. Holbrook. 2006. "Hydraulic design of pine needles: One-dimensional optimization for single-vein leaves." *Plant Cell and Environment* 29: 803–809.

Illustration Credits

Frontispiece—Drawing by Pollyanna von Knorring, based on a photograph taken at the Royal Botanic Gardens, Kew.

Part I opener—Drawing by Pollyanna von Knorring, from the Economic Botany Collection of the Royal Botanic Gardens, Kew.

Pp. 6, 11, 13, 19, 33, 36, 44, 49, 77, 112, 150, 163, 169, 181, 224, 231, 232, 238, 239, 243, 247, 269—Photographs by Peter R. Crane.

Part II opener—Drawing by Pollyanna von Knorring, from photographs in Hori and Hori (2005).

P. 28—Scanning electron micrograph courtesy of Zhou Zhiyan and collaborators.

P. 39—From Siebold and Zuccarini (1835–1870), vol. II, plate 136.

P. 47—Photograph courtesy of Andrew McRobb, © Royal Botanic Gardens, Kew; artwork attributed to Chikusai Kato, 1878, from the Economic Botany Collection at the Royal Botanic Gardens, Kew.

P. 50—Photograph by Elisabeth Wheeler.

Pp. 58, 145—Photographs by Nancy Hines.

P. 69 (top)—Photograph courtesy of Masaya Satoh.

P. 69 (bottom)—Film still from *The Sea in the Seed,* Tokyo Cinema.

P. 71—Photograph courtesy of Marie Stopes International.

Part III opener—Drawing by Pollyanna von Knorring, redrawn from Zhou and Zhang (1989, 1992) and from Zhou (2009).

P. 83—Photograph by Peter R. Crane, from material at the Swedish Museum of Natural History, Stockholm.

Pp. 87, 99, 102, 121—Andrew B. Leslie.

P. 95—Drawing by Pollyanna von Knorring, redrawn from Anderson and Anderson (2003), 288.

P. 105—Photograph courtesy of Meinte Boersma.

P. 109—Photograph by Ghedoghedo, from material at the Paläontologische Museum, Munich.

P. 115—Photograph by Andrew N. Drinnan.

P. 117—Andrew B. Leslie, based on Drinnan and Crane (1989).

P. 126—Image courtesy of Steven Manchester, paleobotanical collection of the Florida Museum of Natural History, University of Florida, Gainesville.

P. 127—Image courtesy of Zhou Zhiyan, from material at the Field Museum, Chicago.

P. 135—Illustration by Francisco Manuel Blanco (1778–1845), from material at the Real Jardín Botánico of Madrid.

Part IV opener—Drawing by Pollyanna von Knorring, based on Hori and Hori (1997, 391).

P. 153—Andrew B. Leslie, compiled and redrawn from Tralau (1967).

P. 166—Photograph by Ernest Henry Wilson (1914), © 2006, President and Fellows of Harvard College, Arnold Arboretum Archives, Jamaica Plain.

P. 172—Drawing by Andrew B. Leslie.

Part V opener—Drawing by Pollyanna von Knorring, based on material from Kyushu Ceramics Museum, Japan.

P. 192—Drawing by Pollyanna von Knorring, based on a model at the National Maritime Museum, Mokpo, Korea.

P. 197—Image from Isaac Titsingh from material at National Library of the Netherlands, The Hague.

P. 202—Photograph by Ashley DuVal from material of the Mertz Library at the New York Botanical Garden, New York.

P. 212—Image by permission of the Linnean Society of London.

Part VI and VII openers—Drawings by Pollyanna von Knorring, based on photographs by Peter R. Crane.

P. 260—Illustration courtesy of Jim Xerogeanes.

P. 274—Photograph courtesy of Andrew McRobb, ©Royal Botanic Gardens, Kew.

Index

Note: Page numbers in *italics* refer to illustrations.

9 Humboldt Redwoods State Park   200 miles North of San Francisco

2. 100,000 different kinds of trees

4 "Suburbia is where the developer bulldozes out the trees and
   then names the streets after them." Bill Vaughan

5. Dawn Redwood and Ginkgo, Wollemi Pine Australia -
       prehistoric

6 William Hooker - Kew Garden (Richmond Garden by
   Capability Brown)   1841 Hooker created Kew collection

8 Chinese Garden in Suzhou - Ginkgo
       Unesco World Heritage site
   South Korea 100 ft tall - 16 ft diameter - 400 years old

Frank Lloyd Wright - Oak Park - large ginkgo tree

2 Ginkgo at Hiroshima survived the bomb
   Kaempfer - ginkgo - 1st Westerner
   Linnaeus - Ginkgo biloba
   Ginkgo leaves are tough and slow to decay
   # of leaves on a ginkgo - 300,000 - 500,000
       giant Ginkgos - 1 million leaves

3 Leaves fall all at once   Howard Nemerov poem

7. short shoots - energy and fill in crown
   long shoots - physical tree growth

8 - short shoots - fan shape deeply divided
   long shoots deep central notch
       Siebold illustration